The Urbanization of the Third World

THE URBANIZATION
OF THE
THIRD WORLD

edited by

Josef Gugler

OXFORD UNIVERSITY PRESS

Oxford University Press, Walton Street, Oxford OX2 6DP

Oxford New York Toronto
Delhi Bombay Calcutta Madras Karachi
Petaling Jaya Singapore Hong Kong Tokyo
Nairobi Dar es Salaam Cape Town
Melbourne Auckland
and associated companies in
Berlin Ibadan

Oxford is a trade mark of Oxford University Press

Published in the United States
by Oxford University Press, New York

First published 1988 in hardback and paperback
Paperback reprinted 1990, 1991, 1992

British Library Cataloguing in Publication Data
The Urbanization of the Third World
1. Urbanization—Developing countries
I. Gugler, Josef
307.7'6'091724   HT149.5
ISBN 0–19–823260–8
ISBN 0–19–823259–4 (Pbk)

Library of Congress Cataloging in Publication Data
The Urbanization of the Third World.
Bibliography: p.
Includes index.
1. Urbanization—Developing countries. 2. Cities
and towns—Developing countries.   I. Gugler, Josef.
HT149.5V75   1988   307.7'6'091724   87–31385
ISBN 0–19–823260–8
ISBN 0–19–823259–4 (Pbk.)

Printed and bound in
Great Britain by Biddles Ltd,
Guildford and King's Lynn

*To the Third World citizens
who have humoured our enquiries*

Contents

Notes on Contributors

Marc Blecher is Professor of Government and Chair of Third World Studies at Oberlin College, USA. He has been a Visiting Fellow at the Institute of Development Studies (at the University of Sussex), the University of Chicago, and the University of California at Berkeley.

Ray Bromley is Professor in the Department of Geography and Planning and in the Department of Latin American and Caribbean Studies at the State University of New York at Albany. He was 13 years on faculty at the University College of Swansea, and has worked as a consultant to various governments, international organizations, and aid agencies.

Manuel Castells is Professor of City and Regional Planning at the University of California, Berkeley. He taught at the University of Paris for 12 years, as well as at the Universities of Montreal, Chile, Wisconsin, Copenhagen, Boston, Mexico, Hong Kong, Southern California, and Madrid.

Abner Cohen, Emeritus Professor of Anthropology in the University of London, carried out field studies in Africa, the Near East, and the United Kingdom, with special focus on the relations between cultural forms and political formations, including religion, ethnicity, marriage and kinship, and symbolism, in both urban and rural areas.

Paul W. Drake is Institute of the Americas Professor of Inter-American Affairs in the Department of Political Science at the University of California, San Diego, where he directs the Center for Iberian and Latin American Studies. He is also President of the Latin American Studies Association.

Susan Eckstein is Professor of Sociology at Boston University. She has written extensively on urban poor in Mexico and on outcomes of revolutions in Latin America. Currently she works on protest and resistance movements in Latin America, and on the political economy of the Cuban revolution.

Alan Gilbert is Reader in Geography at University College and the Institute of Latin American Studies, London. His main research interests focus on urban developments in Third World countries. He has worked extensively on popular housing and service delivery in Latin American cities and is currently working on a project concerned with rental housing in Mexico.

Josef Gugler is Professor of Sociology at the University of Connecticut and Visiting Professor of Development Sociology at the Universität Bayreuth, FR Germany. He served as Director of Sociological Research at the Makerere Institute of Social Research, Uganda. His research on urbanization has taken him to Nigeria, Kenya, Tanzania, Zaïre, Cuba, and India.

Mary Hollnsteiner-Racelis, formerly Professor of Sociology and Anthropology, then Director of the Institute of Philippine Culture, at Ateneo de Manila University, joined the United Nations Children's Fund in 1979. Currently she is Director of UNICEF's Eastern and Southern Africa Regional Office in Nairobi.

Lea Jellinek went to Jakarta in 1971 as a volunteer graduate and was 'adopted' into a poor inner city *kampong*. She has written widely on the lives of the urban poor and worked as a consultant sociologist for the World Bank and the United Nations Development Programme on urban and rural projects.

Michael Johnson has worked at the Universities of Manchester and Khartoum and is currently a Lecturer in Politics at the University of Sussex. While at Manchester he was a member of a research team investigating the politics and society of Lebanon. He has also conducted research on the political economy of India, Sudan, and the Horn of Africa.

Julian Laite is Senior Lecturer in Sociology at the University of Manchester. He has done sociological field research on peasants, industrial workers, and social stratification in Latin America and Canada. His current interests include the changing structures of employment in local labour markets, household structures, and social stratification in Britain and France.

Michael Lipton is Professor of Economics at the University of Sussex, where he has been a Fellow of the Institute of Development Studies since 1967. He has worked on agriculture and rural development, characteristics and differential behaviour of poor and ultra-poor groups, rural–urban relations, employment and labour use, and the impact of modern varieties of cereals upon poor people. His areas of special interest include India, Bangladesh, Sri Lanka, Botswana, Sudan, and Kenya.

Larissa Lomnitz is Professor of Social Anthropology at the National University of Mexico. She has done research on the urban poor as well as the industrial bourgeoisie of Mexico. Kinship and network analysis run through all these studies.

Nici Nelson is currently Head of the Department of Anthropology at Goldsmiths' College, University of London. She lived four years in Nairobi, Kenya. She works in the areas of African urbanization and women and development.

Peter Nientied, a staff member of the Institute of Social Studies, Rotterdam, is presently posted in Delhi, India. He did extensive field-work on government policies and practice of low-income housing in Karachi, Pakistan.

Julius K. Nyerere led the British Trust Territory of Tanganyika to independence as founder president of the Tanganyika African National Union. Prime Minister at independence in 1961, President of Tanganyika Republic in 1962, and President of Tanzania when Tanganyika and Zanzibar united in 1964, he relinquished the Presidency in 1985. He continues as Chairman of the party, since 1977 known as Chama Cha Mapinduzi, the Revolutionary Party.

Samuel Preston is Director of the Population Studies Center and Chairman of the Department of Sociology at the University of Pennsylvania. He was formerly chief of the Population Trends and Structure Section, United Nations Population Division. He is past President of the Population Association of America.

Richard H. Sabot is Professor of Economics at Williams College and Senior Research Fellow at the International Food Policy Research Institute. He had previously been on the staff of the Development Research Department of the World Bank. He has taught at Oxford, Yale, and Columbia Universities. His theoretical and applied research has focused on labour markets in low-income countries.

Jan van der Linden is Senior Lecturer of Development Sociology at the Amsterdam Free University. Since 1971 he has been engaged in a number of studies of low-income housing policy and practice, especially in Karachi, Pakistan.

Peter Ward, University Lecturer in Geography at Cambridge and Fellow of Fitzwilliam College, previously taught for ten years at University College London. He was co-director of a major research project 'Public intervention, housing and land-use in Latin American cities'. He has worked as an adviser on low-income housing policy to the Mexican Government and as a consultant to the United Nations Center for Human Settlements and the World Health Organization.

Martin King Whyte is Professor of Sociology and Associate of the Center for Chinese Studies at the University of Michigan, where he has taught since 1969. His past research has focused on patterns of social change in rural as well as urban China. Currently he is engaged in a collaborative project with Chinese sociologists to compare the changes in the process of mate choice and marital relations in the US and China.

Introduction

MORE than a billion people live in the cities and towns of the Third World. Their numbers are growing rapidly in almost every Third World country, and only in Latin America is there evidence of a slow-down (Table 1). The cumulative effect of this continuous rapid growth is dramatic. Thus Third World cities are outstripping their First World counterparts: Mexico City and São Paulo are about to become the world's largest cities, and by the end of the century seventeen of the twenty-three largest metropolitan areas, with populations over ten million, will be in the Third World (United Nations 1987, 25).

The urban transition constitutes a great human transformation, comparable to the domestication of plants and animals ten thousand years ago that made a sedentary life possible. This second transformation began five thousand years ago, when the first urban settlements were established in the valleys of the Tigris and the Euphrates. However, as recently as the beginning of this century only one in eight people lived in an urban area. The twentieth century is the century of the urban transition: by the end of the century, nearly half the world's population, close to three billion people, will live in urban settlements. And two-thirds of that number will live in the less developed countries of Asia, Oceania, Africa, Latin America, and the Caribbean (United Nations 1987, 8).

The last phase of the urban transition is taking place in the Third World. The magnitude of this transformation, the sheer number of people involved, is without precedent in human history. And the poverty of the Third World makes it a difficult transition. Rapidly growing urban populations have to find employment in urban labour markets characterized by widespread unemployment and underemployment, increase demands for urban housing and services already considered inadequate, and, for better or worse, exacerbate popular pressures on political systems.

Urbanization constitutes a multi-faceted phenomenon, and the Third World is large and diverse. The task of understanding Third World urbanization might be approached by composing a picture from studies

TABLE 1 Demographic Characteristics and Income of Major Third World Countries[a]

Region/country	Population (millions) mid-1985	GNP per capita (dollars) 1985[b]	Urban population as percentage of total population 1985[c]	Urban growth, average annual growth-rate (per cent)		Population growth, average annual growth-rate (per cent)
				1965–80	1980–5	1980–5
East Asia						
Burma	37	190	24	2.8	2.8	2.0
China[d]	1040	310	22	2.6	3.3	1.2
Indonesia	162	530	25	4.7	2.3	2.1
Korea, North	20	930	63	4.6	3.8	2.5
Korea, South	41	2150	64	5.7	2.5	1.5
Malaysia	16	2000	38	4.5	4.0	2.5
Philippines	55	580	39	4.0	3.2	2.5
Taiwan	19	2670	51	5.4	3.1	1.6
Thailand	52	800	18	4.6	3.2	2.1
Vietnam	62	240	20	4.1	3.4	2.6
South Asia						
Bangladesh	101	150	18	8.0	7.9	2.6
India	765	270	25	3.6	3.9	2.2
Nepal	17	160	7	5.1	5.6	2.4
Pakistan	96	380	29	4.3	4.8	3.1
Sri Lanka	16	380	21	2.3	8.4	1.4
Middle East and North Africa						
Afghanistan[c]	17	230	17	5.6	6.2	2.6
Algeria	22	2550	43	3.8	3.7	3.3
Egypt	49	610	46	2.9	3.4	2.8
Iran	45	1690	54	5.5	4.6	2.9
Iraq	16	1930	70	5.3	6.3	3.6
Morocco	22	560	44	4.2	4.2	2.5
Saudi Arabia	12	8850	72	8.5	6.1	4.2
Syria	11	1570	49	4.5	5.5	3.6
Turkey	50	1080	46	4.3	4.4	2.5

Subsaharan Africa

Country					
Cameroon	10	810	42	8.1	7.0	3.2
Ethiopia	42	110	15	6.6	3.7	2.5
Ghana	13	380	32	3.4	3.9	3.3
Ivory Coast	10	660	45	8.7	6.9	3.8
Kenya	20	290	20	9.0	6.3	4.1
Madagascar	10	240	21	5.7	5.3	3.2
Mozambique	14	160	19	11.8	5.3	2.6
Nigeria	100	800	30	4.8	5.2	3.3
South Africa	32	2010	56	2.6	3.3	2.5
Sudan	22	300	21	5.1	4.8	2.7
Tanzania	22	290	14	8.7	8.3	3.5
Uganda	15	_240_	7	4.1	3.0	3.0
Zaïre	31	170	39	7.2	8.4	3.0

Latin America

Country					
Argentina	31	2130	84	2.2	1.9	1.6
Brazil	136	1640	73	4.5	4.0	2.3
Chile	12	1430	83	2.6	2.1	1.7
Colombia	28	1320	67	3.5	2.8	1.9
Cuba	10	_1180_	71	2.7	0.8	0.8
Mexico	79	2080	69	4.5	3.6	2.6
Peru	19	1010	68	4.1	3.8	2.3
Venezuela	17	3080	85	4.5	3.5	2.9

[a] This table includes all Third World countries with a population over 9.5 million in mid-1985.

[b] GNP per capita figures in italics are for 1982 and from a different source; they are not strictly comparable with the other GNP per capita figures.

[c] Urban percentages are based on different national definitions of what is 'urban', and cross-country comparisons should be interpreted with caution.

[d] The data for China include Hong Kong, Macao, and Taiwan.

[e] The demographic data for Afghanistan are for 1983, the growth figures for 1965–73 and 1973–83.

Sources: All data from World Bank (1987: annex Table 1, 27, and 33), except GNP per capita figures in italics (for 1982) from Sivard (1985: statistical annex Table III), demographic data for Taiwan from China (1986: Table 1 and Supplementary Table 3), demographic data for Afghanistan (for 1983) from World Bank (1985: annex Tables 19 and 22).

more limited in scope. A number of fine studies provide broad overviews of urbanization in a region or a major country: Kirkby (1985) and Murphey (1980) for China, Abu-Lughod (1984) for the Arab World, Gugler and Flanagan (1978) for West Africa, Portes (1976) and Roberts (1978) for Latin America. I am preparing a collection, *Patterns of Third World Urbanization*, that will present and contrast such regional/country studies.

The present collection takes a different approach. The aim has been to assemble the most authoritative statements on the varied aspects of urbanization. The authors range across the social sciences, and include one elder statesman of the Third World. If sociologists and anthropologists predominate, demography, economics, geography, history, and political science are also represented. Their studies have been grouped to explore seven aspects of urbanization: the distinctive characteristics of rapid urban growth in the Third World; the relationship between the urban and the rural sector; rural–urban migration as it relates to the urban labour market; the housing question; survival strategies in the city; forms of integration and social control; and the politics played out in the urban arena. Introductions to each of these seven parts put the individual studies into context. Together they provide a composite picture of Third World urbanization today.

A few of the studies are general literature and data reviews. Most, however, are based on research in a specific setting, or on comparative research in a few settings. These settings range across the Third World, but they are by no means randomly distributed. Regional biases arise from the predilection of researchers and the priorities of their sponsors, e.g. the interest of British anthropologists in colonial Africa, of American political scientists in Latin America after the triumph of the Cuban revolution. Not only the extent to which research is pursued in a region, but also the topics it addresses, the theoretical perspective it employs, and the methods it uses, vary across regions. Relatedly, certain issues are more topical in some settings than in others, e.g. government control in China, ethnicity in Africa, and squatter movements in Latin America. If Latin American research preponderates in this collection, this reflects the amount of important research accomplished in the region in the 1970s. This regional bias may be argued to be desirable to the extent that Latin American patterns prove to presage developments elsewhere: as states mature that gained their independence much more recently, as income levels rise in even poorer economies, and as the urban transition proceeds in as yet predominantly rural societies.

Four studies are previously unpublished. The others are reprints of articles or book chapters. Most of these were published within the last ten years; nearly all have been revised for this collection. They thus complement as well as up-date the volume *Cities, Poverty, and Development: Urbanization in the Third World* I co-authored by Alan Gilbert.

Over the years I have accumulated many intellectual debts: debts to colleagues across the social sciences who have made me sensitive to the multiple approaches needed to understand the urbanization process, whose conflicting perspectives have convinced me that we need to assimilate what is valuable in new research, rather than succumbing to the grandiose claims of passing intellectual fashions; and debts to students from three continents who have challenged me to integrate multiple disciplinary approaches and conflicting theoretical perspectives. The selections I have made for this collection, my suggestions to the contributors for revisions, the perspective the chapter introductions provide on the selections, are the outcome of this intellectual journey. But this collection foremost represents the work of the scholars who have agreed to have classic statements reprinted, to revise major writings, to make original contributions. And all of us are profoundly indebted to the men and women in the Third World who were prepared to let us enter their lives, who gave us of their time to answer our queries, some of whom put themselves at risk by responding to us. This collection is dedicated to them.

References

Abu-Lughod, Janet (1984) 'Culture, "Modes of Production", and the Changing Nature of Cities in the Arab World', in John A. Agnew, John Mercer and David E. Sopher (eds.), *The City in Cultural Context* (Boston, London, and Sydney: Allen & Unwin), 94–119.

China, Republic of (1986) *Statistical Yearbook of the Republic of China, 1986* (Taipei: Directorate-General of Budget, Accounting & Statistics).

Gugler, Josef, and Flanagan, William G. (1978) *Urbanization and Social Change in West Africa.* Urbanization in Developing Countries (Cambridge, New York, and Melbourne: Cambridge University Press).

Gilbert, Alan, and Gugler, Josef (1982) *Cities, Poverty, and Development: Urbanization in the Third World* (Oxford and New York: Oxford University Press).

Kirkby, R. J. R. (1985) *Urbanization in China: Town and Country in a Developing Economy 1949–2000 AD* (London and Sydney: Croom Helm; New York: University of Columbia Press).

Murphey, Rhoads (1980) *The Fading of the Maoist Vision: City and Country in China's Development* (New York, London, and Toronto: Methuen).

Portes, Alejandro (1976) 'The Economy of Urban Poverty', in Alejandro Portes and John Walton, *Urban Latin America: The Political Condition from Above and Below* (Austin and London: University of Texas Press), 7–69.

Roberts, Bryan R. (1978) *Cities of Peasants: The Political Economy of Urbanization in the Third World* (London: Edward Arnold; Beverly Hills, Calif.: Sage).

Sivard, Ruth Leger (1985) *World Military and Social Expenditures 1985: An Annual Report of World Priorities* (Washington, DC: World Priorities).

United Nations (1987) *The Prospects of World Urbanization: Revised as of 1984–85*. Population Studies 101. (New York: United Nations).

World Bank (1985) *World Development Report 1985* (New York: Oxford University Press).

—— (1987) *World Development Report 1987* (New York: Oxford University Press).

I

The Urban Transition in the Third World

INTRODUCTION

THE pace of urban growth in the Third World is without precedent. It is fuelled by substantial rural–urban migration: many urban dwellers come from the peasantry. Samuel Preston, in Chapter 1, shows that natural population growth plays an even larger role, accounting for three-fifths of urban growth on average. The rapid natural population growth characteristic of Third World countries since World War II is unique in human history. And the rate of urban growth, the sheer increase in the size of urban populations, is without parallel. However, the rate of urbanization, i.e. the increase in the urban proportion of the population, was similar between 1950 and 1975 to the rate that characterized the urban transition in Europe 75 years earlier. In other words, urban growth is exceptionally rapid in the Third World, not because of an unusually rapid increase in the urban proportion of the population due to rural–urban migration, but because of the rapid increase in the total population to which this proportion is applied.

Preston's emphasis on the role of natural population growth in urban growth has to be qualified to the extent that the coverage of African populations in his data set is quite limited—many African countries have experienced very rapid urbanization since the 1950s when nearly all of them approached independence. In terms of the implications for urban labour markets it has also to be noted that the majority of rural–urban migrants arrive ready to enter the labour force. If they contribute about two-fifths of urban growth, they constitute a considerably higher proportion of the new entrants into the labour force. An analysis of census data for the metropolitan areas of 26 major Third World cities found that in the 1960s net migration accounted on average for 37 per cent of the growth in their population, but for 63 per cent of the growth in the population aged 15–29 (United Nations 1985, 25).

Urbanization and industrialization are frequently assumed to be intimately connected, and urbanization is seen as a prerequisite for development. Such assumptions are problematic. Not only do cities predate industrial manufacture by several millennia, but even today only a small proportion of urban dwellers works in industry. Most Third World cities are foremost centres of public administration and commerce. If the more urbanized countries tend to be more developed—as can be seen in Table 1—this strong correlation does not establish causality. Urbanization does not necessarily cause development, the reverse causal relationship is quite plausible: the more developed countries have greater resources to spend on bureaucracies and public works.

Recurrent concerns about overurbanization expressly deny a functional relationship between urban growth and development. They have been fostered by the high level of non-industrial employment characteristic of Third World cities, whether it is measured in terms of the service, the tertiary, or the informal sector. However, contrary to assumptions commonly made about ever-worsening 'imbalances' in the employment structure, the situation appears quite stable. Preston, in his contribution here, compares regional estimates of labour-force structure with estimates of urban–rural population distribution to assess changes in the ratio of the proportion of the labour force reported in industry to the urban proportion of the population. Between 1950 and 1970, the industrial/urban ratio improved in Middle South Asia, changed little in the rest of Asia and Africa, but declined in Latin America, especially markedly in South America. We will return to this issue in Chapter 5.

A focus on the international position of Third World countries, whether inspired by dependency theory or a world-system perspective, has prompted research on the relationship between dependency or positions in the world system and urbanization. Timberlake and Kentor (1984) report that the greater the level of dependency, as measured by foreign investments, the higher the level of urbanization at a given level of development, and the higher the ratio of services to manufacturing employment. According to Firebaugh (1985), the Third World caught up with Western Europe in the proportion of the non-farm population living in urban areas in 1960—at a much earlier stage in the urban transition. In a particularly ambitious analysis of cross-national data, Bradshaw (1987) finds partial support for dependency theory, as well as the modernization theory it was meant to supplant, and the urban bias argument we will explore in Part II.

Concerns over urban primacy often fused with concerns about over-urbanization, but the two issues have to be distinguished analytically. If the notion of overurbanization entails an evaluation of the size of a country's urban population, urban primacy is an aspect of the size distribution of cities in a given country. It denotes the dominance of one or a few cities *vis-à-vis* the rest of the urban system in a country, a common pattern in the Third World. Smith (1985a; 1985b) discusses various explanations of urban primacy and presents an approach that moves beyond the dependency/world-system perspectives. Preston refutes the common assumption that the largest cities grow most rapidly. His analysis of the growth-rates of 1,212 Third World cities demonstrates that the general relation between city size and city growth-rates is U-shaped: cities in the size classes between

500,000 and four million have the lowest growth-rates. The overall effect of city size on the rate of city growth, in a multivariate analysis, is negative. However, where a country's capital is also the largest city—a common pattern—it will tend to grow disproportionately fast. Finally, Latin American cities grow faster than could be expected on the basis of any of the several variables considered. The same multivariate analysis extended to include 110 Chinese cities produced quite similar results (United Nations 1980, 42–4).

References

Bradshaw, York W. (1987) 'Urbanization and Underdevelopment: A Global Study of Modernization, Urban Bias, and Economic Dependency', *American Sociological Review*, 52, 224–39.

Firebaugh, Glenn (1985) 'Core–Periphery Patterns of Urbanization', in Michael Timberlake (ed.), *Urbanization in the World-Economy* (Orlando, Fla. and London: Academic Press), 293–304.

Smith, Carol A. (1985a) 'Theories and Measures of Urban Primacy: A Critique', in Michael Timberlake (ed.), *Urbanization in the World-Economy* (Orlando, Fla. and London: Academic Press), 87–117.

—— (1985b) 'Class Relations and Urbanization in Guatemala: Toward an Alternative Theory of Urban Primacy', in Michael Timberlake (ed.), *Urbanization in the World-Economy* (Orlando, Fla. and London: Academic Press), 121–67.

Timberlake, Michael, and Kentor, Jeffrey (1984) 'Economic Dependence, Overurbanization, and Economic Growth: A Study of Less Developed Countries', *Sociological Quarterly*, 24, 489–507.

United Nations (1980) *Patterns of Urban and Rural Population Growth*. Population Studies 68. (New York: United Nations).

—— (1985) *Migration, Population Growth and Employment in Metropolitan Areas of Selected Developing Countries*. ST/ESA/SER.R/57. (New York: United Nations).

1

Urban Growth in Developing Countries:
A Demographic Reappraisal*

Samuel H. Preston

GOVERNMENTS and scholars alike have shown rapidly growing concern with issues of population distribution. Among the 116 developing countries that responded to the United Nations' 'Fourth Population Inquiry among Governments' conducted in 1978, only six declared the spatial distribution of their population to be 'acceptable'. Forty-two replied that it was 'unacceptable to some extent', and 68 declared it to be 'highly unacceptable'. To another question addressed specifically to the desirability of current rates of rural–urban migration, only three countries expressed a desire to accelerate such migration. Twenty-three wished to maintain it at present levels, 76 to slow it down, and 14 to reverse it.[1]

Part of the concern with population distribution reflects a belief that current redistributional patterns, particularly net migration from rural to urban areas, are a product of unjustifiable regional and sectoral distortions in patterns of development. Rural–urban migration functions as an indicator of these distortions, at the same time as it may make reversing them more difficult. Another part of the concern on the part of governments grows out of the practical administrative difficulties of planning local public services in the face of unplanned changes in the population of users. Doubtless a third factor in some cases is the belief that dispersed and largely invisible rural masses tend to make fewer demands on the government and to constitute less of an implied threat to social order than do concentrated urbanites, many of whom have made an enormous migratory investment in expectation of economic and social betterment.

In view of the importance being attached to distributional issues, particularly to urban growth, it is useful to examine carefully the demographic processes that are currently responsible for and associated with such growth. Many common views of these processes appear to be seriously misleading and unnecessarily alarmist. In this review, we rely

* Reprinted, in a revised form, from the *Population and Development Review*, 5 (2), 1979, by permission of the Population Council.

primarily upon material developed in the course of a United Nations study of urban and rural population change.[2] This study assembled estimates of urban and rural population and of the population of cities larger than 100,000 from 1950 to the present.[3] The study does not deal with all aspects of population distribution, but only with those demographic aspects that relate to distinctions between urban and rural areas and between places of differing size. Four conclusions of this study are described here, and their bearing on distribution policy is considered.

1. The rate of change in the urban proportion in developing countries is not exceptionally rapid by historical standards; rather it is the growth-rates of urban populations that represent an unprecedented phenomenon.

The most common measure of the rate of urbanization is the annual change in the percentage of the population living in urban areas. According to this measure, urbanization in developing countries did not proceed with unusual speed in the quarter-century from 1950 to 1975. In this period the urban percentage grew from 16.7 to 28.0 in developing countries.[4] While this is a rapid increase, it is very similar to the one that occurred in more-developed countries during the last quarter of the nineteenth century. Between 1875 and 1900, the urban percentage of countries now more developed grew from 17.2 to 26.1.[5] The slight difference from the growth in developing countries 75 years later is well within the margin of error of the estimates. The rates of net rural–urban migration required to achieve the observed increase in the urban percentage may even have been greater in more-developed countries, in view of the higher rates of rural than of urban natural increase that typically prevailed at the time.[6] That is, to achieve a certain increase in the urban percentage, higher rates of net rural–urban migration were required in developed countries than in developing countries, where rural–urban differences in rates of natural increase are far less significant.[7]

Nor does it appear that rates of urbanization or of net rural–urban migration are accelerating in developing countries. Between 1950 and 1960 the urban proportion grew by 5.1 percentage points and between 1960 and 1975, a period 50 per cent longer, by 6.2 percentage points. (These figures include China's uncertain estimates, which show decelerated urbanization.) The pace of urbanization has been accelerating in Africa but decelerating in Latin America. Changes in rates of net migration into urban areas (measured over the base of the urban population) can be computed for 11 countries using intercensal survival techniques applied to adjacent inter-censal periods. These 11 developing countries are the only ones with three post-war censuses, with urban and rural age–sex distributions for each census to permit analysis of growth components, and with definitions of urban areas that are either stable or adjustable to stability. In general, the net in-migration rates showed considerable stability. Among the 11 countries, five had changes in annual net urban in-migration rates of less than 3 per thousand (India, Chile, El Salvador, the Dominican Republic, and Turkey), five countries showed declines larger than that amount: Brazil (−5.7), Ecuador (−4.7), Venezuela (−9.8), Sri Lanka (−3.1), and the Union of South Africa (−6.3), and one country, Panama (+4.4), recorded a rise larger than three per thousand.[8] Geographic coverage is clearly incomplete, but the results support other indications that the pace of urbanization is not quickening.

Many accounts leave the impression that rural–urban migration rates in developing countries, like birth-rates, are high and more or less uniform from country to country. This impression is decidedly false. Net out-migration rates from rural areas in fact have typically been higher in recent years in developed than in developing countries. The mean rural out-migration rate for the most recent intercensal period averaged 18.5 per thousand in the 20 developed countries and 13.7 per thousand in the 29 developing countries for which reasonably reliable measurement is possible. Within developing countries themselves the same tendency is evident: countries more advanced economically have experienced a more rapid flow from rural areas. Thus, Puerto Rico, Argentina, Chile, and Venezuela all had net rural out-migration rates of more than 25 per thousand annually, while Paraguay, Ghana, Guatemala, Bangladesh, India, and Indonesia had rates below eight per thousand. The simple correlation between rate of rural out-migration and gross national product per capita in the 29 developing countries is 0.61. In general, rural out-migration is fastest in countries whose economic performance allows the best possibilities for accommodating the exodus. This view contrasts with one in which absolute deprivation in rural areas, associated in part with rapid rural natural increase, is seen as the motive force driving multitudes to the city. Poorer countries in general have not only more deprived rural areas but also more deprived urban ones. The net effect of poverty seems to be to hold population in rural areas. (During the Great Depression in the United States, reverse net urban–rural migration actually took place.)[9] Obviously, it would be useful to have standardized measures of economic and social performance within urban and rural areas of these 29 countries; regrettably, such information is not available. It is worth noting that constant proportional differences between urban and rural incomes over the sample would suggest that rural–urban migration should be higher in richer countries. Rural–urban migration is reasonably viewed as an investment, the monetary returns from which depend on absolute income differences between the sectors; the same urban–rural income ratio would translate into higher absolute differences in the richer countries.

While the rate of urbanization (the rate of change in the urban proportion) has not been unprecedented in developing countries, the growth-rate of the urban population has been. Between 1875 and 1900, urban populations in now-developed countries grew by 100 per cent and rural populations by 18 per cent. While developing countries were traversing roughly the same range in urban proportions between 1950 and 1975, their urban populations grew by 188 per cent and their rural ones by 49 per cent. The growth factors of both rural and urban populations were much larger simply because rates of natural increase were much faster. Urban growth is currently exceptionally rapid in developing countries, but the explanation is not to be found in unusually rapid changes in the urban proportion produced by rural–urban migration but in the rapid changes in total population to which those proportions are applied.

2. Urban growth through most of the developing world results primarily from the natural increase of urban populations.

This point is readily overlooked in the midst of scholarly and political concern with internal migration. It has been made before by Kingsley Davis, Eduardo Arriaga, Salley Findley, and the United Nations Population Division, and new findings on components of urban growth provide strong confirmation. Of the 29 developing countries whose data support a decomposition of the sources of urban growth during the most recent intercensal period, 24 had faster rates of urban natural increase than of net in-migration (the latter also including area reclassification). The mean percentage of urban growth attributable to natural increase for the 29 countries was 60.7 per cent. Among the largest developing countries the percentage was 67.7 in India (1961–71), 64.3 in Indonesia (1961–71), and 55.1 in Brazil (1960–70). There is apparently a slight tendency for the percentage of urban growth attributable to natural increase to grow over time.

The list of five countries that are an exception to the rule is informative. One is Bangladesh, where international migration and population upheavals were substantial and where the very low urban proportion of 5.2 per cent gives an unstable base for computation. The remaining countries are Puerto Rico, South Korea, Turkey, and Argentina. This group has achieved much higher levels of income per capita than the average developing country and/or has made unusually rapid economic progress. This grouping is consistent with the tendency noted above for the richer developing countries to experience the fastest rates of rural–urban migration. For the remaining countries, where economic performance has been less satisfactory, around two-thirds of urban growth has resulted from natural increase. Thus, natural increase seems to be by far the largest source of urban growth in countries where rapid growth is most problematic. It should be noted that the coverage of African populations in the data set is very poor and that results pertain primarily to Latin America and Asia (except China). Judging from the unusually rapid urban growth in Africa, it is likely that rural–urban migration is a more important source of growth there than is implied by the above account.[10]

3. Among the factors that influence the growth rate of individual cities, national rates of population growth stand out as dominant in inter-city comparisons.

Many factors unique to a particular city have an important influence on its growth-rate: annexation practices, topography and geography, the health of industries in which the city specializes, productivity trends in the rural hinterland, government investment patterns and redistribution policies, rural–urban income and employment disparities, possibilities for accommodating marginal settlements, and so on. Despite the undoubted importance of these individual factors, it is possible to form some solidly based generalizations about more readily measured variables that discriminate among the growth-rates of individual cities.

The analysis reported below is based primarily upon an examination of growth-rates of the 1,212 cities in the world (excluding China) that had reached 100,000 in population at the earliest of the two most recent observations. In most cases, the observations derive from successive national population censuses. Where possible, an agglomeration definition of cities is used in preference to definitions based on administrative boundaries. Since concrete population estimates are used rather than interpolated or extrapolated figures, the dates of estimate vary somewhat from city to city. Typically, results are based upon growth-rates recorded between the 1960 and the 1970 rounds of population censuses. The mean date of the initial observation is 1962.

Four factors reflecting demographic, economic, and political variables are selected for examination of possible correlation with city growth-rates: the size of city and its administrative status; national rate of population growth; national economic level and growth-rates in terms of per capita gross domestic product; and region.

Size of City and its Administrative Status

Much attention has been drawn to the phenomenon of demographic giantism in recent patterns of city growth. Primate cities in developing countries are said to be drawing a disproportionate influx of population from other areas. Their rapid growth is alleged to result from biases in patterns of government expenditure and employment, in part resulting from the undue political influence of these agglomerations. Alan Gilbert points out that large capital cities contain relatively large numbers of government employees, who are often paid above prevailing market rates, and that they often enjoy disproportionate infrastructural investments as well.[11] In some cases the distortions are seen as a legacy of colonial penetration. Thus, Jorge Hardoy points to the extreme coastal concentration of large cities in Latin America as evidence of the distortions resulting from trade relationships with and natural resource exploitation by colonial powers.[12] Graeme Hugo notes that the Dutch had concentrated the administrative bureaucracy in Jakarta and that the centralization tendency was exacerbated under local rule.[13] Extensive treatments of urban bias and colonial exploitation as major factors in urban growth can be found in recent volumes by Michael Lipton and by Janet Abu-Lughod and Richard Hay.[14]

Many economists and regional scientists have emphasized a different set of factors to explain growth patterns by size of city. They point out that firms in large cities enjoy economies of agglomeration: economies of operation that are external to a firm but result from the presence of other firms and of social infrastructure. For example, a firm beginning operation in a large metropolis generally has access to a skilled labour force, banking and credit facilities, networks of buyers and sellers, and a large local market. Consumer agglomerative economies add variety and reduce the cost of consumer goods, while social agglomerative economies reflect efficiencies in providing public services to larger populations. Diseconomies of large size principally occur in the form of congestion and pollution and usually can be sloughed off on the society at large, thus reducing the disadvantages of large city size for firms making locational decisions.[15]

Most but not all of the evidence on agglomerative economies refer to developed countries. It suggests that substantial economies are typically realized by expansion of a city into the range of 100,000–300,000 in population. Beyond that point, social agglomerative economies show sharply diminishing returns except for some vertically integrated services such as water treatment and sewage disposal plants, pipelines, and electrical supply. Agglomerative economies in manufacturing seem to persist throughout the range of observation; productivity is higher in larger cities for reasons not readily explicable in terms of capital per worker or size of enterprise.[16]

Since the desirability of city expansion presumably depends upon the marginal increase in economies of agglomeration, and since there is weak evidence that the marginal gains decrease after a size of 100,000–300,000 is reached, such reasoning suggests that, above this level, city size should be negatively related to population growth. This presumes of course that population growth, and in particular migration, in responsive to the relative economic advantages of places and that these advantages are not totally overridden by biases in patterns of government expenditure and regulation. On the other hand, the political explanation of growth seems to suggest that growth-rates should increase with size of place since larger cities attract more than their 'share' of growth stimulants.

The actual relation between city size and city growth-rates is quite complex and seems to provide some support for both positions, which are of course not mutually exclusive. Table 1.1 displays the relation recently observed between city size and city growth-rates for the world and its major regions.[17] The general relation between city size and city growth-rates in developing regions is U-shaped. Cities in the two size classes between 100,000 and 500,000 are growing at an average of about 3.9 per cent annually. In the three size classes between 500,000 and 4 million, the average growth-rate has declined to 3.1–3.2 per cent. For the cities in developing regions larger than 4 million, growth-rates again reach the level of 3.9 per cent. There are only ten such cities, however—2 per cent of the total number of cities and about 11 per cent of the total urban population of developing countries—so that the predominant relation between city size and city growth-rates is negative. It is also negative, though somewhat irregular, in developed regions. For all cities, the correlation between growth-rates and the log of city size is a modest -0.083. These results are at least consistent with studies showing economic gains from city growth to decline after a size of 100,000–300,000 is reached.

If the slight negative association between city size and city growth comes as a surprise, the reason is probably that so many calculations of urban growth patterns present tabulations based not on individual cities but on size classes. Under the latter format, the set of cities within a particular size class changes over time as cities graduate into and out of the limits that define the class. Under conditions of rapid population growth, it typically happens that many cities graduate into the highest size class, while none devolve out of it. The result is that the highest size class of cities experiences by far the most rapid growth. For example, between 1965 and 1975 it is estimated that the population in cities over 4 million in developing regions grew from 55.9 to 120.6 million, or at the very rapid annual rate of 7.7 per cent. But almost half of this growth resulted from the fact that the number of cities in this class grew from nine to 17, so that 32 million were added to this size class through graduation.

The aggregate results seem to provide support primarily for the arguments regarding agglomerative economies. But the ten largest cities

TABLE 1.1 City Growth-Rates Between Latest Two Censuses as a Function of Size of City at First Census, by Region (Number of Cities in Parentheses)

Average annual intercensal growth-rate

Size of City	World	Developed Regions	Developing Regions	Africa	Latin America	North America	East Asia	South Asia	Europe	Oceania	Soviet Union
4 million +	0.0272	0.0155	0.0389	0.0266	0.0455	0.0170	0.0358	0.0295	0.0097	0.0	0.0085
	(20)	(10)	(10)	(1)	(4)	(3)	(5)	(2)	(4)	(0)	(1)
2–3.999 million	0.0235	0.0182	0.0320	0.0	0.0	0.0191	0.0173	0.0466	0.0174	0.0236	0.0101
	(31)	(19)	(12)	(0)	(0)	(6)	(6)	(6)	(10)	(2)	(1)
1–1.999 million	0.0205	0.0118	0.0308	0.0261	0.0373	0.0153	0.0232	0.0355	0.0074	0.0	0.0214
	(74)	(40)	(34)	(3)	(8)	(9)	(17)	(9)	(22)	(0)	(6)
500–999,000	0.0254	0.0213	0.0320	0.0342	0.0438	0.0270	0.0218	0.0380	0.0150	0.0247	0.0241
	(143)	(88)	(55)	(6)	(12)	(24)	(23)	(15)	(37)	(3)	(23)
250–499,000	0.0271	0.0181	0.0381	0.0445	0.0390	0.0242	0.0367	0.0340	0.0109	0.0319	0.0239
	(288)	(159)	(129)	(20)	(20)	(35)	(48)	(50)	(73)	(2)	(40)
100–249,000	0.0290	0.0223	0.0395	0.0470	0.0360	0.0202	0.0361	0.0370	0.0192	0.0142	0.0271
	(782)	(476)	(306)	(53)	(66)	(96)	(102)	(134)	(203)	(7)	(121)
All cities	0.0276	0.0206	0.0377	0.0445	0.0378	0.0216	0.0329	0.0365	0.0160	0.0203	0.0257
	(1338)	(792)	(546)	(83)	(110)	(173)	(201)	(216)	(349)	(14)	(192)

do seem to have more rapid growth than could be expected on this basis alone, suggesting that political factors may be influential. Procedures more directly tailored to testing the importance of political factors seem to provide stronger support for their influence. Intercensal growth-rates in the 1,212 cities were compared with a variety of demographic, economic, and political indicators pertaining to the initial census date or to the intercensal period. The independent contribution of variables was computed by means of multiple regression, with intercensal population growth rates of cities used as the dependent variable. Results are presented in Table 1.2.

The basic relation between city size and city growth-rates remains negative when other variables are introduced. When initial size of city increases by a factor of four above 100,000, annual city growth-rates decline on average by about three per thousand. However, there is evidence that, apart from the absolute size of a city, its position in a country's urban hierarchy influences its growth-rate. In particular, national capitals grow at an average annual rate of 0.6 per cent, or six per thousand,

TABLE 1.2   Effect of Demographic, Economic, and Political Variables on Intercensal Growth-Rates of 1,212 Cities (in Annual Percentage Growth-Rate)

Variable	Unit of Measurement	Effect of one unit increase in variable on city growth-rate[b]
Demographic		
National intercensal population growth-rate	Annual percentage growth	1.002[a]
Natural log, initial city size	Persons	−0.211[a]
Initial proportion urban	Urban percentage	−0.029[a]
Economic		
Initial level of national GNP per capita	Thousands of US 1964 dollars	0.332[a]
Intercensal growth rate of national GDP per capita	Annual percentage growth	0.239[a]
Political		
Capital city	1 if capital city; 0 otherwise	0.589
Largest city	1 if largest city in country; 0 otherwise	0.292
Regional		
Latin America	1 if in Latin America; 0 otherwise	0.614[a]
Asia	1 if in Asia; 0 otherwise	−0.223
Africa	1 if in Africa; 0 otherwise	−0.025

[a] F value significant at 5 per cent.
[b] Partial regression coefficients. The constant term is 4.119. R^2 is 0.312.

faster than would otherwise be expected. And if a city is the largest in a country, it grows at an annual rate of three per thousand faster than otherwise would be expected. Although neither of these variables is statistically significant, the reason is that the groups of largest cities and of capitals overlap so substantially that the independent contribution of either is quite limited. However, as a *pair* they do make a statistically significant contribution to explaining city growth. To the degree that discrimination between them is possible, being a capital city seems to impart more growth momentum than being the largest city.

These results are thus consistent with the view that spatial patterns of government expenditure bias patterns of city growth towards capital cities and towards the largest city in a country. Whatever economic advantages pertain to size of city should be captured in the variable directly measuring its size. But being the largest city in a country or a capital confers a sizeable additional growth increment. However, it would seem unwise to over-emphasize a political explanation of city growth in view of the small number of cities that fall into these categories. Only 7 per cent of the cities over 100,000 are either capitals or largest cities. Adding the two variables to the equation increases explained variance in city growth-rates by only 1.5 per cent. These cities seem to have attracted an undue amount of attention, perhaps because they are centres of communication and gathering places for intelligentsia. Many other factors can be shown to influence city growth-rates and pertain to a wider set of cities.

National Rate of Population Growth

The rate of population growth in the nation in which a city is located has a powerful effect on a city's growth-rate. The simple zero-order correlation between growth-rates in these 1,212 cities and rates of population growth in their respective nations is +0.516. Thus, of the 31.2 per cent of variance explained in city growth rates by all of the factors introduced, fully 85 per cent is accounted for by national population growth-rates alone $[(0.516)^2/0.312]$.[18] An appreciation for the dominance of this factor can be gained from a simple graph of city growth-rates by city size and by national population growth-rates, presented in Fig. 1.1. Without exception, cities in a particular size class experience faster average growth as their country's population growth-rate increases in increments of 1 per cent. All of the categories in which average city growth-rates exceed 4.5 per cent occur within the group of nations in which population growth exceeds 3 per cent. Within a particular category of national growth-rates, the relation between city size and city growth tends to be flat and somewhat irregular, certainly not as systematic as the relation with national population growth-rates.

Regression results in Table 1.2 suggest that, other things being equal, an

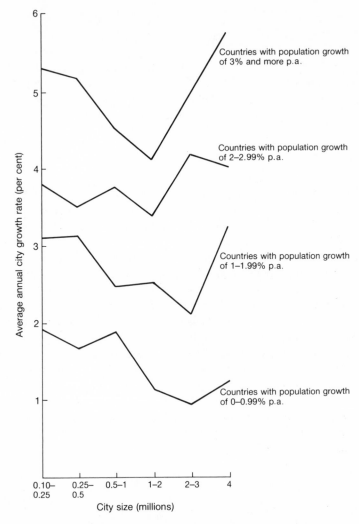

Fig. 1.1 Average annual city growth-rates between latest two censuses as a function of city size

increment of 1 per cent in national population growth-rates is associated with an increment of 1.002 per cent in city growth-rates. Nothing could indicate more clearly that cities draw from the same sources of growth as nations. Although the relationship is hardly surprising, it is worth noting that many mechanisms could have resulted in a different relationship. If high rates of natural increase were propelling rural people to cities, the coefficient of national population growth-rates would be expected to exceed unity. If rapid natural increase in cities made them inefficiently large, the opposite result might be expected. Policies to affect rates of natural increase that were more effective in urban than rural areas would produce a coefficient exceeding unity; if they were more effective in rural areas, it would fall short. Instead, the coefficient is what one would expect if changes in national growth-rates were associated in precisely equal measure with changes in city growth-rates.

National Economic Level and Growth-rates

Other things being equal, nations at higher levels of GNP per capita and with faster rates of economic growth have faster-growing cities. According to Table 1.2, a gain of $1,000 in gross national product per capita is associated with a rise of 3.3 per thousand in annual rates of city growth, and 1 per cent faster annual growth in gross domestic product is associated with a gain of 2.4 per thousand in city growth-rates. These results thus support those cited above—based on a completely different data set and estimation procedure—that suggest rural–urban migration is faster in countries at higher economic levels and with faster economic growth. The positive association between city growth and national economic growth cannot be unambiguously interpreted since city growth could contribute to, as well as result from, more rapid economic growth. However, it is unlikely that the average city is large enough for its population growth to contribute substantially to measured national economic growth during an intercensal period; rather, the lines of causation presumably run predominantly from economic growth to city population growth. Once again, the results imply that city growth is most rapid in the countries with the strongest economies.

Region

The 104 Latin American cities larger than 100,000 in the data set are growing faster, on average, than could be expected on the basis of national population growth-rates or any of the other variables considered. The difference amounts to an average excess in growth of about six per thousand annually. Most other evidence on urban patterns also shows Latin America to be deviant. As indicated in Table 1.1, city growth-rates in Latin America tend to increase with size of city, contrary to relations that generally prevail elsewhere. This pattern is reinforcing a pre-existing tendency for Latin America to have a more top-heavy size distribution of cities than other regions. Furthermore, occupational structures in Latin American urban and rural areas differ from norms established elsewhere. In particular, non-agricultural activities in Latin America are unusually highly concentrated in urban areas. The concentration of manufacturing and service occupations is some 8–14 percentage points higher in urban areas than would be expected based on the percentage of total labour force in agriculture. As a result, rural areas in Latin America are to an unusual extent agricultural enclaves. This concentration may be related to urban growth patterns in the sense that the rising fractions in non-agricultural activity that normally accompany development are disproportionately absorbed by urban areas. Many factors could probably be invoked to account for the shortage of rural non-agricultural activity: land tenancy systems that drain off agricultural profits into cities; better transportation and communication networks than in other developing regions; and so on. The basic point to stress here is that the urbanization process is caused by a multitude of factors operating in each country and each city. Certain of these factors are shared widely enough for them to be identified through the use of global data. Others are evidently widely shared by Latin American cities only; still others—accounting for around 69 per cent of the variance in city growth-rates—can only be identified at a lower level of aggregation.

4. Urban growth in developing countries has typically not been associated with a deterioration in industry/urban ratios.

One of the key arguments underpinning the notion that urbanization in developing countries is abnormal is that their urban populations are 'supported' by an unusually small industrial labour force. This point was solidly established in the 1950s by Bert Hoselitz, who compared current industry–urban relations in developing countries to those in now-developed countries during the late nineteenth and early twentieth centuries.[19] This was an alarming observation to many, largely because industrial activity, producing tangible output, was seen as being in some sense more 'productive' than the service occupations that were tending to substitute for industrial ones. The evident bias against services is a bit odd in view of the fact that the vanguard of most-developed countries has for some time been tracing out a definition of development in which services play the dominant role. It is not altogether clear why developing countries should aspire to the nineteenth-century rather than to the twentieth-century European model. That the nineteenth-century European model was itself anomalous with respect to industry–urban relations is suggested not only by twentieth-century developments in developing and more developed countries alike but also by descriptions of pre-industrial cities as locuses mainly for administrative, religious, military, and commercial activities, rather than for manufacturing.[20]

In any event, it is worthwhile examining whether the 'overurbanization' tendencies as denoted for years around 1950 have persisted since that time. An efficient way to acquire some sense of the trends is to compare regional estimates of labour force structure compiled by the International Labour Organisation with estimates of urban–rural distributions prepared by the United Nations Population Division. Although both sets of estimates are built up from national census publications, they are independent in technique. The comparison will again exclude China and will compare 1950 relations with those in 1970, beyond which date estimates for many countries are more or less arbitrary extrapolations.

Figure 1.2 plots the urban population percentage against the industrial percentage of the labour force for each region in 1950 and 1970. The two observations for a particular region are connected by a line, with the 1950 observation always appearing to the left. It is clear that most of the points fall close to a line through the origin with a slope of 1/2 (each increment in urban percentage being matched by an increment of 1/2 in the industrial percentage). For the world as a whole (excluding China), the ratio of

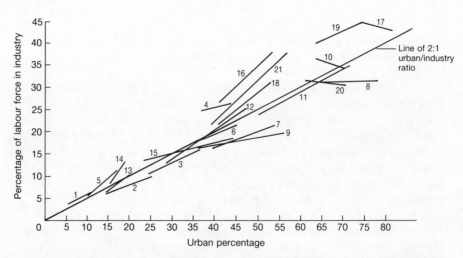

Fig. 1.2 Relationship between urban percentage and percentage of labour force in industry, 1950 and 1970. (Lines connect observations for 1950 and 1970 in a particular region.)

1. Eastern Africa	8. Temperate South America	15. Western South Asia
2. Middle Africa	9. Tropical South America	16. Eastern Europe
3. Northern Africa	10. Northern America	17. Northern Europe
4. Southern Africa	11. Japan	18. Southern Europe
5. Western Africa	12. Other East Asia	19. Western Europe
6. Caribbean	(excluding China)	20. Oceania
7. Middle America	13. Eastern South Asia	21. USSR
	14. Middle South Asia	

industrial to urban percentages was 0.552 in 1950 and 0.578 in 1970. If anything, this small change suggests a reversal of the overurbanization tendency. The degree to which regional trends adhere to the common 1/2 ratio can be inferred from the slope of their lines. Most of the more developed regions display a lower and even negative slope, reflecting the emergence of service-dominated economies. Southern and Eastern Europe and the Soviet Union experienced a sharp rise in the industry/urban ratio during the period, which in the latter two instances is said to reflect a development strategy that attempts to economize on investible resources by restraining the growth of urban populations with their higher consumption requirements.[21]

The largest of the developing regions, Middle South Asia, also experienced a rise (of 29 per cent) in the industry/urban ratio. This region, dominated by India with relatively good censuses, is in large part responsible for the estimated increase in this ratio for the world as a whole. Africa and the rest of Asia show no serious and certainly no consistent departure from the 1 : 2 ratio between 1950 and 1970. But, once again, Latin America appears to be an exception. Each of the four Latin American regions shows a decline in the industry/urban ratio between 1950 and 1970. The decline is especially marked in both temperate and tropical South America. As noted above, the industrial shortage in Latin America is more evident in rural than in urban areas, though it pertains to both.

The foregoing discussion relates to national and international data rather than to labour force structures in urban areas themselves. Data on urban labour force structures provide much less coverage, particularly for trends. Trend data on urban labour force structures have been examined for eight developing countries. Dates of censuses and the change recorded in the percentage of the urban labour force in clerical, sales, and traditional service occupations are the following: Puerto Rico, 1960–70 (−4.3); Costa Rica, 1963–73 (−1.9); Peru, 1961–72 (−1.4); Ecuador, 1962–74 (+0.6); Nicaragua, 1963–71 (0.0); Sri Lanka, 1953–70 (−11.6); Morocco, 1960–71 (−1.8); Thailand, 1954–79 (+5.0).[22] Needless to say, classification of the labour force is inexact, and changes in classification can create bogus trends for a particular country. Relatively uniform upward trends could support notions of service sector inflation; but for these countries, at least, no such tendency is evident. Where the urban service sector is rising in developing countries, it is typically a result of rising fractions of professional, technical, and administrative personnel. Each of the eight countries had a rise in the percentage in professional and managerial occupations, averaging 4 percentage points.

Other research has also questioned the assumption that rapid urban growth leads to an inflated urban service sector in developing countries. Because entry requirements in service jobs are typically less stringent than in industrial ones, it is alleged that the increment in labour supply will tend to be absorbed disproportionately into the service sector. Alan Udall reviews these arguments and finds them unconvincing.[23] Furthermore, he examines a 'natural experiment' in Colombia, where rural disturbances led to a rapid labour flow to Bogotá. The influx of workers, however, did not seem to depress the relative size of the manufacturing sector nor to inflate that of services. Instead, he argues that the distribution of workers among sectors is determined primarily by demand factors related to income growth and government policy. Dipak Mazumdar also questions the prevailing model, particularly the assumption that the service sector plays a predominant role as a point of entry into the labour force for migrants to urban areas.[24]

Discussion

This is not the place for a full-scale consideration of the desirability of reducing rates of urban growth and rural–urban migration. Some of the results reported here would seem to call for moderation of the sometimes frantic tone that such discussions assume. Urban growth has been fastest, other things being equal, where economic levels and economic growth-rates are highest; changes in urban proportions among developing regions are not outpacing historical standards; relations between urban and industrial populations do not seem to have deteriorated in the post-war period; urban growth is partly self-limiting since growth-rates of cities decline as their size increases and as urban proportions grow. Still there is no cause for complacency in these findings, both because the population shifts accompanying urbanization are being superimposed upon what remain very rapid rates of national increase and because the aggregate measures used in the analysis preclude consideration of a wealth of economic, social, and institutional factors that influence and are influenced by this phenomenon. It would be foolish to make general statements about the advantages or disadvantages of slowing urban growth, since these will obviously vary from place to place. In one city, expansion may be very costly for topographical reasons, or it may overtax existing municipal services; in another, expansion may result in economies of agglomeration or facilitate transportation linkages with other cities.

In part, the desirability of slowing urban growth depends on the costs attached to different policies for doing so. In this regard, it is worth emphasizing that urban growth can be strongly influenced by policies affecting rates of natural increase, as well as by policies to influence migration. There is strong evidence that declines in national rates of natural increase tend to be matched one-for-one by declines in rates of urban growth. In most countries, natural increase accounts for the bulk of urban growth. But rates of natural increase or net migration must themselves be dealt with through specific policy measures. A common typology of such measures as they bear on natural increase is also applicable to migration. Migration can be influenced by providing information or services that allow individuals more effectively to exercise their choices; by changing individuals' incentives to move; by 'restructuring' development; or by coercive measures such as identity cards or physical barriers.

The 'free choice' option would seem more likely to hasten than to slow urbanward migration. Urban incomes are practically always higher than rural ones. Migrants to cities in general seem to fare well with respect to acquiring jobs and improving their standards of living. Families that send migrants to the city in Africa and Asia typically enjoy a stream of remittances that enhance their own living standards.[25] Urban standards of public services usually exceed rural ones. Providing information on these matters, along with transportation services to allow people to act on the information, is not likely to slow urban growth. For those whose approach to population policy emphasizes the enhancement of free choice in the prevailing social context, policies to affect natural increase by providing family planning services would seem to provide the most palatable means of reducing urban growth.[26] Oddly, family planning services are rarely seen as a candidate for slowing urban growth, which probably reflects an artificial but well-entrenched distinction between population growth and population distribution policies.

Coercive measures doubtless have been and will continue to be effective in slowing migration to urban areas, and they admit to more effective enforcement than is the case with antinatalist coercion.[27] But most governments find them intrinsically offensive. Tinkering with individuals' incentives to move would seem to be administratively difficult and of questionable effectiveness, apart from making the more fundamental changes that would constitute a restructuring of development patterns.

There are many possible ways of restructuring development to influence urban growth. We have seen at least indirect evidence that government biases towards capital and largest cities promote unusually rapid population growth therein. Redressing the inequities that give rise to these growth imbalances is surely a praiseworthy goal from many points of view. But only a small minority of cities would be affected by policy revisions directed at capitals and largest cities. More general changes likely to affect migration are the promotion of rural development efforts and the provision of basic needs—health, education, food, and so on—to all of the population, rural and urban alike. These efforts are still in their infancy in most parts of the world, and adequate evidence of their impact on rural–urban migration is not available.

While it would seem that improved living standards in rural areas would serve to increase their relative attractiveness and to restrain the flow of rural–urban migration, there are some reasons to doubt that they would act very powerfully in this direction. It is a common observation that rural out-migration probabilities are higher among those with higher educational attainments, no doubt reflecting the greater returns to urban residence for the educated than for the uneducated.[28] However, improving schooling opportunities in rural areas very likely would dampen the flow of individuals (and their families) who migrate to urban areas specifically to acquire more education. Thus, the age pattern of migration may change more than its level as a result of programmes to improve rural education. Rural health advances will almost certainly accelerate rural population growth and, where land tenure systems are not absorbent, they are likely to speed out-migration. Perhaps most important is the impact of advances in agricultural productivity on employment prospects for agricultural labour. Effects on migration are likely to vary somewhat according to whether the advances are labour-saving or labour-using and with elasticities of demand for the product in question. A recent review cites evidence that improvements of labour productivity in the production of basic foodstuffs are likely to accelerate rural out-migration because of deterioration produced in the terms of trade for those products.[29] The most promising avenue of rural development for stemming out-migration seems to be one of increasing the returns to farmers from the production of export crops, either through productivity advances or through elimination of discriminatory agricultural taxes. Focusing on export crops averts the urbanizing influences that arise from inelastic internal demands for agricultural products.

These options do not exhaust the possibilities for rural development activities. Developing small-scale rural industries, opening new lands to agricultural settlement, altering terms of trade between rural and urban areas, improving rural credit and marketing facilities, and many other possibilities exist. That the bulk of population in developing countries resides in rural areas and will continue to do so for at least a generation is surely sufficient reason for focusing development activities and plans on this sector. But the history of developed countries, contemporary relations among developing countries, and evidence on agglomerative economies suggest that success in these development enterprises will ultimately be registered not by rural population retained but by rural population released. If a recession in rates of urban population growth ranks high on the list of development objectives—and its placement requires more careful analysis than it has usually received—then it seems important to recognize the central role of natural increase in current levels of and variations in urban growth-rates. It is conceivable, for example, that many rural development activities will depress urban growth more through their impact on rural fertility than through their impact on rates of rural out-migration.

Notes

1. Data are drawn from United Nations Economic and Social Council, Population Commission, Twentieth Session, *Concise Report on Monitoring of Population Policies*. E/CN.9/338. 22 Dec. 27–8.
2. United Nations, *Patterns of Urban and Rural Population Growth* (New York: United Nations, 1980).
3. Data for all countries were processed in such a way that estimates pertain to a uniform sequence of dates: 1950, 1955 . . . 1975. Such estimates were made by interpolation wherever possible, under the assumption that growth-rate differences between urban and rural areas were constant during the period between two concrete urban estimates. When extrapolation was required, the procedure was to assume constant urban-rural growth differences at a level to that observed in the most nearly adjacent period during 1950–78.
4. This group consists of Africa, Asia except Japan, and Latin America. It does not include Turkey, Cyprus, or Israel.
5. John V. Graumann, 'Orders of Magnitude of the World's Urban and Rural Population in History'. *United Nations Population Bulletin* 8 (New York: United Nations, 1977), 16–33.
6. Kingsley Davis, 'Cities and Mortality', International Union for the Scientific Study of Population, International Population Conference (Liège: IUSSP, 1973), vol. 3, 259–82.
7. *Patterns of Urban and Rural Population Growth*.
8. These migration rates and those cited in the remainder of the paper include the element of reclassification. The population of areas reclassified from rural to urban during an intercensal period is unavoidably included among the urban in-migrants. Those reclassified may represent on average about a quarter of the total growth assigned to migration. (*Patterns of Urban and Rural Population Growth*).
9. Theodore W. Schultz, *Agriculture in an Unstable Economy* (New York: McGraw-Hill, 1945), 90.
10. However, for the three African countries where measurement of components of change was possible, natural increase contributed 57.6 per cent in Ghana (1960–70), 63.0 per cent in Morocco (1960–71), and 74.4 per cent in the Union of South Africa (1960–70). Since the rate of urban growth in Africa was about 4.9 per cent annually during 1960–75 and since a plausible estimate of urban natural increase is about half of that figure, it is reasonable to assume that perhaps half of African urban growth is accounted for by migration and reclassification of areas. Ita Ekanem suggests that migration may have accounted for slightly more than half of urban growth in Nigeria. 'The

Dynamics of Urban Growth: A Case Study of Medium-sized Towns of Nigeria', paper contributed to the Conference on Economic and Demographic Change; Issues for the 1980s, International Union for the Scientific Study of Population, Helsinki, Finland, 28 Aug.–1 Sept. 1978.

11. Alan Gilbert, 'The Argument for Very Large Cities Reconsidered', *Urban Studies*, 13 (1) (Feb. 1976), 27–34.

12. Jorge E. Hardoy, 'Potentials for Urban Absorption: The Latin American Experience', *Food Population, and Employment: The Impact of the Green Revolution* Thomas T. Poleman and Donald K. Freebairn (eds) (New York: Praeger, 1977).

13. Graeme J. Hugo, 'New conceptual approaches to migration in the context of urbanization: A discussion based on the Indonesian experience', in *Population Movements: Their Forms and Functions in Urbanization and Development*, ed. P. A. Morrison (Liège: Ordina Editions, 1983).

14. Michael Lipton, *Why Poor People Stay Poor: Urban Bias in World Development* (Cambridge, Mass.: Harvard University Press, 1977); Janet Abu-Laghod and Richard Hay, jun. (eds), *Third World Urbanization* (Chicago: Maaroufa Press, 1977).

15. For a recent review of agglomerative economies, see Gerald A. Carlino, *Economics of Scale in Manufacturing Location* (Leiden: Martinus Nijhoff, 1978).

16. Among the studies providing evidence on the size of agglomerative economies are Carlino, cited in n. 15; Leo Sveikauskas, 'The Productivity of Cities' *Quarterly Journal of Economics*, 89 (3) (Aug. 1975), 393–413; David Segal, 'Are there Returns to Scale in City Size?' *Review of Economics and Statistics*, 58 (3) (Aug. 1976), 339–50; Yngve Aberg, 'Regional Productivity Differences in Swedish Manufacturing', *Regional and Urban Economics*, 3 (2) (1973), 131–56; Koichi Mera, 'On the Urban Agglomeration and Economic Efficiency', *Economic Development and Cultural Change*, 21 (2) (Jan. 1973), 309–24; Stanford Research Institute, 'Costs of Urban Infrastructure for Industry as related to City Size: India Case Study', *Ekistics*, 20 (Nov. 1969), 316–20; C. A. Rocca, 'Productivity in Brazilian Manufacturing' in *Brazil: Industrialization and Trade Politicies*, ed. J. Bergsmann (London: Oxford University Press, 1970); Werner Z. Hirsch, 'The Supply of Urban Public Services', in *Issues in Urban Economies*, Harvey S. Perloff and Lowdon Wingo, jun. (eds) (Baltimore: Johns Hopkins University Press, 1968); and William A. Howard, 'City-Size and its Relationship to Municipal Efficiency: Some Observations and Questions', *Ekistics*, 20 (Nov. 1969), 312–16.

17. Chinese cities are included in this table, which accounts for the fact that the total number of cities examined is 1,338.

18. The F value of additional variance explained by the national population growth-rate after all other variables are in the equation is an extraordinarily high 105.2, which is three times higher than that of any other variable. A value of 6.64 is significant at 1 per cent.

19. Bert F. Hoselitz, 'Urbanization and Economic Growth in Asia', *Economic Development and Cultural Change*, 6 (Oct. 1957), 42–54; 'The Role of Cities in the Economic Growth of Underdeveloped Countries', *Journal of Political Economy* 61 (1953), 195–208.

20. Gideon Sjoberg, *The Preindustrial City: Past and Present* (Glencoe, Ill.: The Free Press, 1960).
21. Gur Ofer, 'Industrial Structure, Urbanization, and the Growth Strategy of Socialist Countries', *Quarterly Journal of Economics*, 10 (2) (May 1976), 219–44.
22. *Patterns of Urban and Rural Growth*. These data were assembled and processed under the direction of Jean Smith. Turkey is excluded from the present analysis since the proportion of the urban labour force with unknown occupations rose enormously. In the other countries, the absolute change in percentage with unknown occupations averaged 2.5 percentage points. The largest change occurred in Thailand, where unknowns declined by 6.2 percentage points. Interestingly, Thailand is the only country where the service group rose by more than 1 percentage point.
23. Alan T. Udall, 'The Effect of Rapid Increase in Labour Supply on Service Employment in Developing Countries', *Economic Development and Cultural Change*, 24 (4) (July 1976), 765–85.
24. Dipak Mazumdar, 'The Urban Informal Sector', *World Development* (Aug. 1976), 655–79.
25. For good reviews of these matters, see Lorene Y. L. Yap, 'The Attraction of Cities: A Review of Migration Literature', *Journal of Development Economics*, 4 (1971), 239–64; Salley Findley, *Planning for Internal Migration: A Review of Issues and Policies in Developing Countries* (Washington, DC: US Bureau of Census, 1977); Oded Stark, 'Rural-to-Urban Migration and Some Economic Issues: A Review utilizing Findings of Surveys and Empirical Studies covering the 1965–1975 Period', International Labour Organization, World Employment Programme Working Paper WEP/2–21-WP38, May 1976.
26. The most thorough recent review of the effect of family planning programmes on fertility is W. Parker Mauldin and Bernard Berelson, 'Conditions of Fertility Decline in Developing Countries, 1965–75, *Studies in Family Planning*, 9 (5) (May 1978), 89–147. The difficulty of inferring the role of family planning programmes in fertility declines is reflected in the fact that the 11 countries ranked as having the strongest 'programme effort' had an average crude birth-rate in 1965 of only 33.7 per thousand. Such low rates, preceding in most cases vigorous family planning efforts, indicate that strong programme effort is usually conditional upon high levels of pre-existing motivation. Furthermore, the highest figure entering the average crude birth-rate calculation is 42 for North Vietnam, where little is known about the demographic situation.
27. For example, see Brian Berry's discussion of the 'influx control' urban migration policy for Africans in South Africa, in *The Human Consequences of Urbanization* (London: Macmillan, 1973), ch. 4.
28. Derek Byerlee cites evidence from Tanzania and Kenya suggesting that returns to education are much higher in urban than rural areas. 'Rural–urban Migration in Africa: Theory, Policy and Research Implications', *International Migration Review*, 7 (4) (1974), 543–66.
29. Jacques Gaude, 'Causes and Repercussions of Rural Migration in Developing Countries: A Critical Analysis', International Labour Organization, World Employment Programme Working Paper WEP/10–6/WP10, Oct. 1976.

II

The Neglect and Exploitation of the Rural Masses

INTRODUCTION

To understand rural–urban migration—and the rapid urban growth to which it makes a major contribution—we must gain an appreciation of the relationship between the urban and the rural sector within the political economy of Third World countries. In 1967 the leaders of the Tanganyika African National Union (TANU), the country's only political party, assembled in a town by the name of Arusha. They discussed a document that came to be referred to as the Arusha Declaration. Originally published in Swahili, the national language of Tanzania, it appeared in a revised English translation in a collection of essays, *Ujamaa: Essays on Socialism*, by Julius Nyerere, then the country's president. Chapter 2 presents a short statement from the Arusha Declaration warning that the real exploitation in a poor, largely rural country such as Tanzania might become that of the town-dwellers exploiting the peasants. We follow this in Chapter 3 with the introduction from a large volume published ten years later. Michael Lipton entitled it *Why Poor People Stay Poor: Urban Bias in World Development*. In it Lipton set out to demonstrate that the most important class conflict in the poor countries of the world today is between the rural classes and the urban classes.

Lipton argued that 'urban bias', a phrase he coined, provides the main explanation 'why poor people stay poor'. His approach to economic development and social conflict thus specifically relates to the rural–urban divide and, as we shall see, the rural–urban migration engendered by rural–urban inequality and hence the rapid pace of urban growth. Lipton's challenge to the various schools of thought about Third World development provoked a wide range of critical responses, as well as some empirical work to test it. Recently, Lipton (1984) has provided a comprehensive account of these criticisms and answered them in considerable detail. An approach that posits that the most important distinction is between urban and rural dwellers, has to deal in particular with two types of objection. First, any attempt to subsume the diversity of a society into just two categories is altogether too sweeping. Lipton actually qualifies his initial categorization by suggesting that large farmers, or more broadly, rural élites, are favoured by urban-biased policies and align with urban interests (Lipton 1984, 1952–4). He further affirms that much rural–urban migration does not represent a real urban commitment, either because of its temporary character, or in other respects that remain unspecified (Lipton 1977, 216). An argument over what constitutes 'real commitment', is bound to be fruitless. But the suggestion that many migrants thus remain part of the rural sector helps to differentiate Lipton's categorization. Second, other distinctions have commonly been held to be more important in the analysis of class conflict, whether in the classic definition as the opposition between the owners of the means of production and the proletariat, or in a contemporary variant such as an alliance of the indigenous bourgeoisie and foreign interests exploiting the workers and peasants. Ashok Mitra (1977) addressed both concerns when he distinguished the rich peasantry from small farmers and farm workers in agriculture, and the bourgeoisie from the working class outside agriculture. He analysed the allocation of resources in India in terms of an alliance between the industrial bourgeoisie and a rural oligarchy that is able to control much of the rural vote.

Lipton's concern with urban bias grew out of his work in India. However, as he notes, his case is stronger for many other Third World countries. The crisis in African development, where 15 countries recorded declining per capita incomes in the 1970s, prompted the World Bank's report *Accelerated Development in Sub-Saharan Africa*. It found that the growth-rate of agricultural production had begun to decline in the 1960s and was less than the rate of population growth almost everywhere in the 1970s; if food production per capita had been stagnant in the 1960s at best, it had actually fallen in the 1970s. While many factors can be adduced to explain this dismal performance, the report emphasized that there has been a consistent bias against agriculture in government policies: producers of export crops receive only a fraction of the world market price because of export taxes, marketing board levies, excessive marketing costs, and overvalued exchange rates; producers of food crops are affected by policies that fix producer prices below market levels, sell imported food below cost, encourage food imports when domestic food prices rise, and give implicit subsidy to imported foods because of currency overvaluation. For a comprehensive analysis, see Bates (1981).

Socialist countries are expressly committed to foster equality. They invariably reduce income inequality within the urban and the rural sector respectively, but they also appear to find it difficult to deal with inter-sector inequality. In China, personal consumption has been estimated at 574 yuan in urban as against 269 yuan per capita in rural areas in 1982. The ratio of 2.1 : 1 suggests a dramatic improvement over a ratio of 2.5 : 1 in 1981, but is quite similar to that of other developing countries in Asia. The gap in personal income is accompanied by a large gap in collective consumption. The quality of education and health facilities in particular is much higher in urban areas (World Bank 1985, 87 and 90; see also Nolan and White 1984, and Selden 1988). The rural–urban migration that these inequalities can be expected to foster has, however, been restricted—as Blecher spells out in Chapter 7.

How did the peasantry fare in Tanzania where a commitment to socialism had been made and where the Arusha Declaration emphasized the threat of the town-dwellers exploiting the peasants? Urban capitalists and rich farmers alike were virtually eliminated, and the state apparatus was decentralized into small towns and villages. Urban wages were kept down, and industrial prices were controlled. Peasants found, however, that their share of the final sales value of their crops—which they had to sell to parastatal marketing agencies—dropped from 66 per cent in 1970 to 42 per cent in 1980. Ellis (1984) concludes that the increased transfer out of agriculture was required by a proliferating, unproductive state and parastatal bureaucracy.

Confronted with neglect, if not outright exploitation, peasants have three options. They can rebel, like the farmers in the south-west of Nigeria who organized the *agbekoya* ('farmers shall not suffer') rebellion in 1968–9 (Beer 1976). Such rebellions have difficulty reaching across large rural areas, at best obtain limited concessions, and cannot alter the fact that power is urban-based. In the past, when peasants seized power in a revolution, they could not hold on to it, as Emiliano Zapata and his followers found out when they marched into Mexico City in 1915. And in the contemporary world, peasants are no longer in a position to make revolutions—or so we shall argue in Chapter 21. Alternatively, peasants can withdraw from the market and revert to subsistence farming (Hyden 1980 and 1983; but see also Kasfir 1986). This may well force policy-makers to seek an accommodation so as to stimulate food supplies for the cities and the production of export crops—in other words, to reduce urban bias, but it will not erase urban bias. Finally, peasants can opt to join the urban sector by moving to a city. To this option we will turn in Part III.

References

Bates, Robert (1981) *Markets and States in Tropical Africa* (Berkeley, Calif.: University of California Press).

Beer, Christopher (1976) *The Politics of Peasant Groups in Western Nigeria.* Ibadan Social Science Series 7 (Ibadan: Ibadan University Press).

Ellis, Frank (1984) 'Relative Agricultural Prices and the Urban Bias Model: A Comparative Analysis of Tanzania and Fiji', *Journal of Development Studies*, 20 (3), 28–51.

Hyden, Goran (1980) *Beyond Ujamaa in Tanzania: Underdevelopment and an Uncaptured Peasantry* (London: Heinemann; Berkeley, Calif.: University of California Press).

—— (1983) *No Shortcuts to Progress: African Development Management in Perspective* (London: Heinemann; Berkeley, Calif.: University of California Press).

Kasfir, Nelson (1986) Are African Peasants Self-Sufficient? *Development and Change*, 17, 335–57.

Lipton, Michael (1977) *Why Poor People Stay Poor: Urban Bias in World Development* (London: Maurice Temple Smith; Cambridge, Mass.: Harvard University Press).

—— (1984) 'Urban Bias Revisited', *Journal of Development Studies*, 20 (3), 139–66.

Mitra, Ashok (1977) *Terms of Trade and Class Relations: An Essay in Political Economy* (London: Frank Cass; Calcutta, Allahabad, Bombay, and New Delhi: Rupa).

Nolan, Peter, and White, Gordon (1984) 'Urban Bias, Rural Bias or State Bias? Urban–rural Relations in Post-revolutionary China', *Journal of Development Studies*, 20 (3), 52–81.

Selden, Mark (1988) 'Sector, Stratification and the Transformation of China's Social Structure' in Victor Nee, Mark Selden and David Stark (eds), *The Political Economy of Contemporary China* (White Plains, NY: M. E. Sharpe).

World Bank (1981) *Accelerated Development in Sub-Saharan Africa: An Agenda for Action* (Washington, DC: World Bank).

—— (1985) *China: Long-term Development Issues and Options* (Baltimore and London: Johns Hopkins University Press).

2

Let Us Pay Heed to
the Peasant*

Julius K. Nyerere

OUR emphasis on money and industries has made us concentrate on urban development. We recognize that we do not have enough money to bring the kind of development to each village which would benefit everybody. We also know that we cannot establish an industry in each village and through this means effect a rise in the real incomes of the people. For these reasons we spend most of our money in the urban areas and our industries are established in the towns.

Yet the greater part of this money that we spend in the towns comes from loans. Whether it is used to build schools, hospitals, houses, or factories etc., it still has to be repaid. But it is obvious that it cannot be repaid just out of money obtained from urban and industrial development. To repay the loans we have to use foreign currency which is obtained from the sale of our exports. But we do not now sell our industrial products in foreign markets, and indeed it is likely to be a long time before our industries produce for export. The main aim of our new industries is 'import substitution', that is to produce things which up to now we have had to import from foreign countries.

It is therefore obvious that the foreign currency we shall use to pay back the loans used in the development of the urban areas will not come from the towns or the industries. Where, then, shall we get it from? We shall get it from the villages and from agriculture. What does this mean? It means that the people who benefit directly from development which is brought about by borrowed money are not the ones who will repay the loans. The largest proportion of the loans will be spent in, or for, the urban areas, but the largest proportion of the repayment will be made through the efforts of the farmers.

This fact should always be borne in mind, for there are various forms of

* From the *Arusha Declaration*, published by TANU, the national party of Tanzania, in Swahili in 1967, and reprinted from the English version that appeared in Julius K. Nyerere, *Ujamaa: Essays on Socialism* (London, Oxford, and New York: Oxford University Press, 1968). Copyright Julius K. Nyerere, by permission.

nust not forget that people who live in towns can possibly
ʒiters of those who live in the rural areas. All our big
ɪwns and they benefit only a small section of the people of
ve have built them with loans from outside Tanzania, it is
ɪ of the peasants' produce which provides the foreign
ɪyment. Those who do not get the benefit of the hospitals
ɪjor responsibility for paying for them. Tarmac roads, too,
are ɪ.   d in towns and are of especial value to the motor-car
owners. Yet if we have built those roads with loans, it is again the farmer
who produces the goods which will pay for them. What is more, the foreign
exchange with which the car was bought also came from the sale of the
farmers' produce. Again, electric lights, water-pipes, hotels, and other
aspects of modern development are mostly found in towns. Most of them
have been built with loans, and most of them do not benefit the farmer
directly, although they will be paid for by the foreign exchange earned by
the sale of his produce. We should always bear this in mind.

Although when we talk of exploitation we usually think of capitalists, we
should not forget that there are many fish in the sea. They eat each other.
The large ones eat the small ones, and small ones eat those who are even
smaller. There are two possible ways of dividing the people in our country.
We can put the capitalists and feudalists on one side, and the farmers and
workers on the other. But we can also divide the people into urban
dwellers on one side and those who live in the rural areas on the other. If
we are not careful we might get to the position where the real exploitation
in Tanzania is that of the town dwellers exploiting the peasants.

3

Why Poor People Stay Poor:
Urban Bias in World Development*

Michael Lipton

THE most important class conflict in the poor countries of the world today is not between labour and capital. Nor is it between foreign and national interests. It is between the rural classes and the urban classes. The rural sector contains most of the poverty, and most of the low-cost sources of potential advance; but the urban sector contains most of the articulateness, organization, and power. So the urban classes have been able to 'win' most of the rounds of the struggle with the countryside; but in so doing they have made the development process needlessly slow and unfair. Scarce land, which might grow millets and bean sprouts for hungry villagers, instead produces a trickle of costly calories from meat and milk, which few except the urban rich (who have ample protein anyway) can afford. Scarce investment, instead of going into water-pumps to grow rice, is wasted on urban motorways. Scarce human skills design and administer, not clean village-wells and agricultural extension services, but world boxing championships in show-piece stadia. Resource allocations, within the city and the village as well as between them, reflect urban priorities rather than equity or efficiency. The damage has been increased by misguided ideological imports, liberal and Marxian, and by the town's success in buying off part of the rural élite, thus transferring most of the costs of the process to the rural poor.

But is this urban bias really damaging? After all, since 1945 output per person in the poor countries has doubled; and this unprecedented growth has brought genuine development. Production has been made more scientific: in agriculture, by the irrigation of large areas, and more recently by the increasing adoption of fertilizers and of high-yielding varieties of wheat and rice; in industry, by the replacement of fatiguing and repetitive effort by rising levels of technology, specialization and skills. Consumption

* The introductory chapter of Michael Lipton, *Why Poor People Stay Poor: Urban Bias in World Development* (London: Maurice Temple Smith; Cambridge, Mass.: Harvard University Press, 1977), reprinted by permission of the publishers. References to other parts of the book have been deleted.

has also developed, in ways that at once use and underpin the development of production; for poor countries now consume enormously expanded provisions of health and education, roads and electricity, radios and bicycles. Why, then, are so many of those involved in the development of the Third World—politicians and administrators, planners and scholars—miserable about the past and gloomy about the future? Why was the United Nations' 'Development Decade' of the 1960s, in which poor countries as a whole exceeded the growth target,[1] generally written off as a failure? Why is aid, which demonstrably contributes to a development effort apparently so promising in global terms, in accelerating decline and threatened by a 'crisis of will' in donor countries?[2]

The reason is that since 1945 growth and development in most poor countries, have done so little to raise the living standards of the poorest people. It is scant comfort that today's mass-consumption economies, in Europe and North America, also featured near-stagnant mass welfare in the early phases of their economic modernization. Unlike today's poor countries, they carried in their early development the seeds of mass consumption later on. They were massively installing extra capacity to supply their people with simple goods: bread, cloth, and coal, not just luxury housing, poultry, and airports. Also the nineteenth-century 'developing countries', including Russia, were developing not just market requirements but class structures that practically guaranteed subsequent 'trickling down' of benefits. The workers even proved able to raise their *share* of political power and economic welfare. The very preconditions for such trends are absent in most of today's developing countries. The sincere egalitarian rhetoric of, say, Mrs Gandhi or Julius Nyerere was—allowing for differences of style and ideology—closely paralleled in Europe during early industrial development: in Britain, for example, by Brougham and Durham in the 1830s.[3] But the rural masses of India and Tanzania, unlike the urban masses of Melbourne's Britain, lack the power to organize the pressure that alone can turn such rhetoric into distributive action against the pressure of the élite.

Some rather surprising people have taken alarm at the persistently unequal nature of recent development. Aid donors are substantially motivated by foreign-policy concerns for the stability of recipient governments; development banks, by the need to repay depositors and hence to ensure a good return on the projects they support. Both concerns coalesce in the World Bank, which raises and distributes some £3,000 million of aid each year. As a bank it has advocated—and financed—mostly 'bankable' (that is, commercially profitable) projects. As a channel for aid donors, it has concentrated on poor countries that are relatively 'open' to investment, trade, and economic advice from those donors. Yet the effect of stagnant mass welfare in poor countries, on the well-intentioned and perceptive people who administer World Bank aid, has gradually overborne these traditional biases. Between 1971 and 1980 the president of the World Bank, Robert McNamara, in a series of speeches focused attention on the stagnant, or worsening lives of the bottom 40 per cent of people in poor countries.[4] After 1975 this began to affect the World Bank's projects, though its incomplete engagement with the problem of urban bias restricts the impact. For instance, an urban-biased government will prepare rural projects less well than urban projects, will manipulate prices to render rural projects less apparently profitable (and hence less 'bankable') and will tend to cut down its own effort if donors step up theirs. Nevertheless, the World Bank's new concern with the 'bottom 40 per cent' is significant.

These people—between one-quarter and one-fifth of the people of the world—are overwhelmingly rural: landless labourers, or farmers with no more than an acre or two, who must supplement their income by wage labour. Most of these countryfolk rely, as hitherto, on agriculture lacking irrigation or fertilizers or even iron tools. Hence they are so badly fed that they cannot work efficiently, and in many cases are unable to feed their infants well enough to prevent physical stunting and perhaps even brain damage. Apart from the rote-learning of religious texts, few of them receive any schooling. One in four dies before the age of ten. The rest live the same overworked underfed, ignorant, and disease-ridden lives as 30, or 300, or 3,000 years ago. Often they borrow (at 40 per cent or more yearly interest) from the same moneylender families as their ancestors, and surrender half their crops to the same families of landlords. Yet the last 30 years have been the age of unprecedented, accelerating growth and development! Naturally men of goodwill are puzzled and alarmed.

How can accelerated growth and development, in an era of rapidly improving communications and of 'mass politics', produce so little for poor people? It is too simple to blame the familiar scapegoats—foreign exploiters and domestic capitalists. Poor countries where they are relatively unimportant have experienced the paradox just as much as others. Nor, apparently, do the poorest families cause their own difficulties, whether by rapid population growth or by lack of drive. Poor families do tend to have more children than rich families, but principally because their higher death-rates require it, if the ageing parents are to be reasonably sure that a son will grow up, to support them if need be. And it is the structure of rewards and opportunities within poor countries that extracts, as if by force, the young man of ability and energy from his chronically stagnant rural background and lures him to serve, or even to join, the booming urban élite.

The disparity between urban and rural welfare is much greater in poor countries now than it was in rich countries during their early development. This huge welfare gap is demonstrably inefficient, as well as inequitable. It persists mainly because less than 20 per cent of investment for development has gone to the agricultural sector, although over 65 per cent of the people of less-developed countries (LDCs), and over 80 per cent of the really poor who live on $1 a week each or less, depend for a living on agriculture. The proportion of skilled people who support development—doctors, bankers, engineers—going to rural areas has been lower still; and the rural–urban imbalances have in general been even greater than those between agriculture and industry. Moreover, in most LDCs, governments have taken numerous measures with the unhappy side-effect of accentuating rural–urban disparities: their own allocation of public expenditure and taxation; measures raising the price of industrial production relative to farm production, thus encouraging private rural saving to flow into industrial investment because the value of industrial output has been artificially boosted; and educational facilities encouraging bright villagers to train in cities for urban jobs.

Such processes have been extremely inefficient. For instance, the impact on output of $1 of carefully selected investment is in most countries two to three times as high in agriculture as elsewhere, yet public policy and private market power have combined to push domestic savings and foreign aid into non-agricultural uses. The process has also been inequitable. Agriculture starts with about one-third the income per head of the rest of the economy, so that the people who depend on it should in equity receive special attention, not special mulcting. Finally, the misallocation between sectors has created a needless and acute conflict between efficiency and equity. In agriculture the poor farmer with little land is usually efficient in his use of both land and capital, whereas power, construction, and industry often do best in big, capital-intensive units; and rural income and power, while far from equal, are less unequal than in the cities. So concentration on urban development and neglect of agriculture have pushed resources away from activities where they can help growth *and* benefit the poor, and towards activities where they do either of these, if at all, at the expense of the other.

Urban bias also increases inefficiency and inequity *within* the sectors. Poor farmers have little land and much underused family labour. Hence they tend to complement any extra development resources received— pump-sets, fertilizers, virgin land—with much more extra labour than do large farmers. Poor farmers thus tend to get most output from such extra resources (as well as needing the extra income most). But rich farmers (because they sell their extra output to the cities instead of eating it themselves, and because they are likely to use much of their extra income to support urban investment) are naturally favoured by urban-biased policies; it is they, not the efficient small farmers, who get the cheap loans and the fertilizer subsidies. The patterns of allocation and distribution within the cities are damaged too. Farm inputs are produced inefficiently, instead of imported, and the farmer has to pay, even if the price is nominally 'subsidized'. The processing of farm outputs, notably grain milling, is shifted into big urban units and the profits are no longer reinvested in agriculture. And equalization between classes inside the cities becomes more risky, because the investment-starved farm sector might prove unable to deliver the food that a better-off urban mass would seek to buy.

Moreover, income in poor countries is usually more equally distributed within the rural sector than within the urban sector.[5] Since income creates the power to distribute extra income, therefore, a policy that concentrates on raising income in the urban sector will worsen inequalities in two ways: by transferring not only from poor to rich, but also from more equal to less equal. Concentration on urban enrichment is triply inequitable: because countryfolk start poorer; because such concentration allots rural resources largely to the rural rich (who sell food to the cities); and because the great inequality of power *within* the towns renders urban resources especially likely to go to the resident élites.

But am I not hammering at an open door? Certainly the persiflage of allocation has changed recently, under the impact of patently damaging deficiencies in rural output. Development plans are nowadays full of 'top priority for agriculture'.[6] This is reminiscent of the pseudo-egalitarian school where, at mealtimes, Class B children get priority, while Class A children get food.[7] We can see that the new agricultural priority is dubious from the abuse of the 'green revolution' and of the oil crisis (despite its much greater impact on *industrial* costs) as pretexts for lack of emphasis on agriculture: 'We don't need it', and 'We can't afford it', respectively. And the 60 to 80 per cent of people dependent on agriculture are still allocated barely 20 per cent of public resources; even these small shares are seldom achieved; and they have, if anything, tended to diminish. So long as the élite's interests, background, and sympathies remain predominantly urban, the countryside may get the 'priority' but the city will get the resources. The farm sector will continue to be squeezed, both by transfers of resources from it and by prices that are turned against it. Bogus justifications of urban bias will continue to earn the sincere, prestige-conferring, but misguided support of visiting 'experts' from industrialized countries and international agencies. And development will be needlessly painful, inequitable, and slow.

I aim to prove these points: to see how, why, and with what effects the squeeze happens, and to suggest remedies and alternatives. Moral indignation is irrelevant; many members of élites in poor countries struggle to generate equitable development much more unselfishly than did their nineteenth-century European predecessors. The task is to understand the political facts and constraints.

Irrelevant also to this task, but not to my own emphasis, is the fact that (to my own surprise) I first noted urban bias in my analysis of Indian development in the 1960s.[8] Here as elsewhere India 'suffers' for her virtues: relatively good data, honest and first-rate domestic scholarship, and intellectual open-mindedness and curiosity. My work has convinced me that, while Indian development is seriously retarded by urban bias, matters are far worse in most other LDCs. Many of the data in this book, for example those on the allocation of doctors, confirm this.

Three initial objections exist to a theory that urban bias is the mainspring of 'non-disimpoverishing' development. First, does it imply that rural emphasis will solve everything?[9] Development studies have been afflicted by many a misplaced *idée fixe*. Underinvestment, undereducation, and 'underemployment' have in rapid succession been presented as the Cause of All the Trouble, each with its implicit neat cure. It is not my wish to overstate the case for reducing urban bias. Such a reduction is not the *only* thing necessary. But a shift of resources to the rural sector, and within it to the efficient rural poor even if they do very little for urban development, is often, perhaps usually, the *overriding* developmental task.[10] I seek to marshal all the arguments because (for all the easy populist rhetoric of politicians on tour) urban bias is a tough beast: like Belloc, 'I shoot the hippopotamus with bullets made of platinum, because if I use leaden ones his hide is sure to flatten 'em.'

Secondly, does the urban bias thesis imply some conspiracy theory of history? Do people with different interests get together, in reality or 'in effect', and decide on the numerous acts considered in this book, all tending to harm the majority of the population—those who work the land? Do not such flimsy coalitions notoriously split as the interests of their members conflict, and will not the pressures of increasingly articulate mass opinion in the countryside provide natural allies—especially in democracies—for any part of the élite that opposes urban bias? Clearly any conspiracy among several powerful men, representing divergent interests but all opposed to mass interests, is likely to be unstable; and hence any theory of development alleging that *persistent* poverty in many different countries can be explained by such conspiracies is absurd.

However, urban bias does not rest on a conspiracy, but on convergent interests. Industrialists, urban workers, even big farmers, *all* benefit if agriculture gets squeezed, provided its few resources are steered, heavily subsidized, to the big farmer, to produce cheap food and raw materials for the cities. Nobody conspires; all the powerful are satisfied; the labour-intensive small farmer stays efficient, poor, and powerless, and had better shut up. Meanwhile, the economist, often in the blinkers of industrial determinism, congratulates all concerned on resolutely extracting an agricultural surplus to finance industrialization. Conspiracy? Who needs conspiracy?

Thirdly, how far does the urban bias thesis go towards an agricultural or rural emphasis?[11] It was noted (note 10) that there is a rather low limit to the shifts that *can* swiftly be made in allocations of key resources like doctors or savings between huge, structured areas of economic life like agriculture and industry. In the longer run, if the arguments of this book are right, how high do they push the allocations that should go to agriculture in poor countries: from the typical 20 per cent of various sorts of scarce resource (for the poorest two-thirds of the people, who are also those normally using scarce resources more efficiently, as will be shown) up to 50 per cent, or 70 per cent, or (absurdly) 100 per cent? Clearly the answer will differ according to the resource being reallocated, the length of time for the reallocation, and the national situation under review. The optimal extra proportion of doctors for rural India, of investment for rural Peru, and of increase in farm prices for rural Nigeria will naturally differ. However, it remains true that pressures exist to set all these levels far below their optima. To acquire the right to advise against letting children go naked in winter, do I need to prescribe the ideal designs of babies' bonnets?

Linked to the question 'Is there a limit to the share of resources agriculture ought to get?' is a more fundamental question. Does the need for a high share of rural resources last for ever? Does not development imply a move out of agriculture and away from villages? Since all developed countries have a very high proportion of resources outside agriculture, can it make sense for underdeveloped countries to push more resources *into* agriculture? And—a related question—as a poor country develops, does it not approach the British or US style of farming, where it is workers rather than machines or land that are scarce, so that the concentration of farm resources upon big labour-saving farms begins to make more sense?

The best way to look at this question is to posit four stages in the analysis of policy in a developing country towards agriculture. Stage I is to advocate leaving farming alone, allowing it few resources, taxing it heavily if possible, and getting its outputs cheaply to finance industrial development, which has top priority. This belief often rests on such comfortable assumptions as that agricultural growth is ensured by rapid technical change; does not require or cannot absorb investment; and can be directed to the poor while the rich farmers alone are squeezed to provide the surpluses. Such a squeeze on agriculture was overtly Stalin's policy, and in effect (though much more humanely) the policy of the Second Indian Plan

(1956–61) as articulated by Mahalanobis, its chief architect. The bridge between the two was the economic analysis of Preobrazhensky and Feldman. The underlying argument, that it is better to make machines than to make consumer goods, especially if one can make machines to make machines, ignores both the possible case for international specialization, and the decided inefficiency of using scarce resources to do the right thing at the wrong time.[12]

The second stage in policy for rural development usually arises out of the failures of Stage I. In Stage II, policy-makers argue that agriculture cannot be safely neglected if it is adequately to provide workers, materials, markets, and savings to industry. Hence a lot of resources need to be put into those parts of agriculture (mainly big farms, though this is seldom stated openly) that supply industry with raw materials, and industrial workers with food. That is the stage that many poor countries have reached in their official pronouncements, and some in their actual decisions. Stage II is still permeated by urban bias, because the farm sector is allocated resources not mainly to raise economic welfare, but because, and in so far as, it uses the resources to feed urban-industrial growth. Development of the rural sector is advocated, but not for the people who live and work there.

In Stage III, the argument shifts. It is realized that, so long as resources are concentrated on big farmers to provide urban inputs, those resources will neither relieve need nor—because big farmers use little labour per acre—be used very productively. So the sequence is taken one step further back. It is recognized, not only (as in Stage II) that efficient industrialization is unlikely without major growth in rural inputs, but also (and this is the distinctive contribution of Stage III) that such growth cannot be achieved efficiently or equitably—or maybe at all—on the basis of immediately 'extracting surplus'. Stage III therefore involves accepting the need for a transformation of the *mass* rural sector, through major resource inputs, *prior* to substantial industrialization, except in so far as such industrialization is a more efficient way than (say) imports of providing the mass rural sector with farm requirements or processing facilities. For development to 'march on two legs', the best foot must be put forward first.

It is at Stage III that I stop. I do not believe that poor countries should 'stay agricultural' in order to develop, let alone instead of developing. The argument that neither the carrying capacity of the land, nor the market for farm products, is such as to permit the masses in poor countries to reach high levels of living without a major shift to non-farm activities seems conclusive. The existence of a ·Stage IV must be recognized, however, Stage IV is the belief that industrialism degrades; that one should keep rural for ever. This is attractive to some people in poor countries because it marks a total rejection of imitativeness. Neither Western nor Soviet industrialism, but a 'national path', is advocated. Other people, notably in rich countries, argue that environmental factors preclude an industrialized world where all consume at US levels; that there would be too little of one or more key minerals, or that the use of so much energy would disastrously damage the world's air, water, climate, or other aspects of the ecosystem.

The nationalist objections to industry seem to show an unwarranted lack of confidence in the capacity of a great, ancient, localized culture—the Rajasthani or the Yoruba—to preserve or develop its local character in face of changing economic styles and structures. The environmentalist objections are more serious, but most environmentalists themselves recognize that they must be pressed far more strongly on developed than on underdeveloped countries. To do the reverse is a distastefully vicarious form of asceticism (we're rich but you can't afford it). Also such objections rest on a rather static view of technology; in fact, rising mineral and energy prices are already signalling to researchers the need to find new or alternative mineral supplies and to devise ecologically improved paths to growth.[13] For paths to growth there have to be, at least for poor countries.

Growth and development have not so far *sufficed* to raise mass welfare substantially, but are certainly *needed* to provide the resources for that task. 'The wretched of the earth' now know they need no longer live in ill health, hunger, and cultural deprivation. Growth with redistribution appears to offer the only alternative. In my judgement, growth will imply ultimate industrialization; but an incidental advantage of a 'Stage III policy' is that it can offer the ecologically sensitive a wider range of choice. Perhaps, in a few poor countries, a really efficient, egalitarian mass agriculture can offer even a long-run alternative to global industrialization.

In most poor countries, however, the case against urban bias cannot well be made from a Stage IV position. But there is one perfectly valid Stage IV argument for concentrating future agricultural growth in the Third World (most of it has been in rich countries since 1945). Fertilizer and pesticide inputs, per ton of food output, are at much higher levels in rich countries than in poor ones.[14] The increase in food output is less than proportional to the increase in chemical inputs, but the increase in damage to humans from chemical residues is more than in proportion. So an extra ton of agro-chemicals produces *more* environmental damage, for *less* extra output, in rich countries than in poor ones. Apart from that, environmental risks—even if small—are serious enough to warrant insurance policies; and indeed if I am wrong—if the carrying capacity of the land, or the environmental (or human and political) cost of industrialization, proves higher than I anticipate—greater attention to rural development in LDCs will at least have left their options open for a neo-populist solution.

However, the dependence of Stage IV upon such a solution—often backed by a rather idyllic vision of a return to a golden age of happy communal village life—damages it, and sometimes discredits serious advocacy of agricultural development to relieve rural poverty. The traditional village economy, society, and polity are almost always internally unequal, exploitative, and far from idyllic: these features are likely to reassert themselves soon after the initial enthusiasms of a communal revival have evaporated. Even the village in which Mahatma Gandhi settled for ten years lost its cohesive and egalitarian ideals soon after his charismatic leadership was removed.[15]

Both Russian *narodniks* and many Western colonizers confused Stage III and Stage IV. In this study it is accepted that poor countries must grow, develop, and industrialize; and that the three processes are normally locked together. But if countryfolk are to be made richer, happier and more equal by integration into the developing and industrializing national economy, they must first be given—or must take—the chance to reduce the gap in wealth, power, and status that divides them from the cities. The villagers cannot help either themselves or, in the long term, national development if they are neglected (Stage I) or exploited (Stage II). Only on the basis of a tolerable level of living for a mass agriculture of small farmers can most poor countries construct, speedily and efficiently, a modern industrial society. Nor need this mean a world of polluted Tokyos: as Kautsky argued in 1899, it may well be exploitation of the countryside by the city, and not growth or development as such, which bears major blame for the damage to urban (and often rural) environment that has accompanied economic modernization.

This study does not, impertinently, say to those who work in and on poor countries: 'Don't industrialize.' Rather it says: 'A developed mass agriculture is normally needed before you can have widespread successful development in other sectors.' Many reasons for this proposition will be given, but this introduction had better close with the most fundamental. In early development, with labour plentiful and the ability to save scarce, small farming is especially promising, because it is the part of the economy in which a given amount of scarce investible resources will be supported by the most human effort. Thus it is emphasis upon small farming that can most rapidly boost income per head to the levels at which the major sacrifices of consumption, required for heavy industrialization, can be undertaken without intolerable hardship and repression.[16] Except for a country fortunate enough to find gold or oil, poverty is a barrier to *rapid and general* industrialization. To attempt it willy-nilly is to attack a brick wall with one's head. Prior mass agricultural development—building a battering-ram—is a quicker as well as a less painful[17] way to industrialize. The transition point, from mass rural development to industrialization, will signal itself: as good rural projects are used up, so that urban projects begin to 'pay' best even at fair prices; as mass rural demand for urban products emphasizes their new profitability; and as advancing villagers acquire urban skills and create rural labour shortages.

The learning process, needed for modern industrialization, is sometimes long; but it is fallacious for a nation, comprising above all a promising but overwhelmingly underdeveloped agriculture, to conclude that, in order to begin the process of learning, a general attack on numerous branches of industrial activity should be initiated. A far better strategy is to concentrate first upon high-yielding mass rural development, supported (partly for learning's sake) by such selective ancillary industry as rural development makes viable. Rapid industrialization on a broad front, doomed to self-strangulation for want of the wage goods and savings capacity that only a developed agricultural sector can provide, is likely to discredit industrialization itself.

The arguments for rapid general industrialization, prior to or alongside agricultural development, assume against most of the evidence that such a sequence is likely to succeed. But no national self-esteem, no learning-by-doing, no jam tomorrow, can come from a mass of false starts. If you wish for industrialization, prepare to develop agriculture.

Notes

1. The UN target was a 5% yearly rate of 'real' growth (that is allowing for inflation) of total output. The actual rate was slightly higher.
2. Net aid from the donor countries comprising the Development Assistance Committee (DAC) of the Organization for Economic Co-operation and Development (OECD) comprises over 95% of all net aid to less-developed countries (LDCs). It fell steadily from 0.54% of donors' GNP in 1961 to 0.30% in 1973. The real value of aid per person in recipient countries fell by over 20% over the period. M. Lipton, 'Aid Allocation when Aid is Inadequate', in T. Byres (ed.), *Foreign Resources and Economic Development* (Cass. 1972), 158; OECD (DAC), *Development Cooperation* (1974 Review), 116.
3. L. Cooper, *Radical Jack* (Cresset, 1969), esp. 183–97; C. New, *Life of Henry Brougham to 1830* (Clarendon Press, 1961), Preface.
4. See the mounting emphasis in his *Addresses to the Board of Governors*, all published by IBRD, Washington; at Copenhagen in 1970, 20; at Washington in 1971, 6–19, and 1972, 8–15; and at Nairobi in 1973, 10–14 and 19.
5. M. Ahluwalia, 'The Dimensions of the Problem', in H. Chenery *et al.*, *Redistribution with Growth* (Oxford, 1974).
6. See K. Rafferty, *Financial Times*, 10 Apr. 1974, 35, col. 5; M. Lipton, 'Urban Bias and Rural Planning', in P. Streeten and M. Lipton (eds), *The Crisis of Indian Planning* (Oxford, 1968), 85.
7. F. Muir and D. Norden, 'Common Entrance', in P. Sellers, *Songs for Swinging Sellers* (Parlophone PMC 1111, 1958).
8. Streeten and Lipton, *The Crisis of Indian Planning*; Lipton, 'Indian Agricultural

Development: Achievements, Distortions, Ideologies', in *Asian and African Studies* (Israel Oriental Society), Vol. 6, 1970; and 'Transfer of Resources from Agricultural to Non-agricultural Activities: The Case of India', Fifth UN Interregional Seminar on Development Planning (Bangkok, 1969) (UN ST/TAO/SER. C/133), UN, 1971).

9. M. Arnold, '*Porro Unum est Necessarium*', in *Culture and Anarchy* (1869), *Prose Works*, Vol. 5 (University of Michigan Press, 1965). See also G. B. Shaw, *The Doctor's Dilemma*, Act 1, and T. Carlyle, 'Morrison's Pill', in *Past and Present* (1852), *Works*, Vol. 2 (Chapman & Hall, 1891), bk. 1, ch. 4.

10. Nobody who, like myself, has worked in Bangladesh could miss the combination, not rare in Asia or East Africa, of (a) extreme poverty, especially in villages; (b) a responsive peasantry 'raring to go' with improved techniques; (c) a system that steers 70–80% of scarce savings, skills, and political energy into a tiny, inefficient but influential urban sector. Even a total cure for urban bias—given the momentum of outgoing projects—could not efficiently slash this proportion *at once* to, say, 40–55 per cent. But clearly movement towards a cure is the first requirement.

11. More complex and specific objections to the 'urban bias' explanation of persistent poverty—the possible interlocking of rural and urban outlays, the alleged shortage of truly viable rural schemes, the possible need for industrial income as a source of savings—are dealt with below.

12. Alec Nove, *An Economic History of the USSR* (Allen Lane, 1969), 125–7, 132–3; P. C. Mahalanobis, *The Approach of Operational Research to Planning in India* (Asia, 1963), esp. ch. 3; E. Preobrazhensky, *The New Economics* (Clarendon Press, 1965). E. Domar, *Essays in the Theory of Economic Growth* (New York: Oxford University Press, 1957), 223–61; and see below.

13. W. Beckerman, *In Defence of Economic Growth* (Cape, 1974), 220.

14. G. R. Allen, 'Confusion on Fertilizers and the World Food Situation', *European Chemical News*, Large Plants Supplement, 18 Oct. 1974.

15. K. Nair, *Blossoms in the Dust* (Duckworth, 1961), 184–8.

16. P. Streeten, in *The Frontiers of Development Studies* (Macmillan, 1972), 89, tellingly cites Wicksell's discussion.

17. Of course, part of the trouble is that those who benefit from rapid general industrialization are not those who suffer the pain. The Russian finance ministers, Vyshnegradskii and Witte, in the last decades of the nineteenth century, forced near-starving peasants (by taxation in kind, and other means) to produce for foreign markets, thus providing foreign exchange for industrial imports. Vyshnegradskii said 'We must export, though we die'; he meant, as Robert Cassen remarks, 'I shall export, though you die.' T. H. von Laue, *Sergei Witte and the Industrialization of Russia* (New York: Atheneum, 1969), 26–7 and 107.

III

The Urban Labour Market and Migration

INTRODUCTION

UNTIL the last century, many rural populations had little or no connection with urban centres; they lived in quite self-centred societies. By and large, they operated subsistence economies and maintained only limited external contacts. The expansion of the capitalist system, however, under way for half a millennium and accelerated by the Industrial Revolution, has incorporated ever more outlying regions into the emerging world economy. By now the process is virtually complete. All over the world, rural populations have been drawn into the urban nexus. Today, most rural dwellers know what it is like to sell and buy in markets or shops, they have seen what a school certificate can do for the future of a child, and have listened to first-hand accounts of those who work in the city. Some improve their condition while staying where they were born, or by moving to other rural areas as farmers, traders, or artisans. But rural prospects appear dim to many, the urban scene more promising. For a comprehensive discussion of their migration, see Gugler (1986).

The sight of severe and widespread poverty in Third World cities easily leads to the assumption that migrants do not know what to expect, that illusions about the prospects lying ahead bring them to an urban environment in which they find themselves trapped. In fact, most migrants are quite well informed before they move. Many hear the accounts of earlier migrants who have returned to the village on a visit or to stay. Some are able to visit kin or friends in the city before deciding to move. Studies report, time and again, that most migrants consider that they have improved their condition, and that they are satisfied with their move. Thus rural–urban migration continues unabated throughout much of the Third World. 'Why do so many come', the question usually goes, 'when urban unemployment is widespread and underemployment common?' To which a peasant might respond with the counter question: 'Why do so few go, when the rural–urban gap is so unmistakable?—after all, the rural population continues to grow in nearly every Third World country, in spite of rural–urban migration. The answer to both questions is provided by the urban labour market.

The urban labour market in Third World countries, like many markets, is fragmented in a variety of ways, i.e. different categories of people enjoy differential access to earning opportunities. Access is usually largely a function of three criteria: education and training, gender, and patronage. Differential access in turn shapes the composition of the migrant stream. The role of formal education as a prerequisite for access to the more privileged strata motivates parents in rural areas and small towns to relocate with their children or to send their children away to better and more prestigious schools. For those who have ascended the educational ladder, the most attractive career opportunities are in the city. Where new industries have recruited substantial numbers of women, as in South Korea and in Thailand, the proportion of women among rural–urban migrants has increased.

'Credentials' are generally accepted as a screening device, and discrimination on the basis of sex is commonly taken for granted. Patronage, in contrast, is usually frowned upon. It is, however, widespread, sustained as it is by strong interests and effective mechanisms. Most migrants expect and obtain assistance from urban contacts. The urban host, to help the new arrival, to relieve the burden of housing and perhaps even of feeding him, has good reason to find him work. Thus migrants who have secured employment introduce their relatives and other people from 'home' to their firm. Many employers find such 'family brokerage' convenient and even advantageous. They know that skills and knowledge are not as important for many positions as other qualities: dependability, potential for training, persistence, and initiative. Further, in many cases, job advertisements will generate all too many applications for people with similar qualifications. In such circumstances the employer prefers to use a 'broker'. He selects among his employees one or two persons he trusts, and asks them for suitable candidates, whom they will probably have to train. The broker will look at his extended family for suitable candidates, and draw up a short list. He may coach a candidate on how to fill in the application forms, and on how to react at the interview. A close and complex relationship thus arises between the employer, the broker, and the new employee. The broker has increased the socio-economic position of his kin group and his own standing within it, the unemployed has obtained a job, and the employer can exert leverage over his employee through the broker (International Labour Office 1972, 509–10). Because of such particularistic recruitment patterns, migrants of common origin tend to cluster in certain jobs and trades, the labour market is segmented.

The fragmentation of the urban labour market is mirrored in the stream of migrants. Constraints on the participation of women in the urban labour force affect sex selectivity in rural–urban migration (Ferree and Gugler 1983). Migrants with the proper credentials come with a prospect of satisfactory employment. Others have the right connections and come with reasonable assurance that the assistance of their kinsman, fellow villager, or patron will prove effective. Most rural dwellers, however, have no prospects in the city. Some venture forth nevertheless, to succeed or to fail, or to return to the village. But most stay on the land—as long as it supports them.

For many migration is not just a once for all move. Rather they make a number of moves over a lifetime. Such a migratory career is best understood with reference to family and community. Four principal patterns of rural–urban migration in the Third World then stand out:

(a) migrant labour;
(b) long-term migration of men;
(c) family migration to urban areas followed by return migration to the community of origin; and
(d) permanent urban settlement.

These are not fixed statuses. The man who left his family behind may decide to have them join him; the family that expected to return to its community of origin may settle down in the city forever. While changes in migratory status, as perceived by the migrant, typically go in the direction of an increasing commitment to the place of destination, they are clearly affected by changing circumstances in both the urban environment and the area of origin, e.g. deteriorating urban conditions may force men to send their families to the village.

The preponderance of men over women in the cities of South Asia, of much of Subsaharan Africa, and of Oceania (Ferree and Gugler 1985) reflects the tendency of male migrants to leave wife and children in their rural area of origin. If the Industrial Revolution engendered the distinction of work-place and home, the separation of worker and dependants has been drastically magnified for many Third World families. Extended family support typically facilitates such simultaneous involvement in the urban and the rural economy. Indeed, the assistance of male kin in certain tasks, and the protection they afford, frequently appears as a prerequisite for a wife to manage the farm and to hold her own in a male-dominated environment.

The migration of individuals, whether single or separated from their family, has distinct economic advantages: it optimizes labour allocation, and, at least in rural–urban migration, it minimizes the cost of subsistence. Employers save on wages and retirement benefits, and public authorities face less demand for housing and infrastructure. But there are also gains to migrants that motivate them to accept family separation. Living costs in the city are high, while urban earning opportunities for women are usually very limited. Typically wife and children remain on a family farm growing their own food, and perhaps even raising cash crops. Where land is communally controlled and cannot be alienated, as is the case in much of tropical Africa, there is no compensation for those who give up farming it. A wife who comes to town has to abandon an assured source of food and cash to join a husband on low wages.

Family separation has commonly taken the form of migrant labour.

After a period of employment lasting six months perhaps, a couple of years at most, the migrant returns for an extended stay with his family. In the ideal case, the return coincides with peak labour requirements on the farm. In some areas, such migrants go as contract labour, i.e. they are recruited for a fixed period of time at, or close by, their home place, and provided with return transportation. Repetition of the circular movement is common, and many migrants build up extended urban experience. Migrant labour was common in Subsaharan Africa in colonial days when authorities and employers pursued a cheap-labour policy that led to labour shortages. But once urban incomes became rather attractive, relative to rural incomes, the urban labour market was transformed. It became characterized by high unemployment. Today the search for a job takes months and the outcome is aleatory. Under these circumstances long-term migration is the only viable option. The migrant who has secured regular employment now has good reason to cling to it. Instead of extended stays with the family there are short visits, as employment conditions and distance permit. What had been an economic cost to employers—a labour force characterized by high turnover and absenteeism—has become an increase in social costs for workers: more severe strains in their relationships with wife, children, extended family, and village community.

It is tempting to speculate that the temporary migration of men constitutes the initial response of a 'traditional' society to new opportunities to earn wages and acquire manufactured goods, to visualize 'tribesmen' making forays into an alien environment. The facts indicate otherwise. Thus the peasants of highland Peru went to work in the mines already during the Inca period; half a millennium later a widespread pattern of temporary migration persists. Julian Laite, in Chapter 4, reports that the majority of miners and refinery workers are migrants and nearly all these migrants maintain village interests. The most important interest is in village land, even though land has increasingly been transformed into a commodity and brought into the cash arena since the nineteenth century. Three-generation extended families continue to control property and to organize production. The senior generation owns the resources while other members of the family work on them. So, whilst junior members of the household are 'landless', they do have access to land. The migration of one, or several, men provides external resources to meet household needs. During their absence, it is the women who do the work, or, at planning and harvest time, recruit labour. Their task is facilitated by co-operative practices well established in Andean peasant culture.

Rapidly declining urban sex-ratios, especially in Subsaharan Africa, signal a shift to patterns where entire families move to the city. In some settings, however, these families remain socially rooted in the village community which continues to provide ultimate security. I described and analysed this pattern in Eastern Nigeria in the early 1960s and observed its persistence in 1987. Just as the temporary migration of men, such temporary migration of families is sustained by extended kin networks and communal control over land. The insecure economic position of much of the urban population provides powerful motivation to maintain access to the ultimate, however meagre, security the village community offers (Gugler 1971).

Elsewhere families settle permanently in the city. This is the rule in Latin America. The exception presented by Indian communities, such as those of highland Peru, demonstrates the crucial role of rural social structure in the establishment and maintenance of specific patterns of migration. Permanent urban settlement appears to be characterized by a slight predominance of women in the urban population. Such a predominance has been notable in Latin America, the Caribbean, and the Philippines for several generations and is now emerging in rapidly industrializing countries such as South Korea, Taiwan, and Thailand.

In Chapter 5 we argue that Third World countries are characterized by overurbanization: rural–urban migration entails the loss of potential rural output; it brings workers to cities that are unable to fully employ their existing labour force to productive ends; and these additional urban dwellers require more resources for their survival than they would in the countryside. Migrants, however, rationally maximize their benefits. The resolution of the seeming paradox derives from the fact that rural–urban migration has a redistributive effect. Rural–urban migrants lay claim to a share in urban income opportunities, and they gain some access to urban amenities. Rural families send their sons and daughters to the city so that they will be able to partake, however little, of its riches.

Most Third World governments grope for policies to slow down rural–urban migration, whether because they believe that it has negative consequences for economic growth or from concern that it adds to the pressure the urban masses can exert on politicians and bureaucrats alike. Richard Sabot, in Chapter 6, draws on his study of the urban labour market and rural–urban migration in Tanzania systematically to review policies to reduce urban surplus labour.

The most obvious policy—to restrict rural–urban migration through direct controls—constitutes an attempt to keep the have-nots out, to erect a boundary that will shelter urban populations from the competition of migrants, just as affluent countries shelter their citizens by the enactment and enforcement of laws controlling the immigration of foreigners. Boundaries are drawn and barriers erected to protect privilege. Restricting rural–urban migration means closing the remaining escape route to rural masses that have been disenfranchised and neglected. Still, if migration controls are the exception in the Third World, it is not out of concern for equity. Rather, internal boundaries, to a much greater extent than international boundaries, are difficult to police. The countries that regulate internal migration are those that control their populations rather closely; they include South Africa, committed to maintaining inequity between black and white, as well as socialist countries that avow their concern with the 'rural–urban contradiction'. Even where drastic measures are taken to enforce controls on internal migration, however, their effectiveness appears limited. Half of the residents in the African township of Soweto, on the outskirts of Johannesburg, were said to live there illegally even while the pass laws were ruthlessly enforced. In China hundreds of thousands of rusticated middle-school leavers returned to the cities without authorization in the late 1970s.

Contract labour in China, the temporary recruitment of workers from rural areas to work in urban jobs, highlights the dilemmas of equity and efficiency that migration control poses. Marc Blecher, in Chapter 7, provides an account for Shulu, a county in Hebei Province. Between 1964 and 1978 the industrial labour force in Shulu trebled. Three-quarters of the new workers came from rural areas on temporary contracts. By 1978, contract workers comprised over half of the work-force in county-level industry. They received somewhat lower wages than regular workers and had no claim to the fringe benefits enjoyed by regular workers, such as free medical insurance, accident insurance (workers' compensation), pensions, and sick-leave. Most importantly, contract workers were housed in dormitories and had to leave their families in the rural areas. In a period of rapid industrialization, urban growth was thus reined in. Only workers needed in production were authorized to come to urban areas, and most, on contract terms, had to leave their dependants behind. To the extent that the system was effectively implemented, unemployment was avoided and the proportion of dependants kept extremely low in urban areas. At the same time, inequities were created between regular workers and contract workers, as well as between the latter and rural workers. Rural production teams, however, received a share of contract workers' wages; and inasmuch as contract workers and their families remained part of the rural population, the average income of that population was higher than it would have been if they had left altogether. The system can be argued to be more beneficial to the rural masses than if some in their midst were to leave permanently and cut all ties.

There is general agreement that rural development, in order to close the rural–urban gap, constitutes a policy of equity that will truly stem rural–urban migration. Admittedly, initial improvements in education and in communications may encourage migration as long as rural–urban discrepancies remain large. To reduce significantly the gap requires that considerable resources be made available to a huge rural population. Rural incomes have to be raised directly, e.g. through a reduction in taxes or through an increase in the prices agricultural products fetch, and/or indirectly through investments that raise the productivity of the rural labour force whether in agriculture or in small-scale industry. Given the size of the rural population in Third World countries, this would require an enormous reallocation of resources, a reallocation that would be confronted with the determined opposition of urban interests. A long-run rural development policy which in any case would have to focus on labour-intensive investments to retain a growing rural population, might constitute a feasible compromise. Such a policy appears, however, to be beyond the horizon of most governments, preoccupied as they are with their very survival in a much more immediate future.

References

Ferree, Myra Marx, and Gugler, Josef (1983) 'The Participation of Women in the Urban Labour Force and in Rural–Urban Migration in India', *Demography India*, 12, 194–213.
—— —— (1985) 'Sex Differentials in Rural–Urban Migration: Variations across the Third World', Paper presented at the South South Conference, Montreal, May 1985.
Gugler, Josef (1971) 'Life in a Dual System: Eastern Nigerians in Town, 1961', *Cahiers d'Études Africaines*, 11, 400–21.
—— (1986) 'Internal Migration in the Third World', in Michael Pacione (ed.) *Population Geography: Progress & Prospect* (London: Croom Helm), 194–223.
International Labour Office (1972) *Employment, Incomes and Equality: A Strategy for increasing Productive Employment in Kenya* (Geneva: ILO).

4

The Migrant Response in Central Peru*

Julian Laite

IN the highlands of Peru are to be found patterns of circulatory migration between a rural peasant sector and an industrial mining sector, as highlanders establish a migratory response to unstable economic change. One such pattern exists between the mining town of La Oroya and two villages, Ataura and Matahuasi, which lie in the Mantaro Valley. The highland mining sector of Peru was, until 1974, dominated by Cerro de Pasco Corporation. CdeP began its operations in Peru in 1902, and by 1971 owned all three refineries in the country and two of the four smelters. The mining investments of CdeP were mainly in the central highlands, where in the early 1970s the Corporation operated seven major mines, which employed about 6,000 workers. In addition, CdeP operated two major administrative and refining centres in the highlands, located in the towns of Cerro de Pasco and La Oroya. Cerro de Pasco is also a mining location, while Oroya is the administrative heart of highland mining operations, with the main refinery situated there.

La Oroya lies 4,000 metres above sea-level. It is the largest mining town in Peru as well as one of the most important railway centres. Through it passes the main road from the capital to the towns of the highlands. However, it is some five hours' journey to Lima, the capital of Peru, and several hours by road to the highland towns of Junin, Cerro de Pasco, and Jauja. Consequently, La Oroya presents an isolated picture of a crowded jumble of shacks and huts huddled high in the Andes. The town straddles the Mantaro river, which cuts a deep gorge through the mountains. On one side is the municipal township of Old Oroya, a maze of slums and shacks which creep up the sides of the hills. On the other, on the only available flat land, are the long rows of barracks which house the labour force. The town is completely dominated by the vast CdeP refinery. Twenty-six thousand people live in La Oroya. Most of the work-force are directly employed in the refinery, whilst the remainder are dependent upon that labour force for

* Adapted by the author from his *Industrial Development and Migrant Labour in Latin America* (Manchester: Manchester University Press; Austin: University of Texas Press, 1981), especially ch. 6, by permission of Manchester University Press.

their own livelihood. In the refinery there are two major work groups. The first are the refinery's blue-collar workers, or *obreros*, who number 5,300. The second are the refinery's white-collar workers, or *empleados*, who number around 1,200.

The living and working conditions of *empleados* and *obreros* are quite distinct. *Empleados* are mainly office-workers, many inhabiting a large administrative block away from the refinery itself. Even when employed in the refinery complex, *empleados* usually have small offices to themselves. They are employed for long periods and have acceptable redundancy arrangements. After 25 years' service they may retire on full pay. They are paid monthly, in *soles*, between $80 and $400 per month (in 1971). They also receive substantial fringe benefits in the form of subsidized food and schooling, and good medical care. The *empleados*' housing is located away from the refinery and the smoke, and consists of comfortable three- or four-storey flats with two to three bedrooms.

In contrast, the circumstances of the *obreros* are very poor. The *obreros* are employed on a daily basis and paid weekly in *soles* at a rate of between $2 and $4 per day (in 1971). They must work until they are 60 before they can retire on half-pay. If they leave before then, as most do, they receive a small severance sum. Their fringe benefits are fewer than those of the *empleados* and they have no access to the houses of the *empleados*, owing to a weighted points system. The working conditions of the *obreros* are those set by an ageing refinery. The noise in the works is deafening and many *obreros* develop a basic hand language. It is often dark under the ovens and gantries and wet and uneven underfoot. It is very dangerous, the dangers lying not only in accidents but also in illness and disease. Mining has the second highest accident rate of all industries in Peru. Deaths in pit disasters are regular occurrences, whilst fractures, crushings, and bruisings are common. At the same time, La Oroya workers are also subject to lead poisoning, which results in paralysis and death. In 1952 surveys showed that lead in the air in La Oroya departments was 125 times internationally agreed safety margins. Measures were brought in to alleviate the situation, and by 1965 there had been some improvement. However, the *obreros* know that refinery work is dangerous and damaging, and so aim to work there only for short periods.

Ataura and Matahuasi both lie on the left bank of the Mantaro river, some 20 kilometres from each other. As in most villages in the valley, the predominant activity is agriculture, practised on a household-based *minifundia* system. As is the case with other villages, migrants from Ataura and Matahuasi live and work in the mines and in Lima. Of all the valley villages Ataura has the highest percentage of its sons working in the mines, while Matahuasi has the largest number of villagers living and working outside the valley.

Ataura is a small village, with many of its houses perched on the valley sides overlooking other houses on the richer valley floor. Some 1,700 men, women, and children live there,[1] and the village extends over some 600 hectares of land, 400 of which are suitable for crops or pasture. The inhabitants are mainly agriculturalists, four-fifths of them owning or working less than one hectare, while only two people own more than five hectares. All told, there are 107 adult men living and working in the village. Nearly all have some agricultural interest, it being rare to find men who do not own or work some land, but only one-third are solely peasants with no other supplementary occupations. Around half are artisans, with shoemaking, house-building, and driving being the most common occupations. Shopkeepers and professionals each make up one-tenth of the resident males, the former often selling the products of their own fields, while the latter are schoolteachers or retired *empleados*.

Agriculture and semi-skilled trades thus account for most of the population, with construction, transport, and the government sector accounting for one-tenth each. Almost half these men are semi-skilled—often a legacy of migration—and most are able to ply their trades around Ataura, where three-quarters of them work. One-tenth work in Jauja and Huancayo, while a further tenth travel to work in villages in the valley. Only a quarter are dependent labourers, hired by others, and two-thirds are independent workers. Around one-tenth are primarily independent but occasionally hire their services out to others. Of the dependent workers, one-quarter work not for cash but as *partidarios*, splitting the cost and profit of cultivation with the owner of the land. The remainder are employed by small concerns, the government, or private landholders.

Matahuasi is much larger than Ataura, contains some 3,000 inhabitants, and extends over some 2,000 hectares, most of which is cultivable. Matahuasi is a recognized *comunidad* and so, within the village, there exists a group of some 120 households who have usufructory rights to around 60 hectares of communal land. Alongside these there is the private landholding sector. Like Ataura, the system of production is *minifundia*, but unlike Ataura there exists in Matahuasi a large landless group, and a group of large land-holders. Around one-fifth of the Matahuasi population are landless, while another quarter own less than one hectare. Two-fifths, however, own between one and four hectares, while a significant one-sixth own more than four hectares. Thus there is much more land in Matahuasi and more people have access to it.

Again as with Ataura, the predominant occupation is agriculture. A little over half the population are peasant farmers, around one-tenth are shopkeepers, another tenth are craftsmen, and these are followed by teachers, traders, and agricultural workers. Thus, in contrast to Ataura, agriculture is much more of a dominant single occupation for more people.

And although non-agricultural occupational stratification seems similar to that of Ataura, there is in Matahuasi one group of people who have been able to undertake entrepreneurial activities on a large scale, something that has never occurred in Ataura. This group consists of the *transportistas* who own and run lorries from the Mantaro valley to other parts of the country. Overall, socio-economic differentiation is much more pronounced in Matahuasi than it is in Ataura. Broadly speaking, Ataura is comprised of poor peasants, whilst Matahuasi contains both poor and middle peasants.[2]

While industrial development has affected the peasant sector, peasant socio-economic organization in turn has affected the social structures that have emerged along with industrial development. Much of the feedback between the rural and industrial sectors has been transmitted via the process of migration. In the work-place, migrants from particular villages have 'captured' certain departments. In the town, migrants from the same village re-establish relations with one another. In the village, potential migrants know that there are networks in both town and work-place that are prepared to receive them. Thus the migrants are a self-selected group;[3] deciding whether or not to migrate to the mining sector and then attempting to enter one kind of industrial work rather than another. That process of self-selection is grounded in village structures and networks.

The Structure of Contemporary Migration[4]

Migration between the rural and industrial sectors has four major features. It is closely related to household needs and decisions. It occurs within a certain cultural milieu of relationships and expectations that support the move. The organization of the move is around age-cohorts rather than between generations. Finally, migration is not a flight from the land by the landless but rather a means of maintaining work alternatives across a number of economic sectors.

Having to migrate because they owned no land at all was not put forward as an important reason by the migrants. Of course, being 'poor', supporting the family and looking for work opportunities are indicative of a situation in which there is little land and few work opportunities in the village. Yet these do not amount to a situation in which a pauperized peasantry is forced from the land. Migration is embarked upon for a variety of reasons, but not in most cases merely in order to subsist, for this is a condition that can be met in the rural sector (Table 4.1).

The role of the household is of great importance in the decision. Whether to migrate or not is decided in consultation with one's wife or mother and is related to household needs. With the father working away, it is the mother who spells out their family obligations to sons of working age

TABLE 4.1   Seventy-six Ataurinos and Matahuasinos: Reasons for Migration from Village

Reason	% of respondents citing reason
To earn more money	39
To join kin already working	25
No work in the villages	12
Work opportunities presented themselves	11
Dislike of village life	11
To be near the village and family	11
To support nuclear and/or extended family	9
Poor	8
To pay for education	7

as family subsistence and educational requirements are reviewed. The eldest brother often feels that, during the absence of a migrant father, the household responsibilities devolve upon him. Certainly several men are described as the 'father of the family' owing to their assumption of the burden of their younger siblings. Although the eldest moves first, his brothers often join him while the mother handles rural affairs. At her death, or before it, the brothers jointly decide whether to maintain or divide the household lands.

Migrating to be near family and village is a decision taken by two types of people. The first are those working in Lima or some other coastal town who take the opportunity to move to La Oroya to be near ageing parents or to develop landed interests. The second are those who move to the mines from the village so that they can be at hand to maintain kin and village links. That is to say, for both these groups and other migrants, moving to the mining sector does not involve a break with kin or with the village. It is a move contained within a kin and village sphere of influence, supported by social networks. The decision to migrate often turns on news of a job in the corporation from a kinsman already working there. The fact that migrants readily take up these offers shows that migration is a normal part of highlanders' expectations.

Four-fifths of migrants know before moving that there are work opportunities at their destination and half actually have a specific job waiting for them on arrival. Some go chancing their luck—*a la ventura*—but often this will be with a friend to a work-centre that has advertised for labour. Again, nine-tenths go knowing that they can stay with someone for the first few nights: two-thirds had an offer of accommodation with kin and one-fifth of being put up by *paisanos* (compatriots from the same village).

Three-quarters knew about conditions in the mines and in La Oroya either from visits or from information received, and so life in the mining sector did not generally come as a rude shock.

This information and these job opportunities are passed on by kinsmen and *paisanos*, not by fathers. Two sorts of people are involved—peers who have moved a short while previously or established gatekeepers into the industrial labour force. The absence of fathers and the importance of peers in recruitment emphasizes the difference between the mineworkers and a stable labour force in an industrialized country. In the latter, jobs are passed on from father to son, but in peripheral economies industrial employment is so uncertain and so limited that older men are rarely long enough in one job to be able to bequeath it.

The role of 'gatekeepers' is of great importance for village 'capture' and the establishment of village networks in the industrial situation. They are often older men, original village migrants who have reached a position of some importance in the labour force, perhaps as *empleados*. These men can use their position to place migrants from their village in certain departments with CdeP. One such gatekeeper is Arturo.

Arturo was born in Ataura and is now an *empleado* in the research department. He became known in the village as having a good job in a good department and was approached by Ataurinos who wanted to work there too. The Ataurinos who approached him had some claim on him, for they were his kin or neighbours in the village. In return, Arturo could recommend them with some confidence, not only because he knew them but also because family pressure could be brought to bear if they did not work seriously. So Arturo has found work for six men and they in turn have brought in eight more. All 15 are from Ataura and all are related. More than this, all work, or have worked, in this research department, where the duties are not onerous. Indeed, over the years much of CdeP's routine research has been carried out by one family from one small Andean village. It is in this way that 'capture' occurs, and it is replicated by village and department throughout the corporation with Matahuasinos in railways and valley migrants steering clear of the blast furnaces.

These village and kin networks thus establish the possibility of work for migrants and receive them when they move. Migrants travel alone at first, or with a friend, making arrangements about wives and children later. Very few actually 'doss' down wherever they can the first night, since most go direct to a known contact. Kinsmen and *paisanos* find room for the migrant while he looks for work, so he has no rent to pay. They also feed him, although he meets his share of this. The arrangement usually works well: some two-thirds of migrants find work within a week and can then contribute rent and food until they find a place of their own. These successful ones usually have a job waiting for them, or are 'spoken for' by the employee they lodge with. For up to a quarter of migrants, however, the search can take three or four weeks and often these men become tense and desperate as they stretch their familial and financial credit to its limits.

For two-thirds of these valley migrants the first job they get is very often in CdeP itself. Having stayed in the village for a while, they go direct to CdeP. The rest find alternative employment and wait for an opening. Most valley migrants begin with the corporation as *obreros*, but they start as they mean to continue. They steer clear of the toxic areas of the refinery and take work which is not too dirty, noisy, or heavy. They become apprentice mechanics and carpenters. If they have to start near the furnaces they make every effort to get into another section as soon as possible.

The forty-seven Ataurinos and Matahuasinos in La Oroya are not only rural migrants they are also stratified into *obrero*, *empleado*, and commercial groups, as well as being differentiated by age. Consequently there are both similarities *and* differences in their characteristics and the strategies they adopt, depending upon the village they come from, occupation, and age, among other factors.

One feature common to nearly all migrants is their maintenance of village interests. The most important interest is a continuing control over village land, and access to land is through either personal or family ownership. The ownership of land is a family affair, and for individual migrants it is access to land, rather than simple ownership, that is crucial. All the Ataurino and Matahuasino migrants were members of extensive families which embraced both rural and industrial locations. Most of the families owned between half a hectare and two hectares, whilst some owned as much as ten. Some young migrants were 'landless', yet it was they who did the heavy work on their parents' land and they had an equal share in the product. Other young migrants did have a little land, but it was invariably a plot given to them when they married by their parents. The important lands to which these migrants had access were those of the senior household. Overall, three-fifths of the migrants from the villages who were *obreros* owned land, as against five-sixths of the village migrant *empleados* who did. The *empleados'* landholdings were larger than the *obreros'*. Since the *empleados* were older, they had inherited more property, and since they were better paid they had been able to buy more land. It was not the case that *empleados* came from landed families whilst *obreros* came from landless ones, but this is discussed below.

This ownership and access to land by nearly all the migrants had several immediate implications. It meant that land, land prices, food prices, and the state of the weather were a constant preoccupation and topic of conversation. At harvesting and planting times they would be busy journeying back and forth to their village or recruiting labour by sending money to their wives. Some older men rented out their land and were always concerned about whether or not they would be paid. Indeed, such concerns touched all the men, who, although they agreed with the expropriation of the *haciendas*, were afraid that any unworked land of their own might be affected by the agrarian reform laws.

Landed interests are maintained by wives and parents continuing to live in the village. Most migrants travel back to their village every weekend, or once a fortnight, in order to visit their families and look over the household property. When they return they bring cash and consumer goods such as shirts and shoes with them. The cash is either to help the family to buy the things like salt, oil, and meat that it needs, or to recruit labour to help at planting and harvest time. This inflow of cash is more noticeable in Ataura than in Matahuasi. At the weekend Ataura is full of men and the women are able to do their weekly shopping in the small stores. In Matahuasi the shops are larger and there are always enough resident merchants and peasants to keep trade brisk all week.

Whereas both *empleados* and *obreros* in La Oroya maintain village contacts, the *empleados* commercialize those contacts more than do the *obreros*. Having more money, they buy land more. They also purchase houses and shops. Nine-tenths of the *empleados* own houses in the villages, compared to only half the *obreros*, and they spent more on theirs, either purchasing them or building them themselves. The *empleados'* houses were for their retirement. Some have two houses, one in the village and one in Jauja or Huancayo, two highland towns lying at either end of the Mantaro valley, or in Lima. The town house is intended for retirement or for occupation while the children are in education, whereas the country house is kept on for village visiting. A few *empleados* sell their village houses to buy a house in a town.

Some *empleados* also maintain village links by purchasing a shop in the village or running a business there. A shop in the village square in Ataura is run by the wife of an *empleado* in La Oroya. He also has a share in a bus which plies between Jauja and Huancayo. One migrant *empleado* from Matahuasi owns a lorry which his wife manages, as it is used for transport in the valley. These men are the exception, however. Usually, in both Ataura and Matahuasi, it is the wives of *empleados* who may rent a small shop in the village or a stall in the Jauja market, or the *empleados* themselves may have a small sideline in the village, such as bee-keeping or photography. These sidelines flourish more in the small-time economy of Ataura than in the more commercial environment of Matahuasi.

In contrast to these commercializing activities of the *empleados*, the *obreros* rely on non-commercial mechanisms for the maintenance of village links. When *obreros* are house-owners it is more likely that the house will have been inherited rather than bought. When *obreros* recruit labour, they do so through the mechanisms of *al partir* (sharing the labouring on a piece of land and dividing the costs and profits of the work), the *comunidad* (the institution of communal landholding), and the kin network. When *obreros* are landowners it is usually because they have assumed the responsibilities of the rural household, often on the death of the parents.

It is clear that the wider commercial activities of the migrant *empleados* are built up during their working lives rather than being due to the fact that *empleados* come from wealthier families. As the village histories showed, there were wealthy families in both villages and their children became professional people in Jauja, Huancayo, and Lima. The bulk of the peasant population, however, are those owning a half to two hectares, and it is from them that *obreros* and *empleados* are drawn. There was no simple fit between rural and industrial social stratification so far as this mass of peasants was concerned. It was not simply that poorer peasants became *obreros* and richer peasants became *empleados*.

Analysis of the backgrounds of all the 76 migrants resident in La Oroya and the two villages revealed no major predictors of *obrero* or *empleado* status. The parental families of both groups were the same size, and the educational levels of the parents were the same. The fathers of *empleados* and *obreros* held equal amounts of land and were similar in their ownership of houses. Nor did the fathers' occupations differ significantly. The fathers of both *obreros* and *empleados* held similar occupations for similar lengths of time. Around half the fathers had been *obreros* or miners for some years, while one-third had been peasants for many years. There was no correlation between the manual and non-manual status of fathers and the occupational achievement of their sons.

Clearly this lack of precise predictors is not the case for other Departments of Peru. The existence of better educational opportunities in Lima means that significantly more *empleados* are Limeños, while the presence of poor landless labourers on *haciendas* in other Departments may mean that they do not have the resources to gain anything more than elementary education. But the mass of peasants in the Mantaro valley do have a few resources and some education. Consequently their reasons for migrating vary from necessity to the search for education, whilst the fit between industrial and rural socio-economic structures is complex, depending on economic resources and social structures.

In these respects the mechanisms of occupational attainment differ from those found in the proletarianized labour forces of industrialized nations. Among an industrial proletariat the occupational status of a father is a good indicator of his son's occupational status. Manual-worker fathers often have manual-worker sons. Indeed, the links are much closer in that fathers often pass on their job, their experience and their tools to their son. It is fathers who introduce sons into leisure clubs reserved for members of particular occupations. Such a close relation, however, occurs in situations where industrial employment is widespread and there are few alternatives.

In highland Peru industrial development is limited and dependent, so the migrant response of the indigenous peasantry gives rise to a reliance on intra-generational social relations, as in African migration,[5] rather than on inter-generational social relations, as under European industrialization.[6] In African townships young male migrants re-establish similar bonds to those existing in the rural areas. These young men help one another find jobs, share their lodgings, and spend their free time together. It is this cohort pattern of migrant relations that resembles the Peruvian situation, rather than the father–son pattern of European industrial communities.

The similarities in the backgrounds of migrant *obreros* and *empleados* and the differences in their life-histories are revealed by analyses of the vertical and lateral social mobility accompanying migration. In industrialized societies, analyses of proletarian mobility focus mainly on vertical mobility, since there are few alternatives to industrial work. In societies with limited industrial development, however, lateral moves into other sectors, such as agricultural and commerce, must also be considered.

The forty-seven Ataurino and Matahuasino *obreros* and *empleados* in La Oroya were asked to assess whether moving into an industrial occupation had meant a change in their economic status (see Table 4.2). The majority of *obreros* thought that to become an *obrero* was to maintain one's economic status as 'low' or 'middle'. Around one-third thought that the move had improved their status from 'low' to 'middle'. Three quarters of the *empleados* thought that becoming an *empleado* was a rise in status from 'low' to 'middle', however. This confirms the finding that both groups are from similar backgrounds, since becoming an *empleado* is a recognized improvement of social status.

TABLE 4.2   Forty-seven Ataurinos and Matahuasinos in La Oroya: Perception of Status Change by Occupational Group

Occupational group	% maintained status		% changed status	
	Low	Middle	Low to middle	Middle to low
Obreros	30	23	38	8
Empleados		25	75	

Although the backgrounds of the two groups are similar, their careers have differed. Whereas the *empleados* have remained in one sector of the economy, working their way up the industrial occupational ladder, *obreros* have tended to move back and forth from sector to sector. Nine-tenths of *empleados'* working lives have been spent as *obreros* and *empleados*, *obreros* in contrast have spent one-sixth of their working lives as peasants. Here, this low proportion for the *obreros* is due to the poor peasant economy of Ataura, which cannot support much agricultural activity. *Obreros* tend to oscillate between industry and agriculture as industrial and life-cycles turn.

TABLE 4.3 Forty-seven Ataurinos and Matahuasinos Resident in La Oroya: Geographical and Occupational Mobility. Number of Years *Empleados/Obreros* Stay in the Same Department, Province, Village, Occupation, with the Same Employer, in the Same Economic Sector, on Average

	Department	Province	Village	Occupation	Employer	Economic sector
Empleados	16.4	7.0	6.4	6.1	7.4	12.2
Obreros	11.0	6.2	5.2	5.5	5.5	4.9

Table 4.3 shows that, on average, *obreros* change their occupations every five and a half years. The most striking feature of the figures is the high rate of geographic and occupational mobility. Second is the difference in rates between *obreros* and *empleados*. The average sectoral changes for the *obreros*, every 4.9 years, are very rapid and in marked contrast to the sectoral stability of the *empleados*. The different strategies of the two groups are quite clear. The *obreros* oscillate rapidly between jobs, employers, and economic sectors as they establish a survival strategy, working where and when they can. The village column shows how they move often from their natal village to other villages and towns, back to their natal village and then emigrate again. The *empleados*, in contrast, although they may change jobs frequently, do so on account of a promotion within their specific economic sector. Again, the mobility characteristics of the *obreros* may be contrasted with the immobility of the industrial communities created by European industrialization.[7]

The general mobility rates, although differing between *obreros* and *empleados*, do not differ between Ataurinos and Matahuasinos. The differences between the villagers lie in the general availability of more land and economic opportunity in Matahuasi, and the particular strategies established by richer and poorer migrants. However, it is at this point, over differences in migrant strategies, that differences between the groups of villagers re-emerge alongside the *obrero–empleado* differences.

Among the villagers living in La Oroya, the *obreros* did not want to stay long there. Over half of them wanted to stay less than another five years, in contrast to the *empleados*, three-quarters of whom wanted to stay longer than five years. The reasons for the difference are clear. *Obreros'* work is unstable, dangerous and difficult. It is also just one job within an overall migration strategy, so the *obrero* is always looking out for better opportunities. He cannot retire with pay until he is 60 and so he aims to accumulate savings to buy some tools, some land, or a house. *Empleados'* work, however, is well paid and they can retire in their forties. So the *empleados* plan to stay on until they reach retirement age.

On leaving, all want to be independent workers, but whereas both *empleados* and *obreros* want to become peasants and traders in equal proportions, such is not the case for these migrants considered in terms of their status (as orginating from different villages). Half the Matahuasinos want to return to farming, compared to only one-quarter of the Ataurinos. On the other hand, half the Ataurinos want to take up skilled practical trades, compared to only one-third of the Matahuasinos. Similarly, the destinations of different villagers diverged. Although, among both occupational groups, one-half wished to return to their village, one-third to a highland town, and one-tenth to Lima, the village groups differed. Two-thirds of the Matahuasinos wanted to go back to their village, but only two-fifths of the Ataurinos had this aim. This is because there were more economic opportunities in the larger village economy of Matahuasi. Those who did wish to return to Ataura were the older men, on the point of retiring, with an *empleado* or *obrero* pension. The younger Ataurinos opted for taking up jobs in Huancayo and Jauja, realizing that the village economy could not support them.

Whether or not aspirations and strategies are realized cannot be predicted in each case. It is clear that many migrants become enmeshed in family obligations, dependent work, and failing health, and their only hopes for the future lie with their children. Yet the presence of so much circular migration in the region indicates that men do go to the mining sector, do save cash and gain skills, and do then change their work and life-styles. The opportunities for saving are there, through secondary occupations, 'time' money accumulated for years worked, high *empleado* earnings, or subsisting on village products whilst saving the industrial wage. Most of the older men admitted that they had some savings put by, which in the end was a little more than they would have had if they had remained as peasants.

Migration is thus an attempt by highlanders to solve the problems of employment and cash accrual which confront them. The problems arise because of the growth of population on the land, the inability of household lands to provide for all the members of the household, and the change in expectations on the part of peasants who now want to accumulate some capital in the form of savings or consumer goods. The migration occurs within a context of limited and unstable dependent industrial development: it is a strategy of response to that development. The elements of the migratory process are social networks covering different economic sectors, oscillation between sectors, and limited involvement in any one sector. These elements are the characteristics of a migrant labour force, and as such they are in contrast to the characteristics of an industrially involved proletarianized labour force.

Notes

1. 1961 *Censo del Perú*.
2. N. Long and B. R. Roberts *Peasant Co-operation and Development in Peru* (Austin: Texas University Press, 1978).
3. J. Goldthorpe, *The Affluent Worker* (Cambridge: Cambridge University Press, 1968).
4. The following analysis is based on interviews and discussions with migrants from Ataura and Matahuasi in La Oroya and the two villages: A. Laite, 'The Migrant Worker', Ph.D. thesis, Manchester University Press, 1977.
5. P. Mayer, *Townsmen or Tribesmen* (London: Oxford University Press, 1971).
6. P. Willmott and M. Young, *Family and Kinship in East London* (London: Pelican, 1957).
7. J. Goldthorpe, *The Affluent Worker*.

5

Overurbanization Reconsidered*

Josef Gugler

THE proposition that Third World countries are characterized by over-urbanization was widely accepted in the 1950s and into the 1960s. The relationship of level of urbanization and degree of industrialization provided the basis for either a synchronic or a diachronic argument. In cross-section analysis, countries such as Egypt and Korea were shown to deviate from the general pattern of the relationship; in historical comparison, Third World countries were shown to have a degree of industrialization lower than that which characterized First World countries at comparable levels of urbanization in the past. N. V. Sovani's devastating critique of both approaches led to the precipitate retreat of the advocates of the overurbanization thesis; the very notion was all but banned from academic discourse.[1] Unheeded went Sovani's cautionary note that the subject of over-urbanization needed to be investigated further. We do not propose to resuscitate comparative arguments but will focus instead on the economic implications of the rapid urban growth that characterizes most Third World countries.

Third World cities have substantial surplus labour in various guises. Their labour force continues nevertheless to increase, swelled not only by natural population growth but also by rural–urban migration that on average contributes two-fifths of the urban growth in most Third World countries.[2] The process may be labelled 'overurbanization' in so far as (a) rural–urban migration leads to a misallocation of labour between the rural and the urban sectors and (b) rural–urban migration increases the cost of providing for a country's growing population.

Most rural–urban migrants correctly assess that they are improving their life-chances. A paradox arises between the rationality of the individual and

* This is a revised version of an article first published in *Economic Development and Cultural Change*, 31 (1), 1982. It is reprinted here by permission of the University of Chicago Press. Earlier versions were presented to the Migration and Development Seminar at the Center for International Studies, Massachusetts Institute of Technology, Dec. 1978, and to the Conference on Urbanisation in West Bengal at the Centre for Urban Economic Studies, University of Calcutta, Dec. 1980. I wish to thank, without implicating, William G. Flanagan, Alan G. Gilbert, Peter Kilby, and Donald C. Mead for helpful comments.

small group decisions to migrate and the irrationality of the migratory movement when considered at the level of the national economy. This micro/macro paradox is resolved when the migratory movement is seen as a mechanism that allows some of the disadvantaged rural population to partake in a small measure of the resources disproportionately concentrated in urban areas. In the absence of effective policies to redistribute productive resources and/or income across the rural–urban divide, rural–urban migration can be argued to contribute to economic development, defined to include the distributional aspect.

Urban Surplus Labour

Third World cities are characterized by an excess of labour with limited skills. Open unemployment constitutes only one facet of urban surplus labour. A second element is underemployment, that is, the tasks at hand could be satisfactorily carried out by fewer persons. Finally, substantial numbers, while perhaps fully employed, produce goods or provide services that can be judged to contribute little to social welfare; such persons may be labelled 'misemployed'.

Unemployment

Information on open urban unemployment in developing countries is notoriously problematic. First of all, there are few data. The most comprehensive effort at compilation, David Turnham's, provides figures for 12 Asian countries, seven African countries, and 20 countries and dependencies in the western hemisphere; the data are limited to one or a few major cities for several African and nearly all American countries. The rates range from 26.6 per cent in Algeria to 1.6 per cent in India.[3] A. Berry and R. H. Sabot provide more recent data for open unemployment for seven Asian, one African, and five Latin American countries. The rates range from 16.9 per cent for Sri Lanka to 1.3 per cent for Thailand. Regional estimates by the International Labour Office for 1975 put open urban unemployment at 6.9 per cent for Asia, excluding China and other centrally planned economies; 10.8 per cent for Africa; and 6.5 per cent for Latin America.[4] In China, urban unemployment has been limited by controls on rural–urban migration and by 'sending-down' campaigns, most notably the rustication of middle school leavers that took around 17 million youths to the countryside during the Cultural Revolution.[5] When these policies were relaxed or reversed altogether, substantial urban unemployment appeared; a low estimate puts unemployment at 10 per cent of the labour force in the non-agricultural sector in 1979.[6]

Second, there is good reason to doubt how completely urban populations are covered by censuses and how accurately they are represented in surveys. There is probably a systematic bias in that low-income groups tend to go under-reported; in so far as their unemployment rates diverge from the average, the overall employment rates reported are affected.

Third, the extent of unemployment reported is very much a matter of definition.[7] Is it restricted to those actively seeking work or does it cover all who are available for work, including those who have become discouraged about finding work? The distinction is likely to affect in particular the unemployment rate reported for women. This is even more the case for a further definitional issue: Are those searching/available for part-time work to be included? Finally, does part-time work disqualify a person from being considered unemployed?

The unemployed are usually not representative of the most desperate urban living conditions. In countries where very few qualify for unemployment benefits, it is only the not-so-poor family which can support an unemployed member.[8] If unemployment is frequently reported higher among the urban-born than among immigrants, it is because the families of the urban-born are more likely to be already well established in the urban economy. Given family support, an extended search for a satisfactory job can be a rewarding strategy, especially for those with better qualifications. Higher levels of unemployment among the more educated, a common pattern,[9] thus appear as a function of both, the potential rewards for the better educated of an extended job search and the fact that they tend to come from families which are able to support them through a lengthy period of unemployment. In contrast, the poorest, whose relatives and friends cannot help them, and those recent immigrants who have nobody to turn to, are forced to find some livelihood in a hurry; unemployment is a luxury they cannot afford.[10]

Underemployment

We define 'underemployment' as the underutilization of labour.[11] Such underutilization is most conspicuous where labour is idle part of the time. This is a widespread pattern in agriculture. In the urban sector seasonal fluctuations are prominent in industries related to the agricultural production cycle, in construction, and in the tourist trade. Underemployment is not limited to these sectors, however, but is much more pervasive.

Underemployment takes three distinct forms. In one guise it is related to fluctuations in economic activity during the day, for example, at markets; over the week or month, for example, in recreational services; or seasonally. As activity ebbs, casual labour is laid off and many self-employed are without work. Underemployment takes a second form where workers are so numerous that at all times a substantial proportion are less than fully employed, that is, a reduction in the number of workers would not affect aggregate output. In terms of numbers affected, street vendors constitute the most important category in many countries.[12] A third type of underemployment is what may be appropriately called 'hidden unemployment': solidary groups that continue to employ all their members rather than discharging them when there is insufficient work to keep them fully occupied. Such guaranteed employment is typical of family enterprise, but social ties other than kinship proper—for example, common origin or shared religion—can also provide a commitment to maintain every member of the community.[13]

In his study of the urban labour force in Tanzania, R. H. Sabot defined as underemployed those urban wage-earners and own-account workers whose earnings were below average rural income. By this criterion 10 per cent of the urban labour force was underemployed in 1970; another 10 per cent was unemployed—a fifth of the urban labour force could thus be considered surplus labour. Still, in the context of Tanzania, the loss in potential output was limited. Since the urban labour force was rather small—perhaps 6 per cent of the total labour force—the output to be gained by the transfer of the urban surplus labour to the rural sector was unlikely to add more than 1 or 2 per cent to national income.[14] In countries where larger proportions of the labour force are urban, similar levels of urban unemployment and underemployment suggest a much more substantial loss of potential output.

Misemployment

Finally, there is what we propose to call misemployment. Labour may be employed full time, but the tasks performed contribute little to social welfare. Begging is a clear-cut example. More respectable, but also rather unproductive, are the activities of the hangers-on in the entourages of the more powerful and affluent. There is also a wide range of illegal activities. It might be argued that the thief who redistributes resources from the wealthy to his poor family is performing a service not dissimilar to that of many bureaucrats in a welfare state.[15] And, indeed, the productivity of an activity is ultimately socially defined.

The notion of unproductive labour dates back at least as far as Adam Smith's *The Wealth of Nations*.[16] It was part of his polemic against the mercantilist state, whose purpose was to redistribute income from its more productive subjects to the sovereign. The political élite, the religious estate, and the cultural and intellectual superstructure were perceived to have a basically parasitic relationship to the productive classes. Presumably, the larger the surplus income generated by the productive sector, the larger the number of retainers and other parasites that could be supported by the ruling class. Substantial numbers of public administrators in many contemporary societies appear similarly misemployed.[17] As in the courts of yore, their role of hanger-on has become institutionalized.

Much misemployment focuses on getting crumbs from the table of the rich. The member of the local élite or middle class, the foreign technical adviser, or the tourist is begged for a morsel, or made to maintain a company of sycophants, or has his wallet snatched away. The relationship is vividly portrayed by three activities: the army of domestics that cleans and beautifies the environment of the privileged;[18] the prostitutes who submit to the demands of those able to pay, and who in the bargain become outcasts;[19] and the scavengers who subsist on what the more affluent have discarded, who literally live on crumbs from the rich man's table.[20] Admittedly, most forms of misemployment make some contribution to social welfare. In domestic service, for example, the contribution can be substantial where women with qualifications that are in short supply are released from household work. What is at issue here is that large numbers of people are employed in a wasteful manner because their labour is so cheap, relative to the incomes of the élite and the middle class.[21] The point is well demonstrated by the fact that the requirements of middle-class households for domestic help rapidly decrease as domestic wages rise.

The Opportunity Cost of Rural–Urban Migration

An evaluation of the economic implications of rural–urban migration has also to take into account the consequences at the rural end.[22] If our assessment of the marginal productivity of labour in the urban economy has to remain tentative, we now enter into the treacherous realm of counter-factual analysis: What would have happened if some of the migrants had stayed in their rural homes?

For a time the assumption was commonly made that the marginal productivity of rural labour is zero in much of the Third World.[23] Countries such as Bangladesh, India, Pakistan, Egypt, Ruanda, Burundi, and Jamaica and major regions such as Java in Indonesia were seen to be so severely over-populated that labour was redundant in the rural areas. Elsewhere, particularly in much of Latin America, the mass of the rural population has no access to land, either because of institutional barriers— for example large landholdings controlled by absentee owners frequently are farmed in a rather extensive manner—or because of a lack of resources to open up virgin land, particularly in the Amazon basin. The argument gains in strength if it is remembered that the rural exodus is more than compensated for by natural population growth in nearly every country, that is, the rural population continues to increase in absolute numbers in spite of emigration. Admittedly, because of the age selectivity of out-migration, a rural population may increase even while the rural labour force is reduced. Thus the 1960–1 population survey in Burkina Faso found a third of the men away from their homes.[24] But such instances appear to be exceptional, even at the regional level within countries.

In recent years the emphasis has shifted to a recognition that agriculture is characterized by labour bottle-necks during planting and/or harvesting periods, that even where population pressure on land is severe all hands are needed at certain seasonal peaks of labour requirements.[25] Certainly, where additional land can be brought under cultivation, as is the case in nearly every country in Subsaharan Africa, the emigration of able-bodied adults implies a loss of potential output. The argument applies in other areas if institutional obstacles to the more intensive farming of land are not taken for granted, or if the opportunities for opening up virgin land are seriously considered.

In countries that are characterized by severe population pressures, these very pressures may encourage changes in agricultural practice that increase output.[26] These can range from an increase in the frequency with which land is cropped, to irrigation, to higher-yield crops, and to the use of fertilizer. Recognition that agricultural practice is not static and of the need for innovation, especially where the land/man ratio is unfavourable, leads to a full appreciation of the opportunity cost of migration that is selective in terms of age and education. The rural areas lose the young, the more educated, and, we may surmise, the most enterprising.

Effective demand for additional agricultural output is manifest in those Third World countries that have become dependent on food imports as their agriculture proved unable to supply the growing urban population. Certainly, in every country there is scope for improving nutritional standards. Finally, there usually exist opportunities for boosting exports of agricultural products.

We conclude that rural–urban migration entails a loss of potential agricultural output where uncultivated land is still available, where virgin lands could be developed, and where institutional restraints on the intensification of farming could be overcome. We add that the disproportionate loss of the young, the educated, and the enterprising delays innovation where it is most needed, that is, where population pressure on land appears most severe, given present farming methods.

So far we have restricted our argument to agricultural production, but a substantial proportion of the rural population is engaged in non-farm activities. Dennis Anderson and Mark W. Leiserson have presented data for 15 Third World countries showing that from 12 to 49 per cent of the rural population are primarily engaged in non-farm work.[27]

Again, it is the out-migrants who would be best equipped to develop such activities.

A. Berry and R. H. Sabot conclude from their review of the available evidence that labour utilization by farm families is high. Where there is a substantial seasonal element in the labour requirements for agriculture, the average annual labour input into agriculture may be low, for example in parts of Africa, but seasonal labour bottle-necks limit expansion of agricultural production under existing technology, and a considerable amount of time is spent on non-farm activities such as crafts and trading during the off-season.[28]

The Relative Cost of Urban Services and Goods

We have seen that rural–urban migration brings workers to cities that are already burdened with surplus labour, and it now further appears that rural–urban migration entails some loss of potential rural output. If it is plausible that rural–urban migration, at current levels, leads to a misallocation of labour between rural and urban areas, we need to address a second issue that appears to be quite clear-cut: How does the resource cost of providing for a rapidly expanding urban population compare with the task of absorbing such additional numbers in rural areas? Housing, transport, garbage and sewage disposal, provision of fuel, and distribution of staple foods stand out as five amenities that are expensive in urban agglomerations but cheap or not required in rural areas. Whether the rural–urban migrant manages to pay for them, urban hosts provide them, or they are subsidized by public authorities, the higher cost of these amenities will usually more than offset the savings that arise in the provision of other services or goods—for example electricity, pipe-borne water, health care—that come cheaper to urban population concentrations than to a dispersed rural population. This is especially the case if the composition of a low-income budget is taken into account.

The pressure rural–urban migrants exert on existing urban infrastructure, and the cost of new infrastructure required by their presence, have received little attention in the literature on Third World urbanization.[29] This curious silence may be grounded in the tacit assumption that the urban poor remain marginal in that they have neither the political clout nor the market power to make effective demands. Thus, discussions of squatting centre on the diseconomies inherent in such initiatives by the poor to provide housing for themselves but fail to focus on the opportunity cost of land, materials, and labour.[30] Even in shanty towns such costs are substantial. Indeed, the value of the land to be put to alternative uses frequently provides the primary motivation for slum-clearance schemes, opposition to squatting, and demolition of squatter settlements.

W. Arthur Lewis has observed that in international finance in the nineteenth century the distinction between the European lenders and their rich borrowers—Australia, the United States, Canada, Argentina—turned on differences in rates of urbanization. Countries whose urban populations were growing more slowly lent to rich countries that were characterized by more rapid urban growth.

Urbanization is a decisive factor because it is so expensive. The difference between the cost of urban development and rural development does not turn on the difference of capital required for factories and that required for farms. Each of these is a small part of total investment, and the difference per head is not always in favour of industry. The difference turns on infrastructure. Urban housing is much more expensive than rural housing. The proportion of urban children for whom schooling is provided is always much higher, at the stage where less than 60% of children are in school. The town has to mobilize its own hospital service, piped water supplies, bus transportation. In all these respects the towns require more per head in terms of quantity than rural areas, but even if quantities per head were the same, urban facilities would cost more in money terms than rural facilities. Rural people do more for themselves with their own labor in such matters as building houses, or working communally on village roads or irrigation facilities. When they hire construction workers they pay less, both because of a generally lower price level and because they are not faced with powerful construction unions. Rural people also do not hire architects.[31]

To the extent that even marginal migrants have access to better amenities than are available in rural areas, their migration has a redistributive effect, an issue we will address in a moment. This redistributive dimension should not obscure the issue at hand, that is, that the provision of a subsistence minimum is more costly for additions to urban than to rural populations.

A major part of the costs of providing for increases in the urban population are not borne, in market economies, by those who take decisions affecting the location of jobs and people. In centrally planned economies, in contrast, the higher cost of providing for additions to the populations in urban as against rural settings is readily apparent to policy-makers. Numerous studies have shown a clear, regular, positive relation between level of urbanization and per capita GNP. Gur Ofer established the relation for 23 market economies that included all countries with per capita GNP of $280 or more in 1960, but excluded several countries with exceptionally high service shares. When adding the six East European socialist countries, the Soviet Union, and Yugoslavia to the sample, he found a clear negative deviation from the relation for them. The individual residuals were negative for all but the German Democratic Republic. Several policies pursued in these countries appear designed to economize on the costs of urbanization. Socialist countries use a more labour-intensive technique in agriculture and a more capital-intensive technique in manufacturing than do market economies; they encourage a high level of labour participation, focusing especially on women; and they allocate a small number of workers to urban services. The urban labour force required at a given level of industrialization is thus kept relatively low.[32] In addition, substantial numbers of workers are compelled to reside outside the cities and to commute because the provision of housing has typically failed to keep up with the expansion in industrial jobs. If the housing shortage makes it difficult to establish residence in cities, administrative measures expressly restrict in-migration to major cities such as Moscow. György Konrad and Ivan Szelenyi have coined the term 'underurbanization' to characterize this pattern.[33]

In China, after the rapid growth—of the order of 7.9 per cent per year—that characterized the 1950s, the curtailment of rural–urban migration and repeated rustication campaigns resulted in actual declines in the urban population in the early 1960s, followed by slow urban growth—in part a function of a sharp drop in urban fertility—of the order of 2.1 per cent per year between 1965 and 1978.[34] This slow urban growth, combined with new employment opportunities for women in neighbourhood industries, resulted in a remarkable increase in the employment rate from approximately one-third in the mid-1950s to about half of the urban population in the mid-1970s.[35]

The Economic Rationale for Rural–Urban Migration

If rural–urban migration has an opportunity cost of rural output forgone, if it brings workers to cities that are already unable to employ fully their labour force to productive ends, and if additions to the urban population require more resources for survival than their rural counterparts, then the conclusion is warranted that rural–urban migration at current rates is inefficient. The label 'overurbanization' seems in order.[36] In the face of this condition rural–urban migration proceeds at a rapid pace. The majority of migrants are ready to enter the labour force. If they contribute about two-fifths of urban growth, they constitute a considerably higher proportion of the new entrants into the labour force.[37]

A substantial body of research on rural–urban migration has been accumulated over the last two decades, and the evidence is overwhelming: most people move for economic reasons. When migrants are asked why they moved, they usually give the better prospects in the urban economy as their chief reason. Migration streams between regions have been shown to correspond to income differentials between those regions. And over time, as economic conditions at alternative destinations change, migration streams switch accordingly. Finally, studies throughout the Third World report time and again that the great majority of migrants consider that they have improved their condition and that they are satisfied with their move.[38]

If there is a measure of agreement on this basic proposition—that rural–urban migration is largely motivated by economic reasons—two rather different interpretations have evolved. In tropical Africa, analysis focused on migrants coming in search of jobs that offered wages and working conditions regulated by legislation and/or collective bargaining. They would spend several months trying to secure such a job, but, if unsuccessful, eventually return to the village. In the 1950s and 1960s much urban unemployment in tropical Africa appears to have conformed to this pattern. With independence, urban wages rose substantially in many countries. Rural–urban migration surged, the labour shortages that had plagued colonial governments vanished, and urban unemployment appeared. Since much labour migration had been short-term up to this time, recent immigrants faced little competition from entrenched urban workers and their descendants. Furthermore, independence was frequently accompanied by a significant expansion in urban employment. The system of recruiting unskilled labour approximated a random process. Since minimum wages were high relative to rural incomes, even an extended job search was a promising strategy. Joining the urban unemployed, the rural–urban migrant tried his luck at the urban job lottery.[39]

In retrospect it is clear that the urban job lottery pattern occurred in exceptional circumstances. More commonly, labour turnover is low, job creation slow, and recruitment anything but random in Third World countries. Access to the better earning opportunities, and in particular to jobs in the protected sector, is severely restricted. Employers have established criteria for recruitment into different job categories, for example formal educational qualifications, experience, age, sex. And workers who hold protected jobs take advantage of opportunities to assist family members, kin, and friends in joining them.[40] The recent immigrant who has come without the right connections or exceptional qualifications is unlikely to secure a protected job, even if he can afford an extended search. Most migrants have to settle instead for less attractive opportunities; they become part of the marginalized labour force.[41] Still, their earnings are superior to what the rural areas have to offer.

The Distributional Dimension

Our earlier discussion revolved around a concern with the implications of rural–urban migration for aggregate output. The indications that its contribution to urban output is problematic, that it entails a loss of potential rural output, and that it increases demands on scarce resources led us to conclude that rural–urban migration at current levels is inefficient. Now we have argued that rural–urban migration is a rational response to the economic realities ruralites face, that there is advantage to be gained from the move. The resolution of the seeming paradox derives from the fact that rural–urban migration has a redistributive effect. Rural–urban migrants lay claim to a share in urban income opportunities, they gain some access to urban amenities. Rural families send their sons and daughters to the city so that they will be able to partake, however little, and in whatever demeaning way, of its riches. As Larissa Adler Lomnitz put it in her study of a shanty town in Mexico City:

The settlers of Cerrada del Cóndor may be compared to the primitive hunters and gatherers of preagricultural societies. They go out every day to hunt for jobs and gather the uncertain elements for survival. The city is their jungle; it is just as alien and challenging. But their livelihood is based on leftovers: leftover jobs, leftover trades, leftover living space, homes built of leftovers.[42]

The condition of those who stay in the rural areas may improve as migrants remit part of their income to their family, provide villagers with access to urban amenities (e.g. health care), assist village development, and press village interests with officials at the regional or national level.[43] Or, in some cases, rural conditions can be argued to be at least somewhat better than they would have been if the population pressure on land had become even more severe.

There thus appears to be a trade-off between a misallocation of labour and an improved income distribution.[44] In terms of a concept of economic development that includes a distributional dimension, an evaluation of rural–urban migration then depends on an assessment of its consequences for both output and distribution, and on the respective weighting these are given.[45]

Of Policy and Power

There would appear to be an approach that promises a more efficient allocation of labour between the rural and the urban sector as well as a reduction in the extreme inequalities that characterize most Third World countries. It will aim at improving rural living standards by channelling productive resources to the rural areas and/or directing a larger share of income to them.[46] This is not an original policy prescription for a problem that has been recognized for a long time. If policy-makers have rarely given more than token recognition to such advice, their inaction reflects not on the economic and social soundness of the advice but on their priorities. Michael Lipton phrases the issue in terms of 'urban bias'. Echoing the position taken by Julius Nyerere, the president of Tanzania, in the Arusha Declaration,[47] he argues:

The most important class conflict in the poor countries of the world today is not between labour and capital. Nor is it between foreign and national interests. It is between the rural classes and the urban classes. The rural sector contains most of the poverty, and most of the low-cost sources of potential advance; but the urban sector contains most of the articulateness, organization and power. So the urban classes have been able to 'win' most of the rounds of the struggle with the countryside; but in so doing they have made the development process needlessly slow and unfair. Scarce land, which might grow millets and beansprouts for hungry villagers, instead produces a trickle of costly calories from meat and milk, which few except the urban rich (who have ample protein anyway) can afford. Scarce investment, instead of going into water-pumps to grow rice, is wasted on urban motorways. Scarce human skills design and administer, not clean village wells and agricultural extension services, but world boxing championships in showpiece stadia. Resource allocations, within the city and the village as well as between them, reflect urban priorities rather than equity of efficiency. The damage has been increased by misguided ideological imports, liberal and Marxian, and by the town's success in buying off part of the rural elite, thus transferring most of the costs of the process to the rural poor.[48]

Cities are centres of power and privilege. In every Third World country, the urban sector accounts for a disproportionate share of consumption as well as investment.[49] In many cases these are even more concentrated in a primate city, typically the national capital. Certainly, industry and modern transport find advantage in spatial concentration. They require urban infrastructure and highly trained specialists. But it would seem that resources for both consumption and investment are invariably apportioned to urban areas beyond the immediate requirements of industrialization. If this proposition is granted, two characteristics of the contemporary setting may be adduced to explain it. On one hand, nearly all those enjoying high standards of consumption are urban-based. Unlike the feudal order of Europe's Middle Ages, and unlike the hacienda system that characterized much of Latin America in the colonial and early post-colonial days, throughout the Third World today virtually the entire élite is located in cities. Much of the middle class of senior civil servants, substantial traders, managers, and professionals also live there, many out of necessity, some by choice. Finally, the protected sector of the urban labour force enjoys a standard of living that is high when compared to the condition of the rural masses and many urban workers. On the other hand, surplus that was appropriated by private interests under *laissez-faire* capitalism, that was drained abroad in the colonial system, is to a considerable extent controlled by the state. A large part of this surplus goes into public works that are allocated to a hierarchy of places in which the capital city frequently receives the lion's share while rural areas face neglect.[50] Public works are attractive to policy-makers not only because of their visibility but also because their execution usually entails rather unfettered control over large resources.[51] The rationale of their allocation is found in the interests of decision-makers in improving their immediate environment, in assuring the continued collaboration of the middle class, and in placating strategically placed elements of labour. A large part of the state-controlled surplus is thus absorbed in conspicuous investments for the few—in the cities.[52]

Notes

1. N. V. Sovani, 'The Analysis of Over-Urbanization', *Economic Development and Cultural Change*, 12 (1964), 113–22.
2. Samuel H. Preston, 'Urban Growth in Developing Countries: A Demographic Reappraisal', Chapter 1 in this volume.
3. David Turnham, *The Employment Problem in Less Developed Countries: A Review of Evidence* (Paris: OECD Development Centre, 1970).
4. R. Albert Berry and Richard H. Sabot, 'Labour Market Performance in Developing Countries', *World Development*, 6 (1978), 1199–1242.
5. R. J. R. Kirkby, *Urbanisation in China: Town and Country in a Developing Economy 1949–2000 AD* (London: Croom Helm, 1985), 21–53.
6. John Philip Emerson, 'Urban School-Leavers and Unemployment in China', *China Quarterly*, 93 (1983), 1–16.
7. Turnham (pp. 197–201) discusses the Indian data in some detail and suggests that applying more generous definitions of unemployment would roughly double the rate reported in his compilation, leaving it still exceptionally low. Berry and Sabot (1978, 1212), put open urban unemployment in India at 3% in 1971; and J. Kirshnamurty, in 'Some Aspects of Unemployment in Urban India', *Journal of Development Studies*, 11 (1975), 11–19, reinforces the impression that it is indeed remarkably limited.
8. R. Albert Berry, in 'Open Unemployment as a Social Problem in Urban Colombia: Myth and Reality', *Economic Development and Cultural Change*, 23 (1975), 276–91, offers a comprehensive discussion of voluntary unemployment and presents data suggesting its importance. His study, as well as Richard H. Sabot's *Economic Development and Urban Migration: Tanzania, 1900–1971* (Oxford: Clarendon Press, 1979) and Dipak Mazumdar's *The Urban Labour Market and Income Distribution: A Study of Malaysia*, (New York: Oxford University Press, 1981), indicates that the pool of unemployed comprised predominantly the young and married women.
9. Lynn Squire, *Employment Policy in Developing Countries: A Survey of Issues and Evidence* (New York: Oxford University Press, 1981), 71, summarizes data

on rates of unemployment by level of education in ten Third World countries. In eight cases the rate is highest, frequently by a large margin, for those with secondary education. However, Kenya, the only African country represented, shows a regular decline in unemployment as level of education increases; this may reflect an educational system that was only beginning to catch up, in 1970, with manpower requirements. It is noteworthy that unemployment is lower among those with post-secondary than among those with only secondary education in every country, including countries such as the Philippines, Sri Lanka, and India that are notorious for widespread unemployment among college graduates; presumably college students are in a better position to prepare for their transition into the labour market.

10. Janice E. Perlman, in *The Myth of Marginality: Urban Poverty and Politics in Rio de Janeiro* (Berkeley, Calif. University of California Press, 1976), found that among immigrant squatters the best jobs went to those who could afford to take their time and be selective. When men who came from unskilled urban jobs or agriculture were compared with those who previously had skilled employment, the proportion who secured their first job within a month was higher among the less qualified. Also, the proportion who found a job within a month was higher among those who knew no one upon arrival than among those with contacts.

11. Most studies focus on earnings instead, either because the researcher's primary concern is with urban poverty or because earnings data are more easily available and are assumed to reflect productivity.

12. Such underemployment can be a quite stable feature among self-employed who have low overhead costs and are assured of a minimum of customers because of personal ties and/or locational advantages.

13. A comparison with the Japanese permanent employment system springs to mind, but the latter precisely fails to provide for workers outside the major firms and indeed the substantial proportion of casual workers in these firms. The analogy of the commitment to full employment in socialist countries is more apt. In pre-revolutionary Cuba most workers in the sugar fields and the sugar-mills were unemployed for a major part of the year. Since the revolution they are offered employment throughout the year, and major efforts have been directed toward absorbing them in productive activities during the off-season. As in family enterprise, the problem becomes to what extent full employment, while desirable on equity grounds, only hides unemployment instead of employing workers to productive ends.

14. Sabot, *Economic Development and Urban Migration*, 149–77.

15. Illegal activities frequently come to be tolerated as necessary to the very survival of a major part of the urban population. The acceptance of urban squatting is the most salient example.

16. Martin T. Katzmann, *Cities and Frontiers in Brazil: Regional Dimensions of Economic Development* (Cambridge, Mass.: Harvard University Press, 1977), 168–72.

17. More than a million menials work as messengers and guards in the Indian Civil Service. According to one calculation they are on the average usefully employed for 12 minutes a day (Angus Maddison, *Class Structure and*

Economic Growth: India and Pakistan since the Moghuls (New York: W. W. Norton & Co., 1971), 95.) If this estimate appears exaggerated, few would deny that there is a good deal of redundancy in this category as well as in other ranks of the bureaucracy in India—as in many other countries.

18. In Brazil 34% of female labour outside agriculture was in domestic service around 1970. A decade earlier the proportion was 36% in Chile, 45% in Colombia, 31% in Peru, and 32% in Venezuela (Juan C. Elizaga, 'The Participation of Women in the Labour Force of Latin America: Fertility and Other Factors', *International Labour Review*, 109 (1979), 519–38). Almost 40% of working women in Chile were employed as maids, washerwomen, or ironing women in individual homes in 1970 (Michele Mattelart, 'Chile: The Feminine Version of the Coup d'État', in *Sex and Class in Latin America*, June Nash and Helen Icken Safa (eds) (New York: Praeger Publishers, 1976)). In Guatemala not less than 40% of the female labour force outside agriculture were domestics in 1973 (Norma S. Chinchilla, 'Industrialization, Monopoly Capitalism, and Women's Work in Guatemala', *Signs: Journal of Women in Culture and Society* 3 (1977), 38–56). In Mexico 20% of working women were domestic servants in 1969 (Gloria Gonzalez Salazar, 'Participation of Women in the Mexican Labor Force', in Nash and Safa (eds)). The conspicuous use of labour may be argued to be preferable to the conspicuous consumption of imported luxury goods precisely because it provides local employment; in addition, it saves usually scarce foreign exchange. Such an argument takes for granted extreme income inequalities and, in the case of female domestics, sex discrimination in the labour market. The image of the nursemaid who fulfils her charge's every whim while her own children suffer from neglect is all too disturbing.

19. The judgement that some labour is unproductive because its output is ethically undesirable, e.g. prostitution, can also be traced back to Adam Smith. A Scottish professor of moral philosophy, he was less willing than his neo-classical successors to accept the legitimacy of given preferences and was prepared to state that some preferences are better than others. The proportion of urban women who derive their livelihood from prostitution is substantial in most Third World countries; to take a notorious case: more than one million girls and women are said to be working in the sex industry in Thailand.

20. Cigarette-butt collecting constitutes a major street occupation in Jakarta, according to Gustav F. Papanek, 'The Poor of Jakarta', *Economic Development and Cultural Change*, 24 (1975), 1–27.

21. Robert Fiala, 'Inequality and the Service Sector in Less Developed Countries: A Reanalysis and Respecification', *American Sociological Review*, 48 (1983), 421–8, shows that inequality is a significant factor in expanding the service sector. His analysis of 23 less-developed countries found concentration of income in the top 20% of the population in 1960 to be associated with an expanded service sector in 1970.

22. For a comprehensive review of the literature and of research on the rural consequences of rural–urban migration, and an economic analysis, see Henry Rempel and Richard A. Lobdell, 'The Rural Impact of Rural–Urban Migration in Low Income Countries' (manuscript, 1977).

23. Thus the Lewis–Fei–Ranis model assumed a rural sector characterized by zero- or very-low-productivity labour (Arthus W. Lewis, 'Economic Development with Unlimited Supplies of Labour', *Manchester School of Economic and Social Studies*, 22 (1954), 139–91; Gustav Ranis and John C. H. Fei, 'A Theory of Economic Development', *American Economic Review*, 60, (1961), 522–65).

24. Ambroise Songre, Jean-Marie Sawadogo, and George Sanogoh, 'Réalités et effects de l'émigration massive des Voltaïques dans le contexte de l'Afrique Occidentale', in *Modern Migrations in West Africa*, ed. Samir Amin (London: Oxford University Press, 1974); a slightly abridged and revised English version is Ambroise Songre, 'Mass Emigration from Upper Volta: The Facts and Implications', *International Labour Review*, 108 (1973), 209–25.

25. Seasonal migration, where it is geared to the agricultural production cycle, appears optimal from this angle. As long as a substantial part of the established urban labour force is not fully employed, however, the provision of even the most rudimentary urban accommodation for seasonal labour appears to be an unnecessary drain on scarce resources and counterproductive.

26. Ester Boserup launched the contemporary debate on the relationship between population growth and food supply in *The Conditions of Agricultural Growth: The Economics of Agrarian Change under Population Pressure* (London: Allen & Unwin, 1965). She argued that population growth is the prime cause of agricultural change and focused on the more frequent cropping of the available land as the principal mechanism. In *Population and Technological Change: A Study of Long-term Trends* (Chicago: University of Chicago Press, 1981) Boserup emphasizes that investments in rural infrastructure are a prerequisite for the intensification of agriculture. For a review of the available evidence on the relationship between population growth and agricultural change, see David Grigg, 'Ester Boserup's Theory of Agrarian Change: A Critical Review', *Progress in Human Geography*, 3 (1979), 64–84.

27. Dennis Anderson and Mark W. Leiserson, 'Rural Nonfarm Employment in Developing Countries', *Economic Development and Cultural Change*, 28 (1980), 227–48. Apart from the serious difficulties encountered in defining rural areas, it should be noted that these figures include rural residents commuting to urban jobs.

28. Berry and Sabot, 'Labour Market Performance in Developing Countries', 1222–5.

29. For a recent review of research that bears on the cost of garbage and sewage disposal, electricity, pipe-borne water, and health care in urban as compared with rural areas, see Johannes F. Linn, 'The Costs of Urbanization in Developing Countries', *Economic Development and Cultural Change*, 30 (1982), 625–48.

30. See e.g., United Nations, 'Uncontrolled Urban Settlement: Problems and Policies', *International Social Development Review*, 1 (1968), 107–30.

31. W. Arthur Lewis, *The Evolution of the International Economic Order* (Princeton, NJ: Princeton University Press, 1978), 39–40.

32. Gur Ofer, 'Industrial Structure, Urbanization, and the Growth Strategy of Socialist Countries', *Quarterly Journal of Economics*, 90 (1976), 219–44. See Preston, Chapter 1 in this volume, for data on the high proportion of the labour

force in industry, relative to the level of urbanization, in Eastern Europe and the Soviet Union.

33. György Konrad and Ivan Szelenyi, 'Social Conflicts of Underurbanization', in *Urban and Social Economics in Market and Planned Economies: Policy Planning and Development*, Alan A. Brown, Joseph A. Licaro, and Egon Neuberger (eds) (New York: Praeger, 1974).

34. Calculated from Kam Wing Chan and Xueqiang Xu, 'Urban Population Growth and Urbanization Since 1949: Reconstructing a Baseline', *China Quarterly*, 104 (1985), 583–613. Kirkby, *Urbanization in China*, 103–33, gives somewhat different figures, but arrives at similar trends. In Cuba, the growth of the primate city was similarly restrained: Havana grew at 1.4% per year between 1966 and 1976 (Josef Gugler, '"A Minimum of Urbanism and a Maximum of Ruralism": The Cuban Experience', *International Journal of Urban and Regional Research*, 4 (1980), 516–34).

35. Thomas G. Rawski, *Economic Growth and Employment in China* (New York: Oxford University Press, 1978), 29–30.

36. Josef Gugler and William G. Flanagan, 'On the Political Economy of Urbanization in the Third World: The Case of West Africa', *International Journal of Urban and Regional Research*, 1 (1977), 272–92.

37. An analysis of census data for the metropolitan areas of 26 major Third World cities found that in the 1960s net migration accounted on average for 37% of the growth in their population but for 63% of the growth in the population aged 15–29. In all African cities covered, in almost all Asian cities, and in the majority of Latin American cities the contribution of male migrants to the growth of numbers of men in this age-group was larger than the corresponding contribution of female migrants (United Nations, *Migration, Population Growth and Employment in Metropolitan Areas of Selected Development Countries*. ST/ESA/SER. R/57 (New York: United Nations, 1985)).

38. For a comprehensive discussion of rural–urban migration, see Josef Gugler, 'Internal Migration in the Third World', in *Population Geography: Progress & Prospect*, ed. Michael Pacione (London: Croom Helm, 1986).

39. Josef Gugler, 'On the Theory of Rural–Urban Migration: The Case of Subsaharan Africa', in *Migration*, ed. J. A. Jackson (Cambridge: Cambridge University Press, 1969). The proposition that potential migrants take into account not only rural–urban real-income differentials but also the probability of securing urban employment was incorporated into a model by John R. Harris and Michael P. Todaro ('Urban Unemployment in East Africa: An Economic Analysis of Policy Alternatives', *East African Economic Review*, 4 (1968), 17–36; 'Migration, Unemployment and Development: A Two-Sector Analysis', *American Economic Review*, 60 (1970), 126–42). The probability of obtaining urban employment was defined as the proportion of the urban labour force actually employed. The assumptions underlying this definition were problematic even in the early stages of urban unemployment in tropical Africa (Josef Gugler, 'Migrating to Urban Centers of Unemployment in Tropical Africa', in *Internal Migration: The New and the Third Worlds*, ed. Anthony H. Richmond and Daniel Kubat (Beverly Hills, Calif.: Sage Publications, 1976)). For an account of subsequent modifications of the model, see Michael P. Todaro, *Internal Migration in Developing Countries: A Review of Theory, Evidence,*

Methodology and Research Priorities (Geneva: International Labour Office, 1976), 36–45. John R. Harris and Richard H. Sabot have proposed a generalized model of migration and job search in the context of wage dispersion 'and imperfect information, of which the Harris–Todaro model is a special case ('Urban Unemployment in LDCs: Towards a More General Search Model', in *Migration and the Labor Market in Developing Countries*, ed. Richard H. Sabot (Boulder, Colo.: Westview Press, 1982)).

40. Some collective-bargaining agreements in India provide that on a worker's death or retirement his son or other close relative will be employed (Mark Holmström, *Industry and Inequality: The Social Anthropology of Indian Labour* (Cambridge: Cambridge University Press, 1984), 213–14). In China, job inheritance spread during the rustication campaign in the 1970s: parents willing to retire, sometimes early, were able to pick one of their youngsters to succeed them in their job or in some other job in the work unit. This practice was later formally institutionalized (Martin King Whyte and William L. Parish, *Urban Life in Contemporary China* (Chicago: University of Chicago Press, 1984), 40 n. 24).

41. Aníbal Quijano Obrégon, 'The Marginal Pole of the Economy and the Marginalised Labour Force', *Economy and Society*, 3 (1974), 393–428.

42. Larissa Adler Lomnitz, *Networks and Marginality: Life in a Mexican Shantytown* (New York: Academic Press, 1977), 208; first published as *Como sobreviven los marginados* (Mexico City: Siglo Veintiuno Editores, 1975).

43. Josef Gugler and William G. Flanagan, *Urbanization and Social Change in West Africa* (Cambridge: Cambridge University Press, 1978), 64–73.

44. See the reaction to our initial reformulation of the notion of overurbanization by Peter Marris in 'The Political Economy of Urbanization: A Comment', *International Journal of Urban and Regional Research*, 2 (1978), 171–3.

45. For a discussion of poverty-weighted indices of development, see Montek S. Ahluwalia and Hollis Chenery, 'The Economic Framework', in *Redistribution with Growth*, by Hollis Chenery, Montek S. Ahluwalia, C. L. Bell, John H. Duloy, and Richard Jolly (London: Oxford University Press, 1974).

46. The policy options are myriad, from rural infrastructure to subsidizing fertilizer, to credit, to agricultural research, to extension services, to farm mechanization, to small-scale industries; from reducing the tax load of the peasantry, to raising prices that have been depressed by price controls or an inflated foreign exchange rate, to giving rural areas a larger share of health, educational, and other services. The effect of different approaches on the rate of rural–urban migration requires careful enquiry, e.g. mechanization may displace labour and encourage out-migration. The structure of increased demand for agricultural inputs and for consumption goods and services will affect the growth of rural non-farm employment.

47. Julius K. Nyerere, 'The Arusha Declaration', in *Ujamaa—Essays on Socialism* (London: Oxford University Press, 1968); first published in 1967 in Swahili by TANU, the party organization. The relevant section is reprinted as Chapter 2 in the present volume.

48. Michael Lipton, *Why Poor People Stay Poor: Urban Bias in World Development* (Cambridge, Mass.: Harvard University Press, 1977), 13; the introductory chapter of his book is reprinted as Chapter 3 in this volume. Lipton's

proposition has provoked a wide range of critical responses, as well as some empirical work to test it. Recently, Lipton ('Urban Bias Revisited', *Journal of Development Studies* 24 (3) (1984), 139–66) has offered a comprehensive account of these criticisms and answered them in considerable detail. Richard Charles Webb, *Government Policy and the Distribution of Income in Peru* (Cambridge, Mass.: Harvard University Press, 1977) had earlier provided a careful analysis of inter-sectoral transfers in Peru in the 1960s. He concluded that the distributive problem was not so much one of income and savings being drained from rural areas. Rather it arose from a situation where most of the income originated in the modern sector of large-scale, capital-intensive firms—and remained there. Under both the Belaúnde and the Velasco administration distributive measures operated largely within sectors: most income redistribution took place from capital to labour in the modern sector; income and property was redistributed from landowners to rural wage earners and some small farmers. If these measures were progressive, the net effect of fiscal and price policies was to worsen somewhat the position of the rural versus the urban sectors.

49. Even socialist countries, while reducing income inequality within the urban and the rural sector respectively, appear to find it difficult to deal with inter-sector inequality. In China, personal consumption has been estimated at 574 yuan in urban as against 269 yuan per capita in rural areas in 1982. The ratio of 2.1:1 suggests a dramatic improvement over a ratio of 2.5:1 in 1981, but is quite similar to that of other developing countries in Asia. The gap in personal income is accompanied by a large gap in collective consumption. The quality of education and health facilities in particular is much higher in urban areas. In 1979, an infant mortality rate of 12 per thousand in cities compared with 20–30 per thousand in rural areas according to official data, while a Chinese official reported a national average of 56, and outside observers put the rate at 13 for the city of Beijing, 17 for Beijing municipality, and 53–63 for most of the country (World Bank, *China: Long-Term Development Issues and Options* (Baltimore: Johns Hopkins University Press, 1985), 87 and 90; World Bank, *China: Socialist Economic Development: Annex B: Population, Health and Nutrition* (World Bank, East Asia and Pacific Regional Office, 1981), 6 and 97; Ding Chen, 'The Economic Development of China', *Scientific American*, 243 (3) (1980), 152–65). See also Peter Nolan and Gordon White, 'Urban Bias, Rural Bias or State Bias? Urban–Rural Relations in Post-Revolutionary China', *Journal of Development Studies*, 20 (3) (1984), 52–81.

50. For evidence that spatial patterns of government expenditure bias patterns of city growth toward a country's largest city, and particularly the capital city, see Preston, Chapter 1 in this volume.

51. David G. Epstein, *Brasilia, Plan and Reality: A Study of Planned and Spontaneous Urban Development* (Berkeley, Calif.: University of California Press, 1973), 31–41.

52. The allocation of resources produced elsewhere to cities characterized by conspicuous investment and privileged consumption is particularly striking in newly oil-rich countries where it is facilitated by direct state control over oil revenues.

6

Migration and Urban Surplus Labour:
The Policy Options*

Richard H. Sabot

URBAN unemployment is a chronic problem in Tanzania. The problem has its origins in the wide gap between rural and urban incomes for workers of comparable skill and in the responsiveness of rural residents to spatial differences in expected earnings. High, and downwardly sticky urban wages induce more workers to migrate from rural areas than there are urban high-wage jobs. The prospect of eventually finding a high-wage job means that despite unemployment in the short run, the lifetime income of an urban worker may be higher than that of a rural resident fully employed throughout his working life. Some of the excess supply of urban labour finds employment in that relatively small portion of the urban labour market characterized by flexible wages. Incomes in this sector are often below what a worker could earn in the rural areas. Transferring urban surplus labour—the unemployed and those earning less than rural marginal product—to the rural areas would increase total output. However, the output to be gained is unlikely to add more than 1 or 2 per cent to national income.

Nor are the subjective costs of urban surplus labour particularly high: for the open unemployed deprivation is lower than it would be in industrialized countries in the absence of a formal social security system. This is so, not because the unemployed in Tanzania have a greater store of assets on which they can draw, but because the subjective costs are distributed by intra-family transfer from the unemployed to those with jobs and because the agricultural sector is available as an employer of last resort. Similarly the availability of alternative opportunities in rural areas suggests that the subjective costs of employed surplus labour are low. The exception to these conclusions would be surplus workers whose contacts in rural areas are so weak as to limit their rural employment opportunities and restrict

* Reprinted, in a revised form, from Richard H. Sabot, *Economic Development and Urban Migration: Tanzania 1900–1971* (Oxford: Clarendon Press, 1979), 229–48, by permission of the publishers.

their escape from extreme privation in urban areas. However, in Tanzania the number of such workers still seems to be very small.

Low resource and subjective costs do not imply that urban surplus labour is not a social problem worthy of government remedial policy. Tanzania cannot afford to ignore any opportunity to increase national output without increasing the stock of productive resources; the reduction of surplus labour would bring an improved distribution of income to which the Tanzanian government attaches considerable importance. Unless remedial measures are adopted, the problems of surplus labour will remain or even increase. The question is whether the policies at the disposal of planners can substantially reduce surplus labour without imposing counter-vailing political or economic costs.

1. Demand-augmenting Policies

Although the origins of the problem of urban surplus labour in Tanzania are on the supply side of the labour market, this does not mean that the remedy of augmenting demand is inappropriate. Within a certain rather restrictive set of circumstances, an increase in the demand for urban labour would be sufficient to restore the balance of labour supply and demand in a relatively short period. The essence of the intersectoral misallocation model of unemployment that appears to apply in Tanzania is the non-price mechanism for rationing urban jobs by which, given a constant rural–urban income differential, an increase in demand induces the increase in supply necessary to lower the probability of finding a job sufficiently to re-establish equality between rural and urban expected incomes. The interaction, however, will not interfere with attempts to eliminate surplus urban labour where the increase in demand is sufficient to provide an urban job at the prevailing wage for all rural residents who desire one.

To assess the feasibility of a demand-augmenting remedial policy we need estimates of the scope for increases in demand and of the total number of unemployed workers who would volunteer for urban jobs if they were available. While the measurement of the openly employed is relatively straightforward, the measurement of workers in marginal employment, employed surplus labour, presents significant problems. Nevertheless, for Tanzania we have estimated that in addition to the 10 per cent of the urban labour force that is openly unemployed, another 10 per cent is employed surplus labour.[1] In the absence of direct evidence, it is extremely difficult to measure the number of workers currently employed in the rural sector who would migrate to the urban areas if guaranteed an urban job at the prevailing wage. The majority of rural workers have earnings below the urban wage. Both because of the direct and psychic costs of moving involved and because withdrawal of labour from the rural sector would have some positive impact on the incomes of remaining workers, not all of those workers would choose to migrate. Nevertheless, the size of the rural sector relative to the urban suggests that the labour forthcoming would be two or more times the current total of urban wage employment, and that urban surplus labour represents only a small proportion of the total excess demand for jobs in the urban wage sector.

The achievement of significant increases in urban labour demand requires an increase in urban investment or in the labour intensity of production. Immediate and direct employment creation either by government fiat, or, as in the case of Kenya's 1964 'Tripartite Agreement', by some 'voluntary' agreement among employers, trade unions, and government, is unlikely to generate a permanent increase of employment in a system where the aim of employers is to maximize profits. The objective of the Kenya agreement was to create 40,000 jobs in a short period, which entailed 15 per cent and 10 per cent increases in the number of employees in the public and private sectors, respectively. The unions agreed to a one-year moratorium on wage increases and strikes. While in the short run a large number of jobs were created despite the financial difficulties encountered by local authorities and private employers alike, in the longer run it does not appear that any more jobs were created than would have been without the agreement. Short-term job creation in the absence of simultaneous wage reductions could be achieved only at the expense of higher unit costs. Thus the initial effect of 'excess' labour inputs was to reduce the marginal productivity of labour to a level below the wage. Employers neutralized the short-run increase by reducing the rate of expansion of employment relative to output, or by not replacing employees who retired or otherwise left their jobs (Harris and Todaro 1968).

Government fiscal and monetary policies whose aim is to increase the demand for urban labour by stimulating effective demand for the output of goods and services are no more likely to achieve their objectives; they will more probably aggravate domestic difficulties with inflation and balance-of-payments deficits. Structural imbalances may cause significant slack in some industries and in certain circumstances; yet urban surplus labour need not be associated with the sector-wide infinite elasticities of supply of complementary factor inputs of which Keynesian reflationary policies are designed to take advantage.[2]

An increase in the level of investment would require higher domestic savings, greater foreign inflows, or a shift in the distribution of investment in favour of the urban areas. Though the data are sketchy, aggregate savings appear to be already impressively high in Tanzania (van Arkadie in Faber and Seers 1972). The transfer of development resources from rural to urban areas could be increased, but such a course of action is likely to be counter-productive in regard to achievement of urban labour market balance. The urban bias of the government in the 1960s contributed to the stagnation of rural incomes and the emergence of urban surplus labour. To increase urban employment in this way would result in a further widening of the rural–urban differential, in effect increasing the total number of workers desiring urban jobs. An additional squeeze on the agricultural sector would also be contrary to the policy of increased emphasis on rural development in the post-1967 development strategy. There are difficulties in Tanzania both with attracting foreign capital and, given the stated policy of decreasing Tanzania's dependence on foreign decision-makers within the economy, with the hosting of such investment. With the limited development resources available, whether domestic or foreign, the high capital/labour ratio prevailing in urban industry, which implies a high cost per job created, is another constraint on the potential effectiveness of a remedial policy that relies on employment generation.

The hypothesis that the capital intensity of production in the urban industrial sectors of developing countries is due to fixed factor proportions embodied in techniques imported from countries where labour is relatively scarce has not been substantiated. On the contrary, the evidence suggests that there is considerable scope for substitution between capital and labour.[3] The evidence also suggests, however, that, although altering the product mix, eliminating capital-intensive biases in tariff, tax, and wage structures, and providing an environment suitable for the local development of appropriate technology may have a significant impact on the growth rate of employment, such changes are unlikely to generate the quantum leap in urban wage employment necessary to achieve urban labour market balance within a relatively short period.

This brief review of the situation suggests that an employment generation policy alone cannot succeed in eliminating urban surplus labour because the scope for rapid increase in the demand for urban labour in Tanzania is insufficient. On the contrary, feasible increases in labour demand are more likely to aggravate than alleviate the problem of urban surplus labour because urban labour supply is so responsive to changes in labour demand.

Policies focused on the supply side of the labour market must bear the burden of achieving full employment in urban areas. At the risk of over-simplification we can distinguish four types of policy instruments that can be used to regulate migration from rural areas and hence the supply of urban labour: direct controls, urban incomes and prices policies, rural development policies, and education policies.

2. Direct Controls on Mobility

At first glance, direct controls on migration might appear to be the most effective means of reducing the flow of rural residents into the towns. South Africa has successfully used pass laws to this end by requiring a work permit as a condition of urban residence (Hutt 1971); China has reduced urban unemployment by issuing ration tickets for food and clothing that are only convertible into goods at certain locations (Reynolds 1975). But such policies raise difficult questions to which there are at present no definitive answers. How thoroughgoing must the system of controls be and how large and costly the bureaucracy to enforce such a system? To what extent does the implementation of direct controls hinder the achievement of other aspects of the efficient allocation of labour, and of social and distributional goals, thereby imposing political and economic costs in addition to the direct costs of administration?

The government of Kenya has imposed direct controls in the form of housing regulations which exclude the shanties of newly arrived migrants and the unemployed from the city centres, and it has reinforced the regulations by setting fire to such settlements. The Tanzanian government requires urban workers to have an official card certifying their employment, and it periodically 'rounds up' the unemployed and sends them back to the rural areas. Yet urban unemployment has continued to grow in both countries, suggesting that the imposition of additional controls and substantial investment would be required to make their systems work. When controls are sufficient to achieve urban labour market balance, the issue arises of how to ration urban job opportunities among rural workers. Even when a centralized system does not lead to favouritism and corruption, it may be less efficient in matching the skills of workers and the needs of employers than a system in which the employers themselves select employees from all those who come to town looking for employment. Furthermore, though rationing may mean that labour that would otherwise have been idle or marginally employed in the urban area is more productively employed in the rural areas, it does not eliminate the resource costs of other misallocations, particularly in choice of technology, that may arise from high and rigid urban wages. Perhaps most difficult to assess are the political costs associated with the frustration of workers who, despite theoretical freedom to live anywhere within the nation's boundaries, are forbidden to move to places where other workers with the same qualifications earn considerably more, even though equal pay for equal work for people with the same level of needs is a fundamental goal of societies with socialist aspirations.

3. Urban Incomes and Prices Policies

Narrowing the gap between rural and urban incomes is potentially the most effective indirect means of reducing the rate of urban migration and the magnitude of urban surplus labour. Although the effects of decreases in urban incomes and increases in rural incomes are essentially similar and rural and urban policies must be co-ordinated to achieve either, we treat them separately because there is a significant difference in the policy instruments relevant to the different sectors.

Since our analysis implies that the rise in urban wages during the 1960s was an important factor in the emergence of urban surplus labour, a reduction in wages would appear to be justified. Eliminating the gap between rural and urban incomes achieves the desired balance between urban employment opportunities and job seekers. The decline in real wages can be accomplished by lowering money wages or holding them constant while prices rise.

One objection to this policy is its consequences for the distribution of income. Keynesian countercyclical policies are widely accepted in industrialized countries partly because reflation benefits all segments of the community and avoids the difficulties of specifying what constitutes an increase in general welfare. Altering relative factor prices does raise this question. The narrowing of the income difference between urban wage-earners and rural workers is considered beneficial, as is the rise in the income of the previously unemployed group. The functional distribution of income is the focus of concern. The urban wage bill may actually be reduced if the increase in output occasioned by the decline in wages does not offset a reduction of the wage share (Ahluwalia in Chenery *et al*. 1974). Underlying the negative interpretation of this change is the presumption that the increase in the non-wage share of total income all accrues to capitalists as increased income. In fact the increase in profits may induce an increase in investment, either directly by the profiting firms (which may be in the hands of the state rather than privately owned) or by the government, which has taxed away the increment. In this case the implications of a decline in wages for the functional share translate into a choice between relatively high incomes for a relatively small group, urban wage-earners, or lower incomes for that group and higher investment, thus reducing the force of the objection to a wage cut. In China wage stability combined with increases in productivity has contributed to a decline in the unit production costs of manufactured goods and a rise in profit margins and hence in the rate of capital accumulation (Reynolds 1975).

A more serious objection is the possibility that, despite excess supplies of urban labour, higher wages have been the consequence of employers' decisions rather than government policy or the pressure of unions. Though not definitive, there is considerable evidence that because of negative relationships between wages and the rate of turnover of labour in employment and between turnover and productivity, resulting in large part from higher returns to investment in on-the-job training, wage increases actually lowered the cost of labour per unit of output.[4] Thus, simply to eliminate minimum wage and other institutional factors constraining downward flexibility might not in fact lead to a wage decline. Even if the current level of commitment of the labour force to wage employment precluded a return to the former system of high turnover, the smallness of the current wage bill in proportion to total costs in the manufacturing sector would make employers unlikely to risk the possible negative consequences of reduced incentives for productivity. This suggests that the government would have to legislate lower wages and that there is a significant likelihood that the allocative efficiency benefits associated with the elimination of urban surplus labour would be offset, at least in part, by reduced labour productivity. In a situation of declining national income the inadvisability of lowering real wages would have to be qualified. In any case it is doubtful whether this economic option is viable politically.

Either the unions have not been strong enough to resist government actions that severely reduced their power in collective bargaining, or they have come to see that fewer work stoppages and a slower rate of increase in wages is in the national and ultimately in their own interest.[5] Experience during the period 1971–3, to which we refer below, suggests, however, that unions and other organizations of the workers are still strong enough to resist a significant reduction in the real earning capacity of the wage-earners. Though wage reductions may not be feasible, an incomes policy is of the highest priority for the prevention of a further widening of the differential between rural and urban incomes which would be at cross purposes with other measures aimed at reducing urban surplus labour.

Growing inequality of income distribution and a declining rate of growth of wage employment led the government, in 1967, to formulate a national policy on wages and incomes. The background report prepared for the government by the ILO emphasized the importance of such a policy:

The problem of incomes policy is much more urgent for the developing countries than for the advanced economies. Incomes policy involves on the one hand the allocation of resources between current consumption and the investment which is necessary to lay the foundation of future living standards. On the other hand, it involves the structure and distribution of earnings among the people, which determines the allocation of human resources and the incentives to increased production. In an advanced economy with ample resources and high living standards, the delays to growth and the waste of human or material resources which may result from errors in incomes policy are not usually critical. But in a developing economy, with much more limited resources and lower general living standards, the necessity of a sound incomes policy is crucial; it may make the difference between whether or not economic growth occurs at all, and may even involve the political stability of the country concerned. (Turner 1975.)

The specific aims of the incomes policy were to restrict wage increases to a figure corresponding more closely to the actual growth of the economy and to adjust minimum wages only in a close relationship to improvements in peasant farmers' living standards. The Permanent Labour Tribunal was established to ensure the effectiveness of the wage guidelines, which limited annual increases to 5 per cent, with exceptions for genuine incentive schemes.

Whereas in the period 1961–6 wages (urban and rural estates) increased at an average rate of 13.5 per cent a year, in 1967–71 the rate diminished to an average of 3.1 per cent a year. Furthermore wages as a proportion of value added in the manufacturing sector declined from 44.8 per cent in 1968 to 37.8 per cent in 1971. The mission sent by the ILO in 1974 at the request of the government to review the policy concluded that on balance '. . . the Tanzanian incomes policy, from 1966 to 1971 at least, must be reckoned one of the most successful in the world experience; and this is probably very largely due to the comparative unity the Tanzanian people have achieved' (Turner 1975). Since the earlier period witnessed a once for all shift from a low income–high turnover to a high income–low turnover wage sector it is likely that the earlier rate of increase would have slowed even in the absence of an incomes policy. If one considers also the extent to which what happened after 1971 was a consequence of decisions taken during the earlier period, this conclusion seems exaggerated. For the failure to achieve one particular policy goal during this period of general success planted the seeds of the problems which, once they took root, led to the government's virtual abandonment after 1971 of the goals of the 1967 incomes policy.

There was no decrease in the difference between rural and urban incomes during the period 1967–71. Rural incomes continued to stagnate and the gap widened, though at a considerably lower rate than previously. This was largely due to the deterioration of the terms of trade between the agricultural and non-agricultural sectors. Total value added in agriculture at constant prices increased by 30 per cent between 1964 and 1972. While the prices of urban manufactured goods and services rose significantly, however, the price of foodstuffs did not rise. The government, which has primary responsibility for determining the prices peasant farmers receive for their produce, was able to hold the prices of foodstuffs constant, despite increases in the prices of inputs used in their production, as a means of stabilizing urban wages. Apparently the government concluded that a rapidly increasing urban cost of living would make the labour organizations less willing to abide by the terms of the 1967 incomes policy; but reducing pressure on wages in this way, which meant that most of the burden of inflation was borne by the peasant farmers, simply postponed a confrontation with the underlying issues of distribution.

Two consequences of this 'avoidance policy' of artificially low food prices were that supply was reduced and demand increased. There is evidence of an increase in the smuggling of produce into Kenya where prices that farmers received for some products were as much as 50–100 per cent higher than in Tanzania, and of an increase in the production of non-food crops, the prices of which were more closely aligned with world market prices. Not only did higher urban incomes lead to increases in the consumption of food but, in apparent contradiction of 'Engel's Law', the proportion of private urban expenditure devoted to food actually increased because workers at the lower end of the distribution of income benefited the most from the combined rise in wages. In response to the growing shortage of food, producer prices were raised in 1969–70 and organized labour began to press claims for higher wages. The government reacted quite disproportionately by boosting minimum wages 40 per cent in 1972, setting off a classic wage–price spiral, as higher wages increased the demand for food and raised the price of some agricultural inputs, exacerbating the food shortage, which led to further increases in producers' prices and another 40 per cent rise in the minimum wage in 1974.[6]

The wage increases of 1972 and 1974 meant the virtual abandonment of the incomes policy and a reversion to the pre-1967 policy which insulated the incomes of wage-earners but not those of peasants from downturns in national economic activity. The net effect of the oscillation in the real income of wage-earners and peasant farmers appears to have been an even more rapid widening of the rural–urban differential than in 1967–71. If this is the case, we would expect to see it reflected in an increase in rural–urban migration and a rise in urban surplus labour.

One lesson of the Tanzanian experience with incomes policy in 1967–74 is that even during a period of economic expansion the exercise of considerable political will-power is required to achieve the limited goal of holding the rate of increase in urban real wages to that of rural real incomes. A second lesson is that if incomes policy is to help reduce the surplus of urban labour it must be co-ordinated with policies on agricultural prices. In this regard the aim of incomes policy is to stabilize urban wages so that rural incomes can rise and narrow the rural–urban differential. Without any attempt to slow the rise in urban wages in Tanzania, the differential might have widened at an even faster rate than it did. If urban wages are stabilized by measures that effectively constrain rural incomes from rising, however, progress toward the reduction of urban surplus labour can only be achieved if changes in other factors influencing the migration rate are favourable.

4. Rural Development Policies

To reduce or eliminate urban surplus labour significantly in the short run, an incomes policy that stabilizes urban wages without hampering the growth of rural incomes may be necessary, but it is unlikely to be sufficient. In Tanzania rural incomes were stagnant for quite a few years before the freeze of producer prices for foodstuffs. This suggests the need for a positive policy to raise rural incomes as a complement to urban incomes policy.

From 1967 to 1971 government agricultural pricing policy reduced producer incomes below what they would have been without intervention. It is reasonable to suggest that the government might now, in the interests of urban labour market balance, raise prices above market equilibrium as a means of raising rural incomes. During the current period of shortages of domestically produced foodstuffs the world market price for these goods is in effect the equilibrim price. A rise in producer prices above this level is not likely to generate a problem of surplus production, because increases in domestic supply can be used to reduce reliance on imports. Even if a surplus emerged, it could be exported to earn foreign exchange, stockpiled for use in a period of drought, or distributed below market prices in low income areas.

From the point of view of agricultural production and the balance of payments, as well as from that of urban labour market balance, a price support programme would yield significant benefits. The problem with this programme, however, as with the policy of directly lowering urban wages, is the feasibility of government financing of the subsidies involved. The demonstrated ability of urban wage earners to resist erosion of their real income precludes passing the price increase on to consumers. Holding consumer prices constant and raising producer prices, which would reduce the balance of payments benefits of the price support programmes and raise the magnitude of government subsidy required, is not a means of avoiding this difficulty. If the subsidy to farmers is financed out of taxes raised from farmers, as most taxes in Tanzania ultimately are, then in terms of rural–urban differentials the policy does not yield a net gain. Since taxes that lower urban incomes are excluded from consideration and deficit financing would exacerbate inflationary problems, a price support programme appears to be financially unfeasible.

There do not appear to be any rural policies that provide a short-cut solution to the problem of urban labour market imbalance; the only way to narrow rural–urban income differentials is to raise the productivity of the rural labour force. The decision to raise rural productivity as a means of closing the rural–urban income gap, however, poses problems of conceptualization and of implementation. Raising rural productivity significantly is likely to require structural shifts in the underlying organization of the sector and co-ordination of a wide range of policy instruments where predicted effects are highly uncertain.

An assessment of Tanzania's rural development policy is clearly beyond the scope of this study, but let us consider briefly several components in terms of their potential contribution to the achievement of urban labour market balance. Agricultural pricing policy was not the only government-imposed constraint on the growth of rural incomes during the 1960s and 1970s. A net flow of development capital from the rural to the urban areas was accomplished by government expenditure in urban areas in excess of urban taxes, while in rural areas taxes exceeded expenditure.

This urban bias was a consequence of the ease with which agricultural output could be taxed and of the expectation of high social rates of return to investment in urban-industrial projects and infrastructure. The low value added and profitability of the urban manufacturing sector has recently tempered enthusiasm for a strategy of rapid industrialization; domestic food deficits and rural out-migration leading to urban labour surplus have emphasized the cost of neglecting agriculture; and increased urban output from large firms with adequate accounting procedures has eased the difficulty of taxing urban incomes. The urban bias can be reduced by altering the spatial distribution of taxes or expenditure, or of both. While a reduction in rural taxation and constant government expenditure would give a direct boost to rural incomes and thus reduce urban labour market imbalance in the short run, such a policy would once again confront the government with the political obstacles in the way of lowering urban real incomes.

Altering the pattern of government expenditure so as to end the rural–urban capital flow would not mean that the rural and urban sectors would share equally in total investment, or that each sector would have to rely for development capital solely on the surplus it generates. The sectoral share of total investment would also be influenced by the distribution of foreign development capital, which was rurally oriented before Independence but has since been concentrated in urban areas. It would mean, however, that in contrast to most developing countries and most socialist countries (China being an exception (Reynolds 1975)), rural savings would be reinvested exclusively in that sector.

Nor would a policy to increase rural investment by shifting government expenditure, or by any other means, necessarily entail increased investment in agriculture at the expense of industry or social overhead capital. Romantic visions of looms in every home and iron smelters in every village have given a false impression of the nature of rural non-agricultural industry. Some goods can be fabricated efficiently in small-scale plants, and location of agricultural processing near the source of supply can yield significant savings on transport costs when weight is reduced by processing. The relationship between population density and the per capita cost of providing a given level of service may be downward sloping for some types of social overhead capital investments such as water supply. For others, such as electricity and transport, the curve is likely to be U-shaped, with minimum cost at relatively low levels of density. By bringing widely dispersed households together, the villagization programme for rural development can dramatically increase the social returns to rural overhead capital investment, and compares favourably in this respect with similar urban investments.

Nevertheless, the relative abundance of cultivable land in Tanzania and the low level of capitalization of existing farms promise high returns to investment and suggest that a significant portion of any increase in development capital available in the rural areas will be directed into agriculture. During the Industrial Revolution, increases in agricultural productivity both 'freed' workers for employment in the cities and generated the food to feed them. In Tanzania, however, the withdrawal of labour will not provide much of a stimulus to agriculture. Given the high proportion of the labour force currently engaged in the rural sector and the projected rates of population growth, even at the upper limits of projections for urban modern sector employment growth, an absolute decline in the rural labour force would mean a dramatic increase in urban surplus labour. In contrast to historical experience, rural development requires simultaneous increases in productivity and labour absorption.

In Tanzania, where land is still virtually free, the question of labour absorption translates into one of income distribution. Unfortunately the data on rural incomes are not sufficient to shed much light on the consequences for migration of changes in the intra-rural distribution of income. Conceptually, a significant increase in average rural productivity may have limited or even negative impact, if it is achieved by concentrating investments among a small proportion of farmers. Though the effect of the increase in agricultural production on prices may not cause an absolute decline in the income of the poorer farmers, an increased skewing of the rural distribution of income may nevertheless encourage migration because of the influence of the 'demonstration effect' on the rural structure of preferences. The increased ability of the richer farmers to finance higher levels of education for their children and support them through longer periods of urban job search may also increase migration.

This suggests that choice of technology is a policy issue of considerable importance, in relation both to the increasing of rural productivity and to the distribution of income in the rural sector. Factor pricing policies that encourage substitution of capital for labour are likely to impose greater resource costs in the rural than in the urban sectors. Since conceivable increases in development capital available for agriculture will not be sufficient to equip more than a small proportion of farmers with capital-intensive techniques, such policies may also contribute to increased skewness in the rural distribution of income. The identification and development of labour-intensive innovations may prove an essential ingredient of a rural development policy explicitly intended to achieve urban labour market balance.

5. Education Policies

The dramatic increase in educational opportunities and hence in the supply of educated workers in Tanzania since 1955 appears to have reduced significantly expected private returns to investment in education, despite the fact that the rural–urban income differential has widened considerably. School-leavers must now discount the monetary returns to education by the probability of finding a high-wage job and, as a consequence of filtering-down and displacement, even when they find a job, it is generally lower on the occupational and wage scale than it would have been a few years earlier. There is evidence that at the primary school level the excess private demand for education has changed to an excess of school places in consequence.

Expected net private returns to investment in education are in excess of net social returns, both because the government subsidizes costs and because the urban labour market overvalues educational credentials. As a consequence of the recent changes in the urban labour market, net social returns to investment in primary education are probably lower than the social returns to other human capital investments in which the government participates, such as health care, the facilities for which have expanded far less rapidly; they are probably also lower than the returns to investment in physical capital. The uncertain contribution of primary education to rural productivity and the high proportion of school-leavers who return to peasant farming after an unsuccessful period of job search, suggest that the net social returns to investment in education may be very low. Even if these assertions, for which the supporting evidence is all circumstantial, could be substantiated by more rigorous empirical analysis, it may be premature to recommend reducing the rate of investment in primary education which yields substantial non-economic benefits (Edwards and Todaro 1973).

Our analysis suggests that changes in educational policy will have only an indirect impact on urban labour market imbalance. Increasing the social rate of return to investment in education by making it more appropriate for a rural vocation implies, in apparent contrast to the period 1955–70, a kind of schooling that would increase the productivity of at least one segment of the rural labour force and thus contribute to the narrowing of the rural–urban income differential for educated workers. It is doubtful whether a more vocational orientation of the curriculum would accomplish this. An increase in the rural orientation of the school system may however, reduce the propensity of educated workers to migrate, even if the rural–urban income differential remains constant.

Reducing the flow of educated migrants from rural areas would lead to a decline in urban surplus labour in the educated segment of the urban labour force and in the proportion of educated workers in total urban surplus labour, but given the high proportion of rural residents with little or no education, it is unlikely to make a significant impact on the overall rate of urban surplus labour. This is the issue of the influence of a change in the rural distribution of income on migration in another guise.

A decrease in the rate of investment in education rather than an increase in the social rate of return would avoid these distributional complications, if the resources saved were invested in the rural sector in a way that would benefit educated and uneducated alike. Since urban taxes would not have to be raised, such an increase in rural investment would have the additional advantage of avoiding the political constraint of lowering urban real incomes.

Limiting the supply of educational opportunities could be used to achieve urban labour market balance by implementing a system of low level manpower planning. There is virtually no unemployment among high level manpower in Tanzania and few workers with lower secondary education have filtered down into lower level occupations. Manpower planning, practised in Tanzania since the mid-1960s, takes labour demand, projected on the basis of planned rates of expansion of industry sub-groups and of surveys of employers' current skill requirements, as a given and controls supply of high level manpower by the manipulation of only one policy variable, namely the number of entrants into the various post-primary streams of the educational system.

Urban surplus labour could be eliminated by regulating entrance to primary schools and applying a strict educational criterion for hiring in unskilled and semi-skilled jobs. The supply and demand for workers with primary education could be strictly equilibrated and since for workers without the necessary educational credentials the probability of urban wage employment would be zero, by implication the rate of urban surplus labour would also be zero. However, since the annual number of primary school-leavers entering the labour force is already more than four times the total annual increase in non-agricultural employment, such a policy would require a drastic reduction in the size of the school system, a move which would be in obvious conflict with the government's commitment to universal primary education by 1989. Even if politically acceptable, such a policy would simply substitute the rationing of school places for the rationing of urban jobs, would increase the rigidity of the process whereby workers' skills and employers' needs are matched, and would institutionalize the inability of the formal educational system to make a positive contribution to the increased productivity of rural workers.

6. Concluding Remarks

The evidence is quite strong that urban labour market imbalance in Tanzania is due to excessive rates of rural–urban migration which, in turn, are a consequence of high and downwardly inflexible urban wages. Not surprisingly, the implications for remedial policy of the intersectoral misallocation model of urban surplus labour focus on the supply side of the urban labour market. To simplify the policy implications of our analysis of urban labour market imbalance, the advice we would give to the Tanzanian government is, 'intervene in the market so as to narrow the difference in income between rural and urban areas'. To adopt such a policy would improve income distribution and allocative efficiency and thus increase national output and reduce the social problems associated with urban surplus labour. Our review of the alternatives has touched on a number of economic factors that would have to qualify the recommendation of vigorous application of those policy instruments which could achieve this goal relatively quickly. However, the most significant qualification concerns the political feasibility of following a course designed to improve the welfare of the society as a whole when a small segment of the population with disproportionate ability to influence the government would have to bear the transition costs. If there is no alternative to accepting this constraint, there are other policies that can accomplish the task, though over a much longer period.

Notes

1. See Sabot (1979), ch. 6.

2. The fiscal system may still be used to influence factor prices in a way that favours labour-intensive production or at least that eliminates discrimination in favour of the use of capital inputs. See Peacock and Shaw (1972).
3. Winston (1972); Stewart in Edwards (1974).
4. See Sabot (1979), ch. 8.
5. In the period 1958–60 the average annual number of man-days lost through work stoppages was 761,700; in 1973 it was zero. See Knight and Sabot (1983).
6. The serious inflation and balance of payments deficits of the 1970s which followed the moderate price rises and trade surpluses of the 1960s were exacerbated by 'imported' inflation, in particular the rise in the price of oil. However, food constitutes a higher proportion of the increase in total imports than does oil and the period of rapid inflation began prior to the oil price rise of 1973. See Turner (1975).

References

Chenery, H. *et al.* (1974) *Redistribution with Growth: Policies to Improve Income Distribution in Developing Countries in the Context of Economic Growth* (London: Oxford University Press).

Edwards, E. (1974) *Employment in Developing Nations, Report on a Ford Foundation Study* (New York: Columbia University Press).

—— and Todaro, M. P. (1973) 'Educational Demand and Supply in the Context of Growing Unemployment in Less Developed Countries', *World Development* 1 (Mar./Apr.).

Faber, M., and Seers, D. (eds.) (1972) *The Crisis in Planning* (London: Chatto & Windus).

Harris, J. R., and Todaro, M. P. (1968) 'Urban Unemployment in East Africa. An Economic Analysis of Policy Alternatives', *East African Economic Review*, 2 (Dec.)

Hutt, W. M. (1971) *The Economics of the Colour Bar* (London: Merritt & Hatcher).

Knight, J. B., and Sabot, R. H. (1983) 'The Role of the Firm in Wage Determination: An African Case Study', *Oxford Economic Papers*, 35 (1) (Mar.).

Peacock, A. and Shaw, F. (1972) *Fiscal Policy and the Employment Problem in Developing Countries*, Employment Series No. 5 (Paris: Development Centre, OECD).

Reynolds, L. (1975) 'China as a Less Developed Economy', *American Economic Review*, 66 (June).

Sabot, R. H. (1979) *Economic Development and Urban Migration: Tanzania 1900–1971* (Oxford: Clarendon Press).

Turner, H. A. (1975) 'The Past, Present and Future of Incomes Policy in Tanzania', Mimeo (Geneva, ILO).

Winston, G. C. (1972) 'On the Inevitability of Factor Substitution', Research Memorandum No. 46 (Williamstown, Mass., Center for Development Economics, Williams College).

7

Rural Contract Labour in Urban Chinese Industry: Migration Control, Urban–Rural Balance, and Class Relations*

Marc Blecher

TOWNWARD migration has major macro-level consequences for economic development, demographic change, labour market operation, social structure, and political change in developing countries. In micro-analytical terms, it has tremendous impact on the rural and urban people and the communities to and from which they migrate. The general question of townward migration is, therefore, very significant from a number of perspectives. China's rather unique experience in restricting and controlling townward migration through the 1970s is of comparative interest to those concerned with development and urban–rural relations more generally, as are the effects of its reform policies in the 1980s.

Tremendous pressures exist in China for townward migration. In much of rural China, very high levels of population density prevail, leading to underemployment and depressed incomes and standards of living. Other parts of the countryside—vast areas which suffer from poor natural conditions or sheer remoteness—have interrelated problems of very low population density and poverty: the area can support few people, and its poverty is reproduced by the absence of a critical mass of population which could command the political and economic resources or mobilize the labour power to undertake the major investments needed to promote a breakthrough to meaningful development. People tend to want to move away from both kinds of places in China, and they have done so when the opportunity presented itself.[1] Added to the 'push' factors in the countryside are those relating to 'pull' towards the cities: the significantly higher standard of living in urban China, the locus of political power and advantage there, and the relatively greater opportunities for social and

* This is an edited, revised and updated version of my article entitled 'Peasant Labour for Urban Industry: Temporary Contract Labour, Urban–Rural Balance and Class Relations in a Chinese County', originally published in *World Development* 11 (8), 1983. In addition to those whose help was acknowledged there, I would like to thank Josef Gugler for his interest, encouragement, and helpful comments on this revision, and Dorothy Solinger for her special good words.

economic mobility in a country whose already high level of agricultural development but relatively lower level in industry make the urban areas far more dynamic for the foreseeable future in terms of economic growth and occupational upgrading (not to mention social life). Finally, these 'push' and 'pull' factors operate within structural conditions highly conducive to migration: China's historically well-developed commercial and transport networks, the relatively high level of incorporation of rural China into urban-centred political, social, and economic systems,[2] and the generally high level of exchange of information between village and city through official media and informal social networks. In China, then, many villagers have good reasons to want to move to towns and cities, and abundant information, pathways, and other conditions to facilitate the process.

Townward migration has been a matter of deep concern to China's political leaders and economic planners of Maoist and Dengist persuasions alike (and only partly for different reasons). During the Maoist period, raising total food-grain output to keep up with population growth had the very highest priority. Any flight of labour with even the lowest marginal productivity out of agriculture was a source of alarm. (By contrast, the Dengists, who are more concerned with productivity issues, want to encourage movement to labour out of agriculture, but not out of the countryside; they seek expansion of rural sidelines, industries, and commerce.) And both Maoists and Dengists have worried about the danger of further urban crowding. Some of China's cities, especially its larger, more developed (and therefore more attractive) ones, are already very heavily populated and have serious unemployment. Migration of even a small percentage of the massive peasantry to them would cause political, economic, and social problems of potentially explosive proportions. It is not at all surprising, then, that a serious commitment to regulating and controlling townward migration has been evinced by China's political leaders and economic planners of the 'right' and the 'left'; in fact, it is one of the few continuous themes of an otherwise radically shifting Chinese development policy since 1949.

China's leaders have approached this problem with high imagination and broad vision. They have met with success: data indicate that townward migration has probably been lower in China than in any other developing country, despite the tremendous pressures for migration.[3] One set of reasons is the effects, intended or unintended by political leaders and planners, of seemingly unrelated or only indirectly related aspects of Chinese development. For example, rural collectivization has promoted a high level of intra-village economic equality,[4] which Lipton has identified as a major deterrent to rural emigration.[5] But townward migration has also been kept down by the administrative controls preventing 'rural house-holders' from moving to urban areas to live and work.

The Chinese people have been formally classified by their government as 'rural' or 'non-rural householders'. The distinction originated with the start of government regulation of grain distribution in the 1950s, to differentiate those who needed to receive their grain through state agencies in the towns from those who could be provisioned by their rural collective units. Gradually the distinction came to affect the organization of life in other areas too, such as political activity and the distribution of social services. For our purposes it has involved most importantly employment and residence restrictions. In the high collectivist period which ended in 1978,[6] rural householders were employed exclusively by their rural collective units, and were forbidden to reside permanently in a town or city. Even today, when peasants are freed to employ themselves outside their home villages, and when many are engaged in urban–rural trade or in informal sectors of urban economies, they are still legally barred from taking up permanent residence in an urban area.[7] The major purposes of the residence restrictions have been to keep peasants away from already very tight urban labour, housing, and commodity markets.

In some cases, in fact, China has been *too* successful in restricting migration. In many places there has been an excessive build-up of rural labour, and in others a shortage of industrial labour. Shulu County, the subject of this study, is such a place in both respects.[8] There, as in many other parts of China, the sluices of townward migration were opened during high collectivist days, but only in a closely regulated and partial way—partial because urban employment was permitted but change of permanent residence was not. Specifically, peasants have been employed in urban industry and commerce as 'temporary contract workers' (*linshi hetonggong*), while remaining tied to their rural village homes and collective units in fundamental ways. Contract labour of this sort was quite common in China during the high collectivist period, so the situation in Shulu reflects wider Chinese experience.

This discussion describes the practice and terms of temporary contract employment in Shulu County as of 1979, and analyses it in relationship to several problems, some of which have already been mentioned: What has its role been in industrial growth? What effect has it had on the goals of urban–rural equality and development balance? What implications does it have for China's changing class structure and class relations?

Temporary Contract Work: Definition, Rationale, and Magnitude

'Temporary contract workers' were peasants hired by county/state sector industrial enterprises[9] through contracts with their rural collectives. Although 'temporary contract work' was often not very temporary at all—in Shulu most contract workers had been employed as such for many years—the contract workers retained the official status of 'rural house-holders', and maintained their permanent homes in their villages, both formally and actually. Their families remained behind there, and the contract workers visited home periodically—in Shulu usually weekly or monthly—living the rest of the time (indeed, most of the time) in factory dormitories. Only those lucky ones hailing from suburbs of the town in which they were employed could live at home and commute to work daily. The families of contract workers, also, of course, remained 'rural householders', which meant they had no claim on urban employment, housing, or education, as the families of regular county/state sector workers did.

Contract workers also remained members of their rural collective units (*sheyuan*), which retained the ultimate claim over allocation of their labour. In formal terms, then, contract workers were employed by the rural collectives, which assigned them to urban employment under the terms of a contract made with a factory in town. This was no *mere* formality: the collectives had the power to decide which members would be assigned to contract employment. They also controlled part of the contract workers' salaries, paying them like other collective members in the form of work points (which in turn were financed by a part of the workers' salaries paid by the employing unit directly to his/her collective under the terms of the contract).

The Shulu County Planning Commission operated under a very strict quota from higher-level state agencies on the size of the regular industrial labour force (i.e. non-rural householders on the state payroll employed in county/state factories). This had not been revised upward for quite some time, because of the state's unwillingness to increase the number of people depending on state-supplied commodity grain. Yet Shulu industry had been growing and was continuing to do so, and for this it had required and continued to require more workers. Expansion of the contract labour force was the answer. Between 1964 and 1978, the industrial labour force at the county level in Shulu trebled, from 5,150 to 15,034; 7,475 (76 per cent) of these new workers were contract workers. By 1978, contract workers comprised over half of Shulu's county/state industry labour force.

Several factors made contract labour attractive to Chinese policy-makers and economic administrators. First, it permitted expansion of the urban labour force without any concomitant increase in state grain procurements from the countryside to feed the new workers. Temporary contract workers received their allocations of food-grain from their rural collective units directly. Second, contract workers cost less to employ; as we shall see, their wages and benefits were significantly lower than regular workers'. Third, by housing them in dormitories and keeping their families in the village, pressure on scarce urban housing and land was reduced. Fourth, employing enterprises and the planning and administrative agencies which supervised them gained managerial flexibility by using contract labour, which was easier to dismiss (simply by failing to renew the contract) than the regular labour force (which was accustomed to its 'iron rice bowl' and well positioned to threaten the state politically should its security come under attack). Fifth, employment of peasants as contract workers in urban industry provided the state with a way of reducing rural underemployment or unemployment *in a planned way*, avoiding the risks of massive desertion of the rural collectives, which would endanger not only agriculture but also (as it has in the past) China's rural socialist project.[10] Sixth, it provided the rural collective sector with an outlet for excess labour power the use of which also brought revenue into its coffers. The contract labour system, then, had strong attractions for Chinese economic planners, industrial managers and administrators, urban government officials, and rural cadres. Little wonder that it grew so quickly and became so important in Shulu and elsewhere in China.

Perhaps that is also the reason for the firm policy to deny official status as regular workers to the contract workers. Prior to 1972, there was a policy to change the classification of some contract workers to regular workers, probably as a result partly of the obstreperous agitations by contract workers during the Cultural Revolution. In 1972, Shulu factories received a 'general guide-line' to convert their contract workers to regular workers by changing their rural household registrations to non-rural ones. It was implemented in at least one Shulu plant, where it caused so much controversy that it was soon dropped. Since 1972, contract workers have specifically been denied the opportunity to become regular workers. Contract labour continued to be a source of discontent, as we shall see.

Advantages for Urban Areas and for the County/State Industrial Sector

Use of peasant contract labour had several major advantages for Shulu's urban areas and urban county/state industry. First, it reduced wage costs in county/state industry considerably, in 1978 by an estimated Yuan 900,000, 12 per cent of the total payroll in that year. Second, it transferred a large part of the social reproduction costs of the industrial labour force onto the rural collective sector and the individual households that comprised it. Contract workers' housing in town was paid for by the factories that employed them. This could be done relatively inexpensively in dormitories, usually located on factory grounds. Moreover, housing for the workers' families need not be provided at all, since they stayed home in the countryside.[11] This meant in turn that education for the children of these workers also need not be provided by urban factories or city governments. Contract workers were expected to rely for medical care on rural paramedics, doctors, clinics, hospitals, and co-operative health care plans (although they were provided first aid in the factory clinic if necessary). They were paid no pensions and provided no retirees' services. The precise cash value of these savings is difficult to estimate, but certainly it was large, especially when it is remembered that we are talking about all these services for over half of the urban industrial labour force.

Third, the pressure on urban land, institutions, and services was significantly reduced by contract labour. Xinji, the county capital, would have had around twice its 1978 population if the temporary contract system had not prevented the contract workers' families from setting up residence there.[12]

Fourth, urban employers were given the flexibility in labour force matters which they were denied by the various restrictions on hiring and firing of the regular county/state labour force. On the one hand, by 1979 the state had not permitted any expansions of the regular labour force in Shulu for many years. On the other, the 'iron rice bowl' policies had made firing workers almost impossible. While it does not appear that there was much need for industrial employers in Shulu to reduce their labour force in any significant numbers (since industry was expanding apace), the *power* to fire can be a powerful lever in the hands first of employers seeking to enforce greater discipline over their workers, and second of planners seeking to move workers among enterprises. Indeed, the major arguments advanced by proponents of restoring the employers' right to fire workers in China have been the presumed salutary effects on labour discipline and allocative flexibility. Contract workers complained during the Cultural Revolution that almost all the flexibility which factory managers had in labour matters was achieved at the contract workers' expense; this included not only hiring and firing but also transfers, job allocations, and even wage negotiations.[13]

Advantages for Rural Areas and for the Rural Collective Sector

Of course the urban sector's gains in these matters—especially wage differentials and the shunting off of social reproduction costs of the labour force—are the rural sector's losses. The fact is, though, that while the rural collective sector was in effect being forced by the institution of contract labour to bear some of these costs of industrialization in Shulu, it was also benefiting from industry's use of contract labour in some very concrete ways, at least compared with where it would be without it.

First, there is the economic transfer from urban to rural which was effected by contract labour. This included the monies which rural collective units appropriated as a share of the contract workers' salary as well as the remittances which contract workers sent to their families back home. The first can be estimated as amounting roughly to a third of a million *yuan* per year, an amount whose magnitude to the rural collective sector can be appreciated when it is noted that it nearly equalled total state budgetary support for agriculture in 1972 (Yuan 345,000).[14] Remittances are harder to estimate, but if each contract worker were remitting only one-fourth of his or her net income (after paying the fee to the collective) back home— probably a conservative estimate since these workers were getting free housing in town, so their only expenses were self-maintenance and travel— this would still amount to around Yuan 100 per worker per year, or another Yuan 800,000 per year.'[15] Therefore, the total amount of money brought into the rural collective sector by Shulu industrial contract workers in 1978 was, all told, in the vicinity of Yuan 1,000,000. Of course, this does not count the income lost to the rural sector with the departure of the contract workers. This is likely to have been quite small, though.[16]

A second benefit to the rural collective sector from contract labour has to do with the skills and contacts which contract workers brought back with them upon their return to the villages. Given the rapid expansion of the contract labour force in Shulu, it did not appear that many contract workers had been returning home, at least in the aggregate. But the aggregate figures may mask turnover of those who were returning only to be replaced by other peasants, either under some rotation scheme or for demographic reasons (e.g. young women returning home to bear and nurture children or to accompany children who may have been with them in town to start school in the countryside). Even in the absence of a significant return rate, the contacts and information which contract workers could supply to their home villages through visits, phone calls, and letters have been of enormous importance to the rural collective units in helping them identify markets for rural industries and sidelines and obtain needed inputs for industry, sidelines, or even agriculture.

Third, the rural collective sector was spared by the institution of contract labour the very serious losses that can and indeed in recent history have been visited on it by the uncontrolled outflow of peasant labour to towns in search of employment. The experience in the wake of the Great Leap Forward, when peasants deserted many villages *en masse*, may be instructive of the possible effects. Many villages which were poor from the outset, particularly because of low labour densities or lack of infrastructure, lost the very workers whose labour, organized for farmland construction projects under the then-new communes, was the best hope for recovery. A downward spiral was set into motion, in which the desertions of some—and usually it was the 'best and brightest' peasants who left first—only exacerbated the crisis of the rural economy, which increased the pressure on (or, if you will, the incentive for) others to leave, which worsened the crisis further, etc.[17]

This was also a crisis for the rural socialist institutions, whose development was always predicated on a growing rural economy. As skilled or politically influential peasants deserted the countryside, the rural collective units found it difficult to redeem themselves after the disastrous Great Leap Forward and prove their worth and viability by organizing the recovery through collective projects in farmland reconstruction. For a time the only way out was to retreat to less collectivist and even pre-collectivist productive relations. It is certainly no accident that the outflow of labour from the countryside after the Great Leap coincided with the appearance of household contract production and other forms of individual or small collective subcontracting; neither were these purely parallel responses to the crisis brought on by the Great Leap. (It is also not accidental that an essential element of recovery when it did come was the recall of peasant labour to villages which had been victimized by its outflow.)[18] In the absence of the labour power which could be mobilized collectively for recovery, many localities were forced to retreat on the collectivization front. That this was a welcome development to many former rich and upper-middle peasants adds a dimension of class conflict to the matter.[19]

In sum, the uncontrolled outflow of labour after the Great Leap contributed in many localities to developments which can appropriately be seen as struggles between rural collectivist and pre-collectivist modes of production, and between rural classes as well. The contract labour system was born partly as a response to these problems. By placing the flow of labour out of the rural sector under the regulation of state and rural collective sector authorities, the risks posed by uncontrolled outflow to the health of the rural economy, the prospects for socialism, and rural class conflict were reduced.

A fourth benefit from contract labour accrued to a part of the rural areas, namely its poorer part. This has to do with the redistributive effects (among rural *units*, not individuals) of contract labour. Poorer units benefited more than rich ones from the portion of the contract workers' wages which they received, since they paid out to the workers less valuable work points than did the richer ones. It was also claimed in Shulu—and this claim is hardly unique—that in allocating contract labour slots preference was given to poorer collectives.

Contract Labour, Class Structure, and Urban–Rural Cleavage

Temporary contract labour, we have seen, promoted urban–rural *balance* in the areas of financial flows, developmental possibilities, and even, in some respects, income. Yet, at the same time it rested upon and reinforced the structural *cleavage* between urban and rural people, economies and institutions. Contract labour was built around the household registration system, which divides the Chinese people into categories of rural and urban (technically, 'non-rural') householders. People in these categories face entirely different prospects for employment and advancement, structures of income and social services, and social and cultural life. Peasants in Shulu County and much of China have accepted—indeed, they have vied for—urban contract work, despite its more unfavourable terms compared with regular urban labour,[20] precisely because the urban–rural cleavage, and the household registration system which was used to enforce it, posed such severe obstacles to upward and townward mobility in any other way. In this sense, contract labour was based upon the urban–rural cleavage. Moreover, by permitting state-regulated movement of labour from countryside to city in places where it was needed, contract labour prevented the state from having to dismantle the household registration system in order to recruit the labour. In this sense, contract labour actually *reinforced* or at least *preserved the urban–rural cleavage at its point of greatest vulnerability*.

Contract Labour: Five Conclusions

Five main points have emerged in this discussion. First, in Shulu County temporary contract labour was an integral part of industrialization. Shulu had built a solid industrial base by the mid-1960s, but it was only in the latter half of that decade and on into the 1970s that Shulu really grew into the premier industrial county of its area, developing a wide range of industries including some rather modern, sophisticated lines of production. This development was based predominantly on the employment of peasants on a temporary contract basis, to the point that by 1978 these workers made up a *majority* of the labour force in county/state industry. Contract workers also helped subsidize Shulu industry by virtue of their lower wages and benefits compared with the regular urban industrial work-force.

Second, the heavy use of temporary contract labour enabled Shulu to minimize permanent townward migration during its period of rapid industrialization. In other words, urbanization has proceeded at a much slower rate than the county's rapid industrialization would otherwise have caused.

Third, the system of contract labour has helped to promote urban–rural balance in a number of ways, but also set limits to that balance. It enabled rural collective units to appropriate significant amounts of funds which could be used to finance rural development. It also directed significant funds into the rural household sector by paying contract workers far more than they would have made in their villages. By tightly regulating the labour market, the contract system may also have kept wages higher than they would have been if peasants had been permitted to flood the labour market. Whatever the size of the gains in urban–rural balance, they must be weighed against other urban–rural imbalances built into the system. The most glaring of these were the continuing inequalities between contract and regular workers in the areas of wages and benefits. Then there is also the intangible factor of the burden placed on contract workers (and their families) by virtue of separation from family and the rigours of periodic travel home.

Fourth, it helped reinforce the structural class inequality between workers and peasants in China. Most discussions of urban–rural relations address the question of *balance* in the distribution of particular resources or utility values, such as investment funds, knowledge, capital goods, or income. A more *structural* perspective is also needed, paying attention to class relations and differences in terms of incorporation into a particular mode of production, questions about occupational structure and opportunity that flow directly from this, and other matters such as incorporation into a specific structure of political organization, social status, and culture. In this general area, *temporary contract labour has had the effect of reinforcing the worker–peasant cleavage in Chinese society, even while it was redistributing resources in a somewhat more balanced way across that cleavage line*. Thus, while paying contract workers more than they would have made as peasants, it also prevented them and their families from becoming workers and urban dwellers. By putting peasant contract workers in the closest possible proximity to the urban-based working class, it drove home ever more forcefully the class inequality between them.

This leads directly to the fifth point. The contract work system created a social stratum—the contract workers and their families—that could threaten the very basis of the worker–peasant class cleavage which the system was designed, or at least served, to maintain. Contract workers have already expressed their opposition to the class inequalities between themselves and the working class in the form of political protests during the Cultural Revolution (in China generally) and in some, perhaps more measured, form of political action in Shulu County around 1972. While the details of the 1972 dispute remain adumbrated—an indicator of their seriousness, perhaps—what is clear is that the workers lost in their effort to remove the structural barrier dividing them from the working class: their demand for a change in their household registrations to 'non-rural' was not met. The class cleavage still remains, and while its effects have been ameliorated by economic concessions, there is no reason to expect the class conflict that was embodied in, engendered by, and perhaps even reinforced by contract labour to disappear. Indeed, forces operating in the present period of reform may promote greater expression of such conflict.

Urban–Rural Relations and the Post-Mao Reforms

The reforms which have swept over China since 1979 have the most serious implications for urban–rural relations and structures. In terms of economic inequality, macro-economic policy changes such as the upward adjustment of rural procurement prices and some deregulation of cropping patterns by the state have resulted in increased income equality. Average rural income doubled between 1978 and 1983, while urban rose only one-third. In economic terms, there is greater *balance* between city and countryside.

In class *structural* terms there has been much change too. Unlike the Maoist period, where, we have seen, the state—through the household registration system and the strict enforcement of urban–rural labour immobility (ameliorated only in a controlled way through contract labour)—kept peasants and workers segregated spatially, the reforms have had the effect of increasing the spontaneous movement of labour and population from the countryside to the city. Accordingly, temporary contract labour has atrophied in many places. The major changes in political and economic organization which bear upon this are two: first, the reduced political capacity of rural collective units to regulate economic activity in general; and, second, the general freeing up of labour markets in the towns and rural areas.

On the first point, the contract labour system was always enforced partly by the rural collective units, which kept peasants employed in the rural areas and regulated contract arrangements for a select few to undertake urban industrial employment. As the general administrative and political capacities of the rural collective units have declined under the recent reforms, control of rural emigration has become harder to enforce. Specifically, now that rural cadres are not calling upon peasants to show up for work each day on collective fields, peasants are freer to leave the countryside on their own to seek urban employment. And since rural cadres have seen their political power decline under the reforms, there is much less they can do to enforce laws or policies of the state to restrict townward migration.

On the second point, as freer and more informal labour markets have opened peasants have come to towns and cities in great numbers in search of employment. Hence, townward migration has grown, in ways which are less subject than before to state regulation and control. Moreover, with the possible opening up of private and informal housing markets in towns, the increased townward migration may be taking on a broader and more permanent character, as peasants bring all or part of their families with them and stay longer.

For example, in mid-1983, the city government of Wuhan had to issue new regulations to control the influx of peasants by requiring them to have permits (whose number was no doubt to be limited). Permit-holders were not required to be working in the city through the temporary contract arrangements we have described. The permits were to be administered by the Bureau of Public Security (i.e. the police), not the Bureau of Labour. But these regulations were not succeeding in keeping townward migration to the levels desired by the state. Within weeks of the issuance of the new regulations, the Chinese press reported that the 1,400 cadres who had been assigned to monitor the city's street-markets and make sure that all merchants had the proper permits had discovered and banned 1,200 violators. At this same time, the city of Lanzhou, which had issued 200 such permits, discovered no less than 3,000 unregistered immigrants among the street merchants alone.[21]

Simple illegal migration aside, new channels for townward migration are opening up. In Taiyuan, one could dine in 1986 at a small restaurant run as a joint venture by a most unlikely pair of partners: the Ministry of Heavy Industry and a rural production brigade. The brigade was supplying the waitresses, and the Ministry the administrative cover.[22] There are also recidivisms of the most exploitative labour practices of capitalism. Sweat-shops have reappeared: one Zhejiang Province entrepreneur hires peasants for Yuan 1.5 per day—and the days were overlong—in an electrical supplies enterprise which earned him a profit of Yuan 150,000. So have labour gangs, run by bosses who receive over one-third of the workers' wages.[23] This is exactly the sort of practice which had been prevented in part by the temporary labour contracts administered by collectives in which the peasantry had a voice.

The new pattern of deregulated and increased townward migration has produced an explosive situation in terms of social conflict. To be sure, rural migrants provide much that is welcomed by urban residents, including increased and more varied supplies of vegetables and fruit. But they also provide economic competition for urban enterprises and labour, which is greeted much less enthusiastically, at least by the workers. In one case, drivers employed by a state-run bus company in the city of Harbin drove into a private bus owned and operated by a peasant migrant; a month later they beat him up. Opposition to peasant competition with the urban state sector is felt and expressed by state officials too. For example food shops run by peasants are often permitted by urban officials to operate only on out-of-the-way streets in order to help protect state-run competitors.[24]

The present period marks a major change in urban–rural relations and the state's role therein. Whereas in Maoist days the state reinforced the urban–rural cleavage through household registration and restrictions on movement of population and labour, now it has unleashed labour market forces which have promoted such movement much more than the state would like. But at the same time the state continues to maintain the cleavage between urban and rural people. Household registrations have not been abolished, and, as we have seen, the state continues to try to restrict townward migration of rural residents, especially into major cities. But it lacks the ability to regulate the migration as much as it would like, due partly to its own decollectivization policies.

In one sense, then, there has been little change: the state continues to reproduce the urban–rural cleavage through maintenance of household registrations and the differential forms of employment concomitant with them. But beyond this there has been a very important change. While in pre-reform days the state was able to maintain the cleavage by keeping urban and rural residents separate, mixing them only in a closely controlled way that benefited the peasantry at least in relative terms (through contract labour), now *urban and rural residents are brought into much closer contact even as they continue to treat each other, and be treated by the state, as distinct and unequal classes.* The inevitable result is increased exploitation and conflict, between rural residents on the one hand and urban residents and even the state on the other, of the sorts we have seen. The class-like cleavage between rural and urban people remains, then, but now is less subject to state control and more easily expressed in social conflict.

Notes

1. For an example, see Tang Tsou, Marc Blecher, and Mitch Meisner, 'Organization, Growth and Equality in Xiyang County' (Part II), *Modern China*, 5 (Apr. 1979), 140–54 and *passim*.
2. G. William Skinner, 'Marketing and Social Structure in Rural China' (Parts I–III), *Journal of Asian Studies*, 24 (Nov. 1964–May 1965), 3–43, 195–228, and 363–99.
3. Michael Lipton, 'Rural Development and the Retention of the Rural Population in the Countryside of Developing Countries', *Canadian Journal of Development Studies*, 3 (Summer 1982), 18. Nevertheless, 28.52 million Chinese peasants moved to urban areas from 1965 to 1980, which contributed almost as many new urbanites as did natural growth of the urban population (by 32.43 million) over the same period. These figures give some idea of the magnitude of the problem of townward migration which China was facing. The migration control policies adopted by the state were holding back a veritable floodtide. For these data and discussion of them and wider issues of

urbanization in China, see Kam Wing Chan and Xueqiang Xu, 'Urban Population Growth and Urbanization in China Since 1949: Reconstructing a Baseline', *China Quarterly*, 104 (Dec. 1985), 609.

4. Marc Blecher, 'Income Distribution in Small Rural Chinese Communities', *China Quarterly*, 68 (Dec. 1976), 797–816.

5. Lipton, 'Rural Development', 20–2.

6. 'High collectivism' refers to the relations of production that prevailed in Chinese communes from around the early 1960s through the late 1970s. See Marc Blecher, 'The Structure and Contradictions of Productive Relations in Socialist Agrarian "Reform": A Framework for Analysis and the Chinese Case'. *Journal of Development Studies*, 20 (Oct. 1985), 104–26.

7. Dorothy Solinger, ' "Temporary Residence Certificate" Regulations in Wuhan, May 1983', *China Quarterly*, 101 (Mar. 1985), 98–103. The reform period is discussed in further detail in the final section, below.

8. Shulu county is located approximately 70 km. by paved road to the east of Shijiazhuang Municipality, the capital of Hebei Province. It is therefore in a rural rather than suburban area, situated well out into the North China Plain. It is about average for its area in agricultural development, but relatively advanced industrially; in fact, it is the most industrially advanced county in the prefecture. Though industrial output value made up 55% of total 1978 output value by official count (and actually 65% if the output of brigade-level industry is included), still only 4% of the county's population was classified as 'non-rural householders'. With 6.94 persons per cultivated hectare, in 1978 it was densely populated. For further description of the county, see Marc Blecher *et al.*, *Town and Country in a Developing Chinese County: Government, Economy and Society in Shulu Xian*, forthcoming; and Marc Blecher, 'Balance and Cleavage in Urban–Rural Relations', in William Parish (ed.), *Chinese Rural Development: The Great Transformation* (Armonk, NY: M. E. Sharpe, 1985).

9. Urban county/state industry includes all state-run enterprises as well as collective enterprises administered by the county government. See Blecher *et al.*, *Town and Country in a Developing Chinese County*.

10. Outflow rural labour has been characterized by some Chinese as part of the country's class struggle. See Tsou, Blecher, and Meisner, 'Organization, Growth and Equality', 149.

11. The only exception was that pre-school children of mothers with contract work could remain with them in town, housed in their mothers' dormitories and cared for during the day in factory crèches. But when the children reached school age they had to return to their parents' home villages. (Often this was also the time for their mothers to resign from contract work to move home too.)

12. In 1978 Xinji had a population of 27,269, not including the 9,994 contract workers employed in industry and commerce. If they and their dependants (assuming the average Shulu household size of 3.98) all moved to Xinji, its population would have reached around 67,000.

13. Hong Yung Lee, *The Politics of the Chinese Cultural Revolution* (Berkeley, Calif.: University of California Press, 1978), 132.

14. See Blecher, 'Peasant Labour', 744, n. 24.

15. Ibid., n. 25.
16. Ibid., 738–9.
17. Tsou, Blecher, and Meisner, 'Organization, Growth and Equality', 140–54.
18. Ibid.
19. Ann Thurston, 'The Revival of Classes in Rural Kwangtung: Production Team Politics in a Period of Crisis', paper presented to the Workshop on the Pursuit of Political Interest in the People's Republic of China, Center for Chinese Studies, University of Michigan, Aug. 1977.
20. Contract workers' salaries averaged annual Yuan 116 (20%) lower than regular urban workers. They also received health and retirement benefits.
21. Solinger, ' "Temporary Residence Certificate" Regulations'.
22. Personal communication from a visiting Western scholar whose name must remain anonymous to protect the confidentiality of Chinese informants.
23. Daniel Kelliher, 'State-Peasant Relations in Post-Reform China', paper presented at the 1986 Annual Meeting of the Association for Asian Studies, Chicago.
24. David Zweig, 'Up from the Village into the City: Reforming Urban–Rural Relations in China', paper presented to the Thirty-second North American Meeting of the Regional Science Association, Philadelphia, Nov. 1985.

IV

Housing for the urban masses

INTRODUCTION

SHANTY towns and run-down tenements dramatically demonstrate the poverty of much of the urban population in the Third World. Many urban dwellers lack adequate protection from rain and flooding, from heat and cold. Still, their health, and indeed their life, is more directly affected by inadequate nutrition and hunger, by contaminated water, and inadequate sanitation. Also, of course, all of these, as well as shelter, are a function of income. Having examined the urban labour market, we shall explore how people earn an income in the urban setting in Part V.

Three considerations argue for interrupting our focus on income to direct our attention to urban shelter. The cost of housing absorbs a major part of the budget of most urban dwellers; housing construction takes up a large share of national investment and constitutes long-term investment at that; and construction entails a long-term commitment in the allocation of space. Not surprisingly then, it is in this dimension of urbanization, more than in any other, that governments are deeply involved, that political change has immediate repercussions on policy. Lozano (1975) discusses the changes in housing policy when Salvador Allende was elected president in Chile and established his Unidad Popular government in 1970; Hamberg (1986) details the major changes implemented as soon as Fidel Castro took over in Cuba in 1959, as well as several shifts in housing policy since—indeed, the legalization of private real estate transactions she describes in 1985 was reversed the following year; and Kirkby (1985, 164–79) describes the policy reversal in China after the Cultural Revolution: between 1958 and 1976 urban housing had deteriorated in general, and living-space per capita in particular had diminished, but in recent years as much as a quarter of all basic construction investment has gone to housing.

The cost of urban housing is staggering. Regimes that can exert a measure of control over migration reduce the demand for urban housing by tailoring the urban population to the needs of the urban labour market, as we have seen in Part III. Only workers needed for the urban economy are allowed into the cities, and many of them are recruited on contract; they will have to leave the city when no longer needed, they cannot bring their dependants to the city. Furthermore, attempts are made to force excess labour out of the cities, as in the 'up to the mountains and down to the villages' campaign in China in the 1960s and 1970s. In spite of such restrictions on urban growth, socialist countries invariably have severe housing shortages. The state limits investment in the housing construction it controls, and neglects maintenance, because such expenditures are seen as unproductive.

The cost of urban land constitutes a major element in the cost of urban housing. Speculation may inflate land values, but in any case proximity to central locations constitutes a value that is revealed in the costs of transport and time that more distant locations entail. But more than distance from central locations is involved in land values. Alan Gilbert and Peter Ward, in Chapter 8, describe the stratification of urban space in Bogotá, Mexico City, and Valencia. They emphasize that land values reflect not only the inherent but also the acquired characteristics of land. The rich may choose to live in an attractive location at considerable distance from their place of work, but an expressway is likely to shorten the distance.

Squatting in the face of repression demonstrates that the cost of land constitutes a crucial element in the provision of urban shelter. Many squatters choose quite unattractive locations—land unsuitable for building, or land on the distant outskirts—so as to minimize the risk of being displaced. They are prepared to endure the hardships of squatting, and they take the risks it entails, because even such unattractive land would be costly to purchase. Indeed, once squatters have secured a measure of recognition of their claim to the land they occupy, it commands a price on the market. Whether such land thus becomes part of the general urban land market, or two land markets need to be distinguished, is an issue further pursued by Gilbert and Ward here. A more detailed examination of the land supply question in Mexico is provided by Ward (1986).

Self-help holds out the promise of lowering the cost of urban housing. Over the last two decades the movement in favour of self-help housing has found widespread support. The most important achievement of its advocates has been to dissuade governments from indiscriminate eradication of squatter settlements, unauthorized structures, and slum areas. Turner played the leading role in this movement. Peter Nientied and Jan van der Linden, in Chapter 9, present his arguments, detail the critique by Burgess, and demonstrate that any debate between Marxist and non-Marxist approaches is impeded by epistemological differences. Nientied and van der Linden further show how these theories diverge from practice in exaggerating the importance of housing policy, in erroneously identifying the housing policy of the World Bank with the self-help approach, and by a structuralist interpretation of the state that can accommodate any empirical reality. Nientied and van der Linden argue that, in spite of these difficulties, common ground can be found between Marxists and non-Marxists.

We will have occasion to return to the housing question. In Part V Nelson includes rental housing in her account of income-earning opportunities in a squatter settlement and Lea Jellinek reports the experience of squatters resettled in government housing. In Chapter 19 Castells addresses the political role played by squatter movements.

References

Hamberg, Jill (1986) 'The Dynamics of Cuban Housing Policy, in Rachel G. Bratt, Chester Hartman and Ann Meyerson (eds) *Critical Perspectives on Housing* (Philadelphia: Temple University Press), 586–624.

Kirkby, R. J. R. (1985) *Urbanisation in China: Town and Country in a Developing Economy 1949–2000 AD* (London and Sydney: Croom Helm; New York: Columbia University Press).

Lozano, Eduardo E. (1975) 'Housing the Urban Poor in Chile: Contrasting Experiences under "Christian Democracy" and "Unidad Popular" ', in Wayne A. Cornelius and Felicity M. Trueblood (eds.) *Urbanization and Inequality: The Political Economy of Urban and Rural Development in Latin America*. Latin American Urban Research 5 (Beverly Hills, Calif. and London: Sage Publications), 177–94.

Ward, Peter (1986) *Welfare Politics in Mexico: Papering over the Cracks*. The London Research Series in Geography 9 (London, Winchester, Mass., and Sydney: Allen & Unwin).

8

Land for the Rich, Land for the Poor*

Alan Gilbert and Peter M. Ward

It is clear that in any society the process of land allocation is highly competitive between groups. There has to be some mechanism by which particular groups gain priority over certain areas of land before others. In most capitalist societies it is the market which essentially allocates land. Those who can afford to pay more or, according to economic theory, are less indifferent to location, acquire the more desirable areas. According to one's ideological perspective the process is efficient and equitable, or inefficient and inequitable, or somewhere in between. In all capitalist societies the process is one which creates residential segregation; the rich tend to live in one part of a city and the poor in another. Clearly, the greater the inequality of income and wealth, then the greater the degree of residential segregation and the greater the exclusion of poor people from access to the better land or, at times, any land at all. Thus, in societies with wide disparities in income and wealth, such as Colombia, Mexico, and Venezuela, the degree of residential and land-use segregation is likely to be very great. This segregation is not only a reflection of the *inherent* characteristics of land, that is to say its height, soils, etc., but also of its *acquired* characteristics, its servicing, location relative to other activities, social character, etc.

In our three cities the poor occupy the worst land in terms of acquired characteristics; the areas with the worst pollution, the areas with least services and worst transportation. More often than not they also occupy the worst land in terms of inherent characteristics, the areas most liable to flooding, areas subject to subsidence, the areas with the poorest soils. Where rich areas develop on inherently poor land, the difference is that negative elements are swiftly remedied, indeed they may be turned to good advantage. In Bogotá, Valencia, and Mexico City the process of residential segregation is not only brought about by differential powers to bid for well-

* Reprinted, in a revised form, from Alan Gilbert and Peter M. Ward, *Housing, the State and the Poor: Policy and Practice in Three Latin American Cities*, Cambridge Latin American Studies 50 (Cambridge, Cambridge University Press, 1985), 62–71, by permission of the publishers.

located land but also by the mechanisms which determine whether the land is serviced or not.

As Map 8.1 shows, the poor in Bogotá tend to occupy broad swathes of land to the south-east, south, south-west, and north-west. By contrast, the major affluent groups live in clearly demarcated zones in the north of the city. Although the demarcation is by no means as clear as it was in the Bogotá of the 1940s, homogeneous class areas are still found across broad areas of the city. Needless to say, the affluent suburbs are well planned, with green zones, good services, paved roads, and street lighting. The poor settlements are partially serviced and are located closest to the centres of pollution, flooding, and heavy traffic. The middle-income groups are broadly distributed between the areas of high-income and low-income settlement. Broadly, these areas correlate closely with the inherent qualities of the land; the micro-climate of the north is superior to that of the south and the land most liable to flooding is occupied by the poor (Amato 1969).

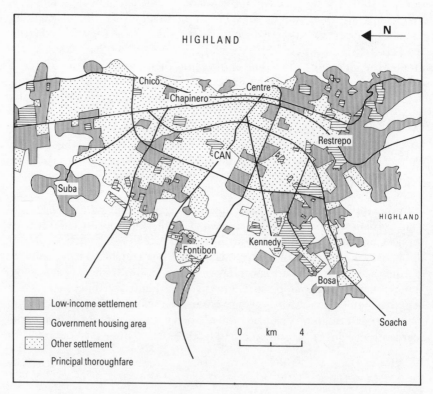

Map 8.1 Distribution of settlement by housing submarkets in Bogotá

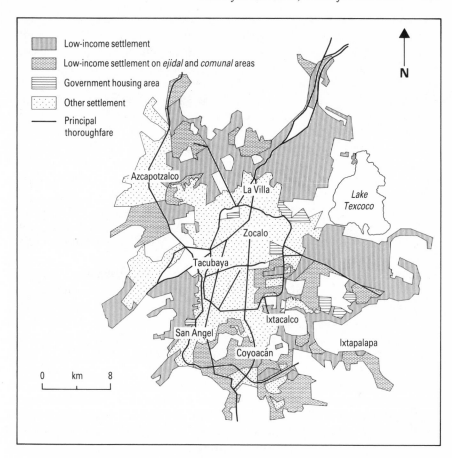

Map 8.2   Distribution of settlement by housing submarkets in Mexico City

In Mexico, residential segregation is more difficult to describe because of the size and the complexity of the city. But, broadly, the poor live in the east, north and north-east, while the rich occupy land in the west and south-west (Map 8.2). Middle-income groups eschew the poorer zones and live as near to their would-be peers as possible, either in well-serviced subdivisions or in housing recently vacated by the rich. However, there are large pockets of low-income housing close to more affluent areas, mainly owing to the occupation of state or *ejidal* land. Where possible the rich have opted for hilly, wooded districts. Although the topography is often broken, the rich can afford to landscape their neighbourhoods and adapt house styles to the terrain to make the most of the views. The western hillsides suffer less pollution than elsewhere and enjoy easier road access to the rest of the city. With the major exception already noted, the poor occupy the two broad zones which are not higi.ly desired by more affluent residential groups. The first is the area to the east and north of the Zócalo, where most rent accommodation, live in cramped conditions, and suffer from deteriorating services and high levels of pollution. The second is the dessicated bed of Lake Texcoco where the soils are sterile, become waterlogged in winter, only to dry out in the summer to become subject to dust storms. The pattern of residential segregation has become more complex through time as the city has become larger and has absorbed more peripheral areas, as high-income groups have established new areas of preference, and as infrastructural developments have transformed particular zones of the city.

In Valencia, the rich and middle-income groups live north of the city centre. The higher-income estates take advantage of certain physical features such as access to Lake Guataparo and the higher land but the main feature is their separation from lower-income areas and their good access to main roads. With one major exception, the area of Naguanagua to the north-west, the poor occupy the south of the city, the low-lying land liable to floods and closest to the industrial areas (Map 8.3). As in the other cities, the high-income areas are well laid out and serviced, the low-income areas are not. The pattern has changed little over the past thirty years except for the creation of the industrial parks in the south-east, the associated development of the Isabelica public-housing complex, and the general expansion of the city.

How can we best generalize about the similarities in the patterns of residential land use that have emerged in the three cities? Clearly, the processes determining land use are very complex. During the development of urban areas, agricultural land is converted into residential, commercial, industrial, institutional or recreational land. Which it becomes depends upon the location of the land with respect to existing uses, upon the investment decisions of private and public institutions, and upon the planning decisions of the authorities. Over time, these initial land uses will change, low-income residential areas may be upgraded, high-income areas lose social standing, and residential land close to commercial and industrial areas may go through a process of urban renewal. All these changes influence the price of land. In developed societies most land is exchanged at commercial values in a legally binding process. In Latin America, however, a considerable proportion of land transactions are illegal in the sense that the land has been occupied against the wishes of the original owner, there are doubts about the legal documentation, or because the settlement, or the buildings themselves, do not meet the planning regulations. It is frequently suggested, or at least is implicit in most discussions of irregular and spontaneous housing, that land allocation in

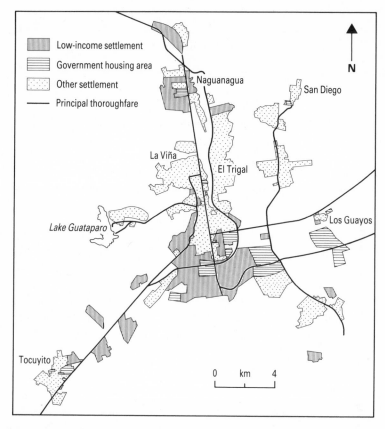

Map 8.3   Distribution of settlement by housing submarkets in Valencia

these illegally developed areas lies outside the market mechanism. Prices are irrelevant because the land is invaded by groups or individuals or because the prices are so much lower than the 'commercial value' as a result of the illegality of transactions. According to this argument there are effectively two land markets, the one legal and guided by the normal processes of supply and demand and subject to government intervention as planner, the other guided by non-market and usually political considerations. While there is some truth in this assertion, there are certain dangers implicit in it. In general, we would argue that the difference between the two sets of areas relates less to the method of valuation of the land, which is generally similar, than to the method of lot distribution. All areas, legal or illegal, have commercial values; the differences in price reflect differences in location, services, prestige, legality and transferability. If some illegal areas are initially allocated through non-market mechanisms such as invasion, their occupation gradually transforms the land into a marketable commodity. Thenceforth the value of individual lots is determined by the forces of supply and demand and influenced by servicing decisions and patterns of urban expansion.

In so far as there are *initially* two land markets, the key question to be answered is why some land is subject to legal market transactions and other land is not? To some extent, of course, historical factors can explain this apparent separation. In Latin America, Spanish law established certain areas as reserves for indigenous peoples.[1] While many of these areas have been incorporated into the private land market in ways which we shall consider in a moment, there is no doubt that in some cities large areas are controlled by the public sector. In Valencia, for example, much of the land in the south of the city belongs to the municipality for this reason. In Mexico, most communal land was sold to private owners during the latter half of the nineteenth century but after the Revolution was reconstituted in the form of the *ejidos*.[2] There is no doubt that the presence of this land has an important influence upon the operation of the land market, notably upon the incidence of invasions.

Nevertheless, even in the case of *ejidal* and common land, we would submit that market mechanisms have intruded in a variety of ways to help determine the social distribution of land. First, the fact that it is often municipal land that has been invaded in cities such as Lima, Caracas, Valencia, and Barranquilla should be considered carefully. In general, invasions have occupied only those areas of public land that would fetch low prices on the market. For example, most land invasions in Lima have occurred on desert lands distant from the city centre (Collier 1976), the invasions of Guayaquil have occupied swampland (Moser 1982), those of Caracas the unstable hillsides (Marchand 1966), and those of Barranquilla the land close to the river estuary (Foster 1975). A few exceptions, such as the *favelas* in Copacabana or the Policarpa Salavarrieta invasion in Bogotá, have occupied central land and have been vigorously opposed and sometimes eradicated. In these cases opposition has been generated by the fact that the land would have a high value if placed on the formal land market. Generally, valuable public land has not been invaded.

One reason is that the more valuable public land is sometimes held by the state for commercial or speculative reasons. The Beneficencia of Cundinamarca in Bogotá, for example, has held land since 1938 in a prime spot for urban development two miles from the city centre. Another reason is that the interests of private landowners may be threatened even by the invasion of public land. The occupation of land close to high-income residential areas, for example, would have a powerful negative effect on land prices in those areas. Private landowners are likely to persuade the state to oppose the invasion of land in those areas of the public domain which are continguous to those of the private sector. In addition, it has been a common practice in many Latin American cities for the private sector to alienate the more desirable public land. Collier (1976) notes how the more valuable public land around Lima has been used for agriculture by private interests. In Valencia, the council sold off up to 10 million square metres of land in the late 1950s and early 1960s for industrial use (CEU 1977 39). Ford alone bought 415,950 square metres for *bs*. 2 per square metre, a price far below its potential market value (CEU 1977 36).[3] If the market did not determine the sale price of this land, it is clear that the attractiveness of the land to the industrial purchasers was not unconnected with their view of the real market value of that land.

It can also be argued that much public land was acquired by the state for communal and public use because of its poor quality and low market value. Much of the *ejidal* land created in Mexico was of dubious agricultural value, its primary purpose. Similarly, the *ejidos* in Valencia occupy the low-lying land liable to floods in the south of the city, the least desirable land in colonial times. It was the absence of alternative uses that dictated the incorporation of land into the 'non-market' sector.

Finally, once public, or indeed private, land has been invaded, it rapidly becomes subject to market transactions. The fact that low-income areas begin illegally does not mean that they continue to be illegal. Indeed, in most cities such areas are gradually incorporated both functionally and legally into the urban fabric. This is true when the land has been sold illegally and when it has been the subject of invasion. As soon as the tenure of a settlement is assured, lots start to change hands. Invaders sell out to new occupants, as do speculators and the purchasers of lots who can no longer afford to build a house or whom other circumstances force to move. The price of these lots is determined by the commercial land market. The price, of course, is lower than in formal residential areas, but it is lower not because the land is outside the market, nor wholly because of the inherent characteristics of the land, but because of planning decisions. In short, the price is low because the land is outside the urban perimeter, or because it is not serviced, or because it is still technically illegal and therefore not serviceable or creditworthy. A favourable planning decision to service the land would automatically raise the price of the land.

We would submit, therefore, that there is some clear connection between the way in which public land is used and the potential price of that land on the commercial market. While some land is kept out of the 'formal' land market despite its commercial value, for example certain *ejidal* lands in Mexico, much of the public land in our cities is indirectly subject to market forces. If this is true of public land, it is still more true of illegal subdivisions and other kinds of private land. Indeed, we believe it is misleading to stress the distinctiveness of the two land markets. Clearly, the mechanisms by which purchase and servicing occur in legal and illegal settlements sometimes differ widely, but whether land is allocated to low-income or high-income groups is substantially determined by the market. High-income, commercial, and institutional users bid for specific, well-located, areas of land. That which is left over becomes vulnerable to illegal land transactions, it is invaded, or is sold without servicing and planning permission. Indeed, there are relatively few differences between the present pattern of residential location in Latin American cities and what would have developed had all transactions been subject to legal sales on the 'formal' market. In every city, low-income residential uses are found on the low-value land. Government housing projects are relegated to distant, unserviced land because they cannot afford the higher value sites more accessible to service lines. In Bogotá, the annual reports of the National Housing Agency (ICT) since 1965 have consistently complained that the agency has been forced to purchase peripheral land which is expensive to service. In Valencia, the main public housing scheme, Isabelica, is located far from the higher-income areas. In Mexico City, too, housing developments by state agencies during the 1970s were, almost exclusively, located in the less accessible and initially unserviced periphery. We would argue that the market effectively determines land use whether the land is public or private, whether the use is public or private, whether the transaction is legal or illegal. In the process each bid affects the price of the land, further raising the price of high-value land and lowering that of low-value land. Of course, this does not go on for ever as some high-income areas become downgraded, some low-income areas, upgraded, but in general it maintains social segregation. In addition, land-use planning in Latin American cities generally favours high-income uses. It often ignores low-income areas beyond making sure that low-income housing is systematically excluded from high-quality residential areas. These areas are not outside the market, nor in another part of the market, but form an integral part of it. Planning decisions merely change the price of land. The desirable land is generally occupied by the more affluent, the least valuable land by the poor.

We must be careful, however, in our use of the terms desirable and valuable, for the value of, and the desire for, land does not depend only on its inherent qualities. In urban areas, the price of land depends upon externalities, neighbouring uses, government decisions about zoning, and so on. Thus some land becomes expensive not because of its inherent qualities but because of decisions made with respect to it. Thus suburban development has often turned land of little inherent quality for, say, agriculture and with little suitability for other uses into high-value property. Where private or public investment has occurred it has raised land values, where it has not occurred it has reduced relative values, thereby increasing the probability of invasion. The process of urban development, and the decisions which influence it, modify and redistribute land values.

This fact must be taken into consideration when discussing the ways in which individual members of the urban poor obtain land. The form in which land is acquired by the poor in our three cities varies considerably. But, despite this, the poor inevitably occupy the poorest land, both in terms of its inherent and acquired characteristics. This process of residential segregation has not occurred by chance. It has been the outcome of market forces, planning decisions, and servicing policy.

Notes

1. The Spanish crown reserved certain areas of land for the use of the indigenous Indian population. These are often known as *resguardos indígenas*.
2. The *ejidos* in Mexico were established after the Revolution, usually through the break-up of extensive *haciendas*, and the lands made over in usufruct to be held by the community.
3. The Council did this in a campaign to attract industry to the city. Between 1959 and 1968, 43% of Valencia's municipal land was sold to private industrialists (Lovera 1978, 150).

References

Amato, P. W. (1969) Environmental Quality and Locational Behaviour in a Latin American City', *Urban Affairs Quarterly*, 83–101.

CEU (Centro de Estudios Urbanos) (1977) 'La intervención del estado y el problema de la vivienda: Valencia' (Caracas CEU).

Collier, D. (1976). *Squatters and Oligarchs: Authoritarian Rule and Policy Change in Peru* (Baltimore: Johns Hopkins Press).

Foster, D. W. (1975) 'Survival Strategies of Low-income Households in a Colombian City', Doctoral dissertation, University of Illinois, Urbana.

Lovera, A. (1978). 'Desarrollo urbano y renta del suelo en Valencia' (Caracas: Universidad Católica Andrés Bello, Facultad de Ciencias Económicas y Sociales).

Marchand, B. (1966) 'Les ranchos de Caracas, contribution à l'étude des bidonvilles', *Cahiers d'Outre-Mer*, 19, 104–43.

Moser, C. O. N. (1982) 'A Home of One's Own: Squatter Housing Strategies in Guayaquil, Ecuador', in A. G. Gilbert *et al.* (eds.), *Urbanization in Contemporary Latin America* (Chichester and New York: John Wiley) 159–90.

9

Approaches to Low-Income Housing in the Third World*

Peter Nientied and Jan van der Linden

FROM the mid-1960s onwards, several authors took a fresh look at low-income urban housing in the Third World. One of their main policy recommendations was that governments should stop trying to provide standard housing for the poor, and instead should use their human potential by permitting and enabling them to house themselves.[1]

In practice, implementation of this recommendation implies the execution of two main types of projects: upgrading of existing squatter settlements and creation of so-called 'sites and services projects'. Ideally, the two approaches are executed simultaneously: squatment upgrading intends to deal with existing, but poor and illegal housing; sites and services are intended to increase the housing stock and thereby remove the need of the poor to resort to squatting.

Upgrading projects of existing squatments mostly consist of two components. First is legalization of housing, implying that the residents become legal owners or lessees of the land on which they dwell. It is assumed that this security of tenure stimulates the residents to improve their houses by themselves. The second component, improvement, usually signifies that the government brings in those parts of the infrastructure which the inhabitants cannot provide for themselves: access to drinking-water, sewerage, electricity, etc.

Sites and services projects show an enormous variety. The principle is that the government provides cheap plots of land with access to basic infrastructure. The construction of houses is left to the participants, individually or in groups.

From the mid-1970s onwards, Marxists have severely attacked the self-help housing recommendations and the arguments upon which they are based. However, there has not been much debate between the two viewpoints. This is the more regrettable because some of the self-help

* Reprinted, in a revised form, from the *International Journal of Urban and Regional Research* 9 (3), 1985, by permission of Edward Arnold Ltd.

recommendations were taken up by governments and international organizations, especially the World Bank.

From 1972 onwards, the World Bank has very actively pursued the new approach to urban low-income housing. In the four fiscal years 1980–3, the Bank approved loans for urban projects in 28 different countries, totalling $US1,778.9 million (World Bank 1980–3). The major part of these loans is absorbed by sites-and-services, squatment upgrading and integrated urban development projects—this latter category usually includes a sites-and-services and/or squatment upgrading component. For the period up to 1986–7, a lending programme amounting to $US4 billion, spread over some 90 projects, is under way. By now, there are hardly any Third World countries which do not have some low-income housing projects in which the influence of the new recommendations is clear. Also, previous policies of slum eradication and the provision of standard low-income houses have often been abandoned.

Here we review the debate over self-help housing, especially the Marxist critique, with the aim of opening up a debate about housing and self-help policies (cf. Ward 1982b). First, the main arguments of the 'liberal' and the Marxist approaches will be recapitulated.[2] Second, we relate these arguments to the divergence between theories and practice.

1. The Arguments of the 'Liberal' and Marxist Approaches[3]

One of the best-known promotors of the liberal approach is John Turner (1967; 1968; 1976; 1982a; Turner and Fichter 1972). Turner's key ideas may be summarized as follows:

1. The concept of 'housing' should be viewed as a verb, rather than as a noun: housing is not just shelter, it is a process, an activity.

2. Consequently, the house should not be seen simply in terms of its physical characteristics ('what it is'), but in terms of its meaning for those who use it ('what it does'). By implication, the material values of housing should be substituted for by human use values, of which the material value is only one amongst many indicators.

3. Since housing needs change, e.g. according to the family cycle or according to stages in the migrant's life in the city, and since there is an endless variety of individual needs, priorities, and possibilities among the users, large organizations can never adequately cater for all these. Large organizations, such as the state or municipality, always have to standardize procedures and products and thus fail to respond adequately to the majority of individual and changing needs and priorities. In other words, the main components of the housing process have to be left to the users. This does not necessarily imply that dwellers should build their own

houses, but that they are the ones who should judge and decide about housing, individually or through decentralized local institutions.

4. All this does not do away with a role for government, either at the municipal or national level. Only the government can enable the users to become involved in housing activities, e.g. planning, organizing building and maintenance. First, certain elements, such as main roads or sewage treatment plants, obviously cannot and should not be planned, built, and maintained by the community, let alone individually. Second, the government has to formulate the *proscriptive* laws that define the limits to what people and local institutions may do (rather than *prescriptive* laws that tell them what they should do). Third, the government has to provide, and actively protect, access to the elements of the housing process for the users. These elements include land, laws, building materials, tools, credit, know-how, etc.

Many of Turner's ideas are based on what he has observed in autonomous settlements, where—informally or illegally—the dwellers control large parts of the housing process. However, often these dwellers' efforts were frustrated by governments who saw squatters as trespassers who had to be controlled and preferably rehoused in what the authorities considered to be adequate housing. Often, governments demolished the self-built squatter dwellings and even more often the threat of such government action put limits on what the squatters dared to invest in housing. What Turner proposes is that, instead of threatening the existing housing systems, the government should respect and support them. He considers his main task to be convincing governments that there are better solutions to the low-income housing problem than the ones applied thus far. In his view, his proposed solutions are beneficial both to the low-income groups and to governments.

Turner's propositions came at a time when there was a widespread disappointment about the development strategies followed in the 1950s and 1960s. Rethinking the concept of development led, on the one hand, to target-group strategies, aiming at the poor more directly than had been done previously. Turner's ideas fitted very well into such strategies and were soon embraced by several development agencies.

On the other hand, rethinking of the concept of development also led to investigations into the basic causes of poverty, especially by Marxists. To scholars who view the capitalist mode of production as the basic cause of underdevelopment, Turner's proposals are of little appeal, since they do not question the economic and social structure shaped by capitalism.

One of Turner's main Marxist critics is Rod Burgess, who aims to demonstrate that what Turner proposes in fact boils down to 'an economic and ideological means necessary for the maintenance of the status quo and the general conditions for capitalist development' (Burgess 1978, 1107). Burgess's argument (1978; 1982b) may be summarized as follows:

1. By separating and opposing use value and market value, Turner fails to appreciate their dialectic interrelations. Because commercial products (or waste materials, valorized through labour) and labour are invested in the self-help house, they cannot fail to assume a value on the capitalist market. Turner does not deny this, but wrongly assumes that use values dominate. This cannot be—or remain—true since constant expansion of the sphere of commodity production for capitalist exchange is a condition of capitalist development.

2. By blaming industrialism (with its hierarchical organization, large-scale, and fixed standards), Turner seems to overlook that industrialism is just an aspect of capitalism. The squatter does not build outside the sphere of capital circulation, rather he is part of the specific circuit called 'petty commodity production', which is an integral part of the capitalist system. The fact that this specific circuit is allowed to survive can be understood by realizing that in this way one of the essential conditions for cheap reproduction of labour is fulfilled.

3. Turner holds the basically bourgeois view of a well-meaning but misinformed government. In his ideas on the government's role, there is no reference to imperialism (which a class-based analysis would contain). Turner also overlooks the interests of politicians, administrators and financiers, and their manipulation of the squatters financially and politically. In short, Turner depoliticizes both the housing problem and the state. 'Access to the elements' which Turner advocates is a function not of law and administration, but of the capitalist market. The government's function of maintaining the general conditions for the reproduction of capital will always prevent it from intervening against the interests of capital; it cannot reasonably be expected to legislate against the interests on which it depends and which it serves.

A number of practical results can be derived from these considerations. On the one hand, Turner's proposals will be welcomed by the state as the protector of capitalist interests. First, because labour is enabled to reproduce and maintain itself cheaply, thus having a lowering effect on wages. Second, self-help housing functions as an ideological means to quieten the fundamental political demands of the poor, especially because this individualistic approach will result in a growth of social inequality and therefore blunt collective consciousness. Third, by legalizing autonomous housing, the whole housing process is incorporated in the capitalist market. The penetration and dominance of industrial, financial, and landed capital is greatly facilitated by the incorporation of the thus far illegal housing process. Housing will assume its manifest commodity form. The ultimate result of this will be the expulsion of the poor from their self-help settlements (turned into a commodity). Renters in those settlements—to whom Turner pays very little attention—will be the first victims.

On the other hand, there are also some far-reaching implications of Turner's proposals, should they be implemented on a significant scale. Guaranteeing access to the elements of housing, if taken seriously, would, in its implications, shake the foundations of capitalist society. The strongly redistributive policy which is needed to guarantee access to the element 'land', is only one example of this. Therefore, it is unrealistic to expect any government to implement Turner's approach beyond a limited scale, i.e. governments would either seek a compromise and take up only some convenient components of the Turner strategy, or limit themselves to a few projects.

This apparent paradox in Burgess's critique of Turner in its turn gave rise to further comments. In Burgess's view, on the one hand, Turner appears to be a bourgeois liberal, whose proposals further the interests of capitalist development; on the other hand, no government can be expected to implement these proposals on a large scale, since in their implications, they run counter to these capitalist interests.

In this brief presentation of the views of Turner and Burgess, some stereotyping has been inevitable. Turner and Burgess have been treated as single opponents,[4] and as representatives of the self-help and the Marxist views respectively; however, both qualifications hold true only to a certain extent.

In the introduction we stated that a real discussion between the two viewpoints has not taken place. Although it has been suggested that Turner does reply to the Marxist critique (Gilbert 1982; cf. Lea 1979), Turner (1978, 1135), in fact, never went any further than to *suggest* a 'creative dialogue' between, what he calls, 'conservative anarchists and radical authoritarians'. But this dialogue could start only after the development of 'commonly recognized terms of reference', since, 'unless these issues and positions are recognized, the discussion of all significant points will be at cross-purposes—each party will be using a different set of definitions and will, therefore, be discussing different questions' (ibid.). Not surprisingly, Turner's proposals for these terms have never obtained the mutual agreement required. Thus, in this respect, Turner and Burgess are not really opponents, but, rather, authors exemplifying different approaches, chosen for the present discussion to demonstrate an opposition of viewpoints more clearly, and to suggest the distinction between a Marxist and a liberal approach.[5] In the literature about the low-income housing question the views of the various authors show, however, more diversity and more gradation. There are not just two groups of authors around Turner and Burgess; in fact, some points of view have other starting-points, and do not share the theoretical notions of Turner or Burgess. Sometimes, they do not have a theoretical justification for their policy proposals at all.

In the remainder of this paper we aim to explore various aspects of the housing issue. In the first place an epistemological difference between Marxist and non-Marxist approaches will be examined, raising the question whether a debate between Turner and Burgess can take place at all. Second, we will argue that the significance of the housing issue tends to be overstressed, a consequence of a drifting apart of theory and practice. A somewhat different, but related aspect of the distance between theory and practice is illuminated by comparing the World Bank's views with those of John Turner. The housing policy proposed by the World Bank is very different from Turner's self-help approach, despite the World Bank's adoption of Turner's jargon. In a fourth comment we discuss a characteristic of the Marxist writings on Third World urban housing, viz. their structuralist interpretation. By examining the concept of the relative autonomy of the state, we aim to show why practice and theory are so divergent. In a concluding section we propose some themes of the low-income housing question which may be a fertile starting-point for both Marxist and non-Marxist approaches at a practical level.

II. The Divergence of Theories and Practice

1. A Different Epistemology

We stated above that the suggested debate between Burgess and Turner never took place, because of a lack of agreement on the 'mutually recognized terms of reference'. We think that this point can be related to a general problem of the 'debates' between Marxists and non-Marxists, namely their difference in scientific method. The approaches have a different interpretation of the empirical situation which leads to differing definitions of the empirical situation.

Marxism applies a distinction between social appearance and social reality. Implicit in the critique of the Marxists is the view that Turner, as all other non-Marxists, deals with the phenomenal forms of reality. What can be observed in the everyday experience of daily life—in the cities of the Third World as anywhere else—are phenomenal forms. This is the realm of appearance, the reality of scholars like Turner. But, according to Marxists, this 'reality' is deceptive, and 'a true scientific approach should try to penetrate the world of phenomenal form to a deeper underlying reality' (Basset and Short 1980, 160). This penetration can only be achieved through the use of certain theoretical concepts (e.g. abstract labour and value) that can be used to explain the generation of phenomenal forms.

Saunders (1981, 15) describes this Marxist method as follows: the material world exists prior to our conceptions of it, and our ideas about the world bear some relation to what the world is actually like, but reality is not directly reflected in our consciousness. 'Consciousness is thus not a reflection of material reality but the mediation of it, and it follows from this that the existential reality that science attempts to discover may be obscured by the phenomenal forms through which the reality is represented in our everyday experience. . . .'.

This epistemological difference between the two approaches is pertinent to the nature of the publications that have appeared on low-income housing in Third World cities. In fact this difference between the approaches 'really' renders the term *discussion* a misnomer since the different views on what 'reality' is and how it must be studied does not just separate the various participants, but makes them 'discuss' different things.

Turner's definition of 'housing' was described above; the process, the activity of housing and its use to the dweller ('what it does') were the focus. Burgess (1982b, 59–61) criticizes Turner's conception in three respects: Turner fails to use the right concepts of 'product', 'utility', and 'market value' related to housing, and Burgess infers that it 'merely reflects the fact that Turner remains on the level of appearance of the housing object and hence remains mystified by its reified form. Taken together with the statement that the "opposition" between use value and market value can be traced to an a priori condition of scarcity, the ideological nature of this concept of housing becomes more obvious' (1982b, 59).

Thus, Burgess reproaches Turner's mode of analysis for remaining on the level of appearance, while the Marxist method delves into the underlying reality. To give two other illustrations; Burgess (1978, 1112) speaks of 'the real relationship' between systems and, regarding Turner's concept of the state, he puts it as follows: 'so too is the concept of the state which informs his work, deprived of any "real" content: class contradictions and interests are either denied or considered of little relevance' (1982b, 75).

The different principles of scientific analysis lead to diverging views on the purpose of studying urban housing problems and policies of the cities in the Third World. An observation that Harloe made of the debate between Marxist and Weberian approaches of urban sociology in the developed societies can be repeated with equal validity for the low-income housing question: 'In fact argument and criticism between Marxist and non-Marxist theories, which often remain overtly at the level of argument over particular theoretical explanations which are assumed to be comparable, sometimes ignores the fact that the questions which each side seeks to answer are also different' (Harloe 1977, 13).

Applying Harloe's remark to the subject of Third World urban housing leads to an identification of the starting-points of the two approaches: the advocates of self-help try to find—and do find—field situations and evidence for their stand and their policy recommendations whereas Marxists aim to analyse housing in relation to the wider urban situation and to capitalism. Thus, two different types of questions are posed.

It must be said that both approaches are internally quite heterogeneous in their starting-points, therefore it does not seem possible to construct a continuum of 'liberal to Marxist' views, or of 'orthodox-Marxist to neo-Marxist' views. Rather, on the basis of an epistemological criterion, a more or less clear dividing line between Marxist and non-Marxist approaches can be drawn, although this difference is clearer to Marxists than to non-Marxists, so it often seems. Just applying the Marxist jargon does not result in a Marxist analysis.[6]

Our tentative conclusion is that the two different approaches of the study of Third World urban housing problems have a different method of analysis, they start with different questions and refer to different principles. Hence, comparison of the two approaches and 'cross-fertilizations' meet with serious difficulties. We think that the recent call for the joint 'application' of different theories at the same time—Marxist in conjunction with non-Marxist theories—(e.g. Goldsmith 1980; Johnston 1982), must, because of these difficulties, be responded to with a certain reservation when low-income housing and self-help in the Third World are concerned.

2. The Limited Significance of the Housing Issue

A second point that needs some discussion is the significance of housing and housing policy. In the third world housing literature, it appears as if housing policy is the 'most important' policy issue for Third World cities. To a certain extent this may be due to the nature of Turner's writings. He started to write in the 1960s about the problems and policies of uncontrolled settlements in Latin America. The concept of housing, which had to be seen as a verb rather than a noun, as was described earlier, also included the environment and services, and referred to local community organization, to small-scale activities, and so on. Thus, the term 'housing' was taken in a very broad sense and hence it became an issue of overwhelming importance.

There are, however, some reasons for believing that the importance of housing—and especially when it is defined in such a broad sense—has been exaggerated. First, housing and improvement of houses often do not have a high priority among squatters, or even the homeless pavement dwellers (Peattie 1979, 1017). Economic conditions are considered more important than housing. Because an informal housing delivery system with many options exists in nearly all Third World cities, constraints on one's economic position rather than government policy will determine the individual's housing situation. Moreover, popular demands addressed to governments are often for services, and not for houses, house-building loans, etc. Second, in the literature there is an emphasis on the problem of urban housing, especially on the different forms of irregular settlement. However, in the rural areas housing conditions are worse and also the availability of and access to basic infrastructure are less (Gilbert and Gugler 1982, 96). High visibility seems to be the prime reason why the urban situation attracts so much attention. Third, assuming that local government policies are, to a certain extent, responsive to the consumption demands of the working classes in the cities, the needs of the poor are not, or are hardly, met by the actual implementation of housing policies. It is true that many countries have some kind of low-income housing plan, but cities where those plans are being implemented on a significant scale are exceptions. Dwyer (1975) characterizes the policy situation as one of apathy towards the problem, and housing schemes are more symbolic than realistic, while Rothenberg (1981, 48) asks the interesting question 'Why have a few (and only a few) governments opted for policies which do focus on housing needs of the poor?' Drakakis-Smith (1981, 23) also indicates that 'it is generally true that government investment in urban low-cost housing is limited'; he uses the term *laissez-faire* for the meagre investment policies of most governments in the Third World!

From these three points we conclude that the significance of the housing issue has been somewhat exaggerated. This conclusion is not without implications for theory. Marxist-oriented critics have tried to explain low-income housing policy as a logic, as a response of the capitalist system, but what are they explaining? If, as Marxists suggest, self-help housing policy benefits the dominant classes, and is functional to the capitalist system in so many ways, then why is it not implemented? Or, as Gilbert and Gugler (1982, 113) put it: 'If spontaneous settlements have been so useful to the rich, why have the World Bank and the United Nations had to work so hard to sell the idea of sites-and-services programmes?' It seems as if the structural critique of the self-help approach is a theoretical reaction, not an empirically grounded one.

Observations like these make it painfully clear how wide the gap between theory and practice has become. This is not to say that theories about housing are irrelevant because governments are not interested. Rather, theory should not drift apart from practice and should recognize the state's reaction in the real world.

3. The World Bank and Housing[7]

In this section, we will show that the World Bank, which seems to have a theoretically based policy, has in fact a very particular view. The Bank presents its thinking as a mixture of Turner's recommendations and response to the empirical situation. But, when we look in detail, the Bank's view is rather different from that of Turner. It bears more relation to economic theory than to Turner's position.

The World Bank recognizes that the provision of conventional permanent housing is not possible given the limited resources available. The housing deficit is then explained in market terms: evidently, there is sufficient demand for housing, but numerous constraints make for weaknesses on the supply side. As long as the supply side keeps providing conventional permanent housing only, it is not properly geared to the enormous existing demand. The only viable method of correcting this is to bring the supply cost down, so that housing and services become accessible to large parts of the population presently excluded from the formal housing market.

A number of devices are proposed to bring the supply cost down. The main components of the supply viz. land, services, and finance, need restructuring. In all three instances, the main bottle-necks are traced to 'institutional constraints'. Regarding land, although it is admitted that 'the will is not always there', the problems of its provision are viewed as 'for the most part an institutional problem'. As for services, costs can be substantially brought down by reducing standards. Standards, both for houses and services, are often unrealistically high. Reduction of such standards (including density standards) will remove important barriers preventing the poor from building permanent legal houses. Again, constraints in housing finance are 'largely institutional'. In short, correction of imperfections in the supply of urban land and services and development of the 'typically underdeveloped finance institutions' would go a long way to rendering housing accessible to the existing demand.

Even with substantial cost reductions, it is important that public expenditure be minimized. In view of scarce public resources and enormous demand, reliance on subsidies would not only be uneconomic, it would also be sure to limit replicability of any such undertakings. Replicability, in its turn, is possible only through sound pricing policies and appropriate standards. A second device to economize on the use of public resources is by shifting the financial burden from the public sector towards the private sector and the urban population itself. Partly, of course, this can be achieved by 'sound pricing policies', but an active mobilization of the energies and resources of the community is also required. Such resources can be tapped by letting the users build their own houses, as is much emphasized in early documents, and/or by the beneficiaries taking contractual and managerial responsibilities, as tends to be stressed more heavily in recent documents. In either case, the principle of 'progressive development' (or 'incremental improvement', or 'phasing') is advocated. It is assumed to have a cost-reducing effect on housing, especially since it relieves the beneficiaries from the burden of substantial down payments. It should be noted finally that all these devices have been chosen on the basis of an analysis of the market forces of supply and demand. Therefore, other possible approaches, such as building low-cost houses and prefabrication, have a low priority but are not a priori excluded.

With regard to the goals behind the adoption of this approach, the central focus is on fighting poverty. One explicit reason for this is that massive urban poverty may pose a political threat. 'Historically, violence and civil upheaval are more common in cities that in the countryside. Frustrations that fester among the urban poor are readily exploited by political extremists. If cities do not begin to deal more constructively with poverty, poverty may well begin to deal more destructively with cities' (McNamara 1975, 20).

Another goal is contained in the proposition that 'housing is a tool for macro-economic development'; it has 'substantial multiplier linkages, through the economy', it 'can lead to higher national productivity, making productive underutilized labour, material and financial resources'. Housing, then, is viewed as one means of attacking poverty by increasing the productivity of the poor. One way of increasing productivity is the creation of employment in house construction which is emphasized in several World Bank documents on the subject. But, more generally, the harnessing of underutilized resources by investing them in housing will yield a flow of income. Thus, making housing accessible to the poor will enhance equity, while, as one document remarks, the participation of the poor in legal housing better integrates them in society. It is mostly for other, more pragmatic reasons that participation of the beneficiaries is stressed: better cost recovery, maintenance, etc. are the assumed results of participation.

A third and related goal is to enhance the development of 'economical urban patterns'. In this connection, site-and-services programmes will provide 'orderly, more efficient alternatives to squatter invasions'. In a wider context, the aim is 'to produce more rational patterns of urban growth by harnessing market forces'. Clearly, housing programmes alone cannot achieve this goal. Rather, they should be seen as integral parts of an approach in which other components, such as land use and transportation, have their place as well. Regarding housing, squatter settlement upgrading programmes, replacing earlier policies of eradication, and site-and-services programmes are the two complementary components of the policy to be pursued.

The World Bank has no pretentions to be 'solving the housing problem'.

Rather, the approach suggests that the housing problem should be capable of solving itself. In view of the logic of the approach, there is an implicit—and sometimes explicit—assumption that the single root cause of the low-income housing problem is the lack of understanding on the part of local or national housing agencies and/or government. The Bank's role, then, is more of a catalytic nature. First, this role consists of demonstrating that there are low-cost affordable and user-acceptable solutions to the problem of shelter. The principle of replicability implies that governments can continue applying this approach once they are convinced that it works. Second, in view of the paramount importance attached to 'institutional constraints' as a root cause of poor housing supply, institution building is a major aim of World Bank involvement in shelter programmes.

Many similarities between John Turner's and the Bank's thinking can be detected. However, it is only fair to note some quite fundamental differences between the two views. More importantly, this exercise can also provide a useful example of the drifting apart of theory and practice, and of the functioning—or dysfunctioning—of theory in practice.

1. To start with, both views regard housing as a tool for development (Turner 1967, 13; 1976, 91) and as a source of employment (Turner 1976, 50), and both believe that the solutions proposed will provide 'economical urban patterns' and 'aesthetically satisfying and culturally meaningful environments' (Turner 1976, 136) as an alternative to unplanned, direct action by squatters. Both views also consider the housing problems as a result in the first place of institutional problems (Turner 1967, 15; 1976).

2. The starting-point for both views is also that conventional attempts to solve the housing problem do not work. However, the World Bank does not reject the possibility of conventional solutions. Only because of its analysis of the market the choice is made of the twin approach, upgrading/site-and-services, to tackling the housing problem of low-income groups in Third World countries. Should market conditions change a recourse to 'conventional solutions' would be fully acceptable. But Turner believes that a proper division of tasks between the public, private and community sectors—in which the last sector has the major say in the planning, construction and management of housing—is the only solution to the housing of any class of people in any place of the world (cf. e.g. Turner 1982b).

3. Although Turner views housing also in terms of supply and demand, in the solution which he proposes, market values do not play a dominant role: the material values of housing are only one indicator of the human use values which according to Turner should take the place of materially determined market values. In the World Bank's view, 'harnessing of the market forces' is the key to the solution of housing problems. 'Bringing down supply costs' by lowering standards is a logical consequence of this

line of thought, whereas Turner stresses the users' autonomy which—amongst other things—results in adopting more realistic standards.

4. Both views emphasize the need to economize. However, while one of the main emphases of the World Bank is bringing down the cost to the public sector and shifting them to the private and community sectors, Turner's emphasis is on the diseconomies of large-scale organizations which cannot possibly cater for the variety of changing housing needs. 'Mobilizing the popular sector's resources', then, has rather different meanings in the two views. In the World Bank's approach, it is a matter of shifting the burden from the one sector to the other, while in Turner's thinking it is the setting free of frustrated human aspirations. In the same sense, 'progressive development' is a means of economizing in the World Bank's eyes, whereas Turner sees it as the logical, natural expression of the recognition that housing is a process. So economizing is one amongst the beneficial results, not the primary aim.

5. In line with this, the result of the World Banks strategy can be—and often is—that categories of people are provided with 'classified bundles of housing goods and services' (Turner 1976, 119). This is precisely what Turner does not want. To him, the essential thing is that every individual user is guaranteed optimal access to the elements required for every individual user's housing. In the same vein, mobilizing popular resources should—in Turner's view—not be misinterpreted as incorporating the resources of the popular sector in centrally administered programmes of housing action (Turner 1976, 131). If this and similar principles are not adopted, Turner fears that site-and-services programmes 'may prove to be more effective instruments in the hands of oppressive governments' (Turner 1976, 119). To the World Bank, centrally administered large-scale housing programmes are by no means tabooed, as long as they are 'efficient'. Also, in the World Bank's documents, the main argument for people's participation is that it increases the project's efficiency, cost recovery, etc., while in Turner's view it is the starting-point and cornerstone of a successful housing process.

Thus, in many basic respects, the World Bank—with its emphasis on what is supposed to be good for governments (political order, economizing on budgets, rational patterns of urban growth, etc.) and Turner—with his emphasis on satisfaction of individual human users' needs—have different views.

4. The Concept of the State and Low-Income Housing

Marxists, following the work of Althusser as elaborated by, for example, Castells (1977) give persuasive interpretation of an empirical situation, but quite often, the same interpretations are valid for the opposite situation as well. This is in part due to the structuralist-Marxist view of the nature of the state.

Regarding the functions of the state, authors like Pradilla (1979; 1981) and Burgess (1982a) believe that the state's role includes maintaining class domination, guaranteeing the reproduction of the capitalist regime of production, assuring the solution of political conflicts tied to the provision of housing and services, and conciliating the secondary contradictions between the fractions of the bourgeoisie. Thus, not only is the state's role quite comprehensive, its actions may be contradictory in their effect. For instance actions for the maintenance of political and ideological dominance may conflict with the state's role of increasing the productivity of land.

In interpreting such conflicting actions, the concept of the 'relative autonomy of the state' (e.g. Pradilla 1981) is important. This concept implies that within an ultimately dominant economic structure, the state—notwithstanding its dependence on the mode of production—has a certain freedom from the dominant classes. Because the dominant classes are internally heterogeneous and have conflicting interests the state needs this relative freedom to serve the long-term interests of these classes.

By applying the concept of the state's relative autonomy in empirical situations, it can be shown that in one particular situation the government directly protects the short-term interests of, e.g., the landowning class at the expense of, e.g., industrial capitalists. At another time, the government may even seem to be responsive to popular demands and to benefit the working class. In the latter situations, other aspects of the state's role than the protection of direct class interests dominate. For example ideological aspects, when the state implements some policy with the purpose of legitimizing certain values and norms. However, the interests of the dominant class are always served, albeit sometimes indirectly and in the long term.

A problem with these interpretations is that they are 'capable of explaining both the grinding down of the working class by authoritarian governments and the improvement of the conditions for the poor. Since nothing is precluded, nothing is explained' (Gilbert and Ward 1982, 118). In fact, the concept of the relative autonomy of the state appears to be tautological and immune from critical evaluation or falsification (Saunders 1981, 279; Sandbrook 1982, 78; cf. also Saunders 1982; Peattie 1979; Nientied and van der Linden 1983).

This concept and the theory behind it can only establish their value if specific conditions can be identified under which specific interests dominate; in other words, when the theory develops a predictive power rather than only being capable of explaining situations *post hoc*. Thus, academic debates can—and indeed do—proceed irrespective of what is happening in practice; consequently, the practical relevance of such debates is quite limited, to say the least.

III. A Quest for Common Ground

Throughout our paper, we have tried to illustrate the considerable gap between theories and practice of low-income housing in the Third World. What we considered as a main problem in the theoretical discussion between Marxist authors and their non-Marxist colleagues, was that their scientific methodologies are so dissimilar that meaningful discussion is impossible. What adds to this problem is the identification of World Bank projects with Turner's theory—a misunderstanding we hope to have made explicit.

We will conclude with a brief exposition of an approach that promises to overcome some of the theoretical difficulties encountered, and propose an area of research in which different theoretical perspectives can complement one another. Here, we follow an argument developed in a different context by McDougall.

Following Marx, McDougall (1982, 264) states that theory develops from systematic reflection on practice, but also directly contributes to practice: 'if the principle of praxis is adhered to, it is logically impossible for theory to be practically irrelevant'. He criticizes the view that sees the state primarily as a functional response to the needs of capital, and advocates the recognition of the contradictory nature of the state: the (welfare) state represents both the achievements of past working-class struggles and the requirements of capitalist production.

From this point of departure, it is possible to identify areas in government policies in the Third World which embody positive elements for the working classes, in the sense that they represent a defence or promotion of their interests. Such elements may be conceived of as achievements of working-class pressure, or as the result of influence which progressive workers in the state have been able to exert. Such 'positive elements' may serve a dual purpose in relation to our above discussion.

Firstly, we believe that the positive evaluation of such elements provides common ground for progressive scholars whether they belong to Marxist or to non-Marxist schools of thought. Secondly, these positive elements in state policy may constitute a practical object of research for the different groups of scholars, and, as such, a common point of reference and discussion. Of particular importance is that this common research object is a practical one. Therefore, theoretical debates can take place on this basis rather than about an empirically irrelevant body of theory.

Two topics of research meet the requirements of serving as a shared object of study particularly well. Firstly, certain government policies, such as squatter settlement upgrading, embody positive elements for the urban working population: low-income groups gain access to urban land and basic services. Admittedly such policies may enable governments to evade their responsibilities to provide proper housing. However, upgrading programmes which, contrary to conventional public housing, can be implemented on a large scale, may lead to community organization and to increased popular pressures on public decision-making. Thus, upgrading programmes may serve several, often conflicting, interests. It is important to investigate the conditions under which positive elements dominate, and, starting at the community level, to explore which routes can be followed to overcome the forces opposing squatter settlement dwellers.

Secondly, the potential power of workers in the state and of the recipients of state services demand enquiry. How can these workers and recipients act to serve their class interests? For recipients of state services this often implies a struggle against bureaucratic values and procedures. An understanding of the working of the bureaucracy is needed to inform the struggle for rules and procedures which do not obstruct lower-class interest, but, on the contrary, allow popular pressure and demand making.

Research on these issues promises to be fruitful in two ways. On the one hand, both Marxists and non-Marxists can study the links between theory and practice. At the same time, progressive researchers can play an important role in supporting popular classes in and outside the state in the defence and promotion of their interests. They can assist the popular classes by, for instance, uncovering the nature of rules and procedures, and by exploring the relationships between positive elements in government housing policies and increasing popular influence on public decision-making.

Notes

1. This recommendation for autonomous housing, often labelled 'self-help', has wide, and sometimes ill-defined implications. When in this paper we use the term 'self-help', we mean the practice of users autonomously taking decisions on planning, building, and maintaining their houses, irrespective of whether they construct houses by themselves or not.
2. When, for lack of a better expression, we use the word 'liberal' in this context, we do not mean to imply that this approach advocates a *laissez-faire* policy. As spelled out in section I, the 'liberal' approach is critical of the traditional state interventions in housing and advocates other interventions instead.
3. This section is partly adopted and slightly revised from Nientied *et al.* (1982). Elsewhere these approaches have been reviewed in greater detail; see Drakakis-Smith (1981) and Grose (1979).

4. This has been done in much of the literature on the subject.
5. Although Burgess adopts Pradilla's argument to a large extent, Burgess's articles are discussed here since they are more easily accessible.
6. A practice called 'eclecticism' by some authors, cf. Gilbert (1982). Burgess (1981, 57) speaks about geography's eclecticism as 'shameless relativism, that sees no particular theoretical difficulty in the espousal of mutually contradictory theories'.
7. World Bank documents on which this section is based include World Bank (1972; 1974; 1975; 1979; and 1980), Grimes (1976), McNamara (1975), and TUE (1982).

References

Basset, K. and Short, J. (1980) *Housing and Residential Structure, Alternative Approaches* (London: Routledge and Kegan Paul).

Burgess, R. (1978) 'Petty Commodity Housing or Dweller Control? A Critique of John Turner's Views on Housing Policy', *World Development* 6, 1105–33.

—— (1981) 'Ideology and Urban Residential Theory in Latin America, in D. T. Herbert and R. J. Johnston (eds), *Geography and the Urban Environment: Progress in Research and Applications*, Vol. IV (Chichester: John Wiley, 57–114).

—— (1982a) 'The Politics of Urban Residence in Latin America', *International Journal of Urban and Regional Research*, 6, 465–80.

—— (1982b) 'Self-help Housing Advocacy: A Curious Form of Radicalism. A Critique of the Work of John F. C. Turner', in P. Ward (ed.) (1982a), 55–97.

Castells, M. (1977) *The Urban Question* (London: Edward Arnold).

Drakakis-Smith, D. (1981) *Urbanisation, Housing and the Development Process* (London: Croom Helm).

Dwyer, D. J. (1975) *People and Housing in Third World Cities: Perspectives on the Problems of Spontaneous Settlements* (London: Longman).

Gilbert, A. (1982) Introduction, in A. Gilbert, J. E. Hardoy, and R. Ramirez (eds) (1982), 1–18.

—— and Gugler, J. (1982) *Cities, Poverty and Development, Urbanization in the Third World* (Oxford: Oxford University Press).

—— Hardoy, J. E. and Ramirez, R. (eds) (1982) *Urbanization in Latin America, Critical Approaches to the Analysis of Urban Issues* (Chichester: John Wiley).

—— and Ward, P. (1982) 'Low-income Housing and the State'. In A. Gilbert, J. E. Hardoy, and R. Ramirez, (eds) (1982), 79–127.

Goldsmith, M. (1980) *Politics, Planning and the City* (London: Hutchinson).

Grimes, C. F. (1976) *Housing for Low-income Urban Families* (Baltimore: Johns Hopkins University Press).

Grose, N. R. (1979) *Squatting and the Geography of Class Conflict, Limits to Housing Autonomy in Jamaica*. International Studies in Planning (New York: Cornell University).

Harloe, M. (1977) Introduction, in M. Harloe, *Captive Cities* (Chichester: John Wiley), 1–47.

Johnston, R. J. (1982) *Geography and the State: An Essay in Political Geography* (London: Macmillan).

Lea, J. P. (1979) 'The Politicization of Housing Policy in the Third World'. Paper presented to Waigani Seminar, Port Moresby.

McDougall, G. (1982) 'Theory and Practice: A Critique of the Political Economy Approach to Planning', in P. Healey, G. McDougall, and M. J. Thomas, (eds) *Planning Theory, Prospects for the 1980s* (Oxford: Pergamon Press), 258–71.

McNamara, R. S. (1975) *Speech by the President of the World Bank to the Annual Meeting* (Washington: World Bank).

Nientied, P. and van der Linden, J. (1983) 'The Limits of Engels's "The Housing Question" for the Explanation of Third World Slum Upgrading', *International Journal of Urban and Regional Research*, 7, 263–6.

—— Meijer, E., and van der Linden, J. (1982) *Karachi Squatter Settlement Upgrading: Improvement and Displacement*. GPI, bijdragen tot de sociale geografie en planologie 5 (Amsterdam: Free University).

Peattie, L. (1979) 'Housing Policy in the Developing Countries: Two Puzzles', *World Development*, 7, 1017–22.

Pradilla, E. (1979) *Bourgeois Ideology and the Housing Problem: A Critique of Two Ideological Theories* (Rotterdam: BIE). (First published in 1976 in *Ideologica y Sociedad*, 19, translated by R. Burgess.)

—— (1981) 'Notes on Housing Policies of Latin America', in M. Carmona, P. ter Weel, and A. Falu (eds), *De stedelijke crisis in de derde wereld* (Delft: TH) 109–38. (First published in 1976 in *Ideologica y Sociedad*, 16, trans. by R. Burgess.)

Rothenberg, I. F. (1981) ' "Symbolic Schemes" and Housing Policy for the Poor: Lessons from Colombia, Mexico, Chile and Hong Kong', *Comparative Urban Research*, 7, 48–75.

Sandbrook, R. (1982) *The Politics of Basic Needs, Urban Aspects of Assaulting Poverty in Africa* (London: Heinemann).

Saunders, P. (1981) *Social Theory and the Urban Question* (London: Hutchinson).

'—— (1982) The Relevance of Weberian Sociology for Urban Political Analysis', in A. Kirby and S. Pinch, *Public Provision and Urban Politics, Papers from the IBG Annual Conference* (Department of Geography, University of Reading, Geographical Papers 80), 1–24.

TUE (The Urban Edge), (1982) No. 4. Periodical publication (Washington: World Bank Council for International Urban Liaison).

Turner, J. F. C. (1967) 'Uncontrolled Urban Settlements', in UN, Economic and Social Council, Committee on housing, building and planning, *Progress Report of the Secretary General on the activities of the Centre for Housing, Building and Planning, 5th session* (Geneva).

—— (1968) 'Uncontrolled Urban Settlement: Problems and Policies', in G. Breeze (ed.), *The City in Newly Developing Countries* (Englewood Cliffs, NJ: Prentice Hall), 507–34.

—— (1976) *Housing by People. Towards Autonomy in Building Environments* (London: Marion Boyars).

—— (1978) 'Housing in Three Dimensions: Terms of Reference for the Housing Question Redefined', *World Development*, 6 1135–45.

—— (1982a) 'Issues in Self-help and Self-managed Housing', in P. Ward (1982a), 99–113.

—— (1982b) 'Who shall do What about Housing?', in A. Gilbert, J. E. Hardoy, and R. Ramirez (1982), 191–204.

—— and Fichter, R. (eds) (1972) *Freedom to Build* (New York: Collier–Macmillan).

Ward, P. (ed.) (1982a:) *Self-help Housing: A Critique* (London:Mansell).

—— (1982b) 'Introduction and Purpose', in P. Ward, (1982a), 1–13.

World Bank (1972) *Urbanization*. Sector working paper. (Washington).

—— (1974) *Sites and Services Projects* (Washington).

—— (1975) *Housing*. Sector policy paper (Washington).

—— (1979) *Report on the 6th Annual Conference on Monitoring and Evaluation of Shelter Programs for the Urban Poor* (Ottawa and Washington: Urban and Regional Economics Division).

—— (1980) *Shelter*. Poverty and Basic Needs Series (Washington).

—— (1980–3) *The World Bank Annual Report* (4 vols.) (Washington).

V

Making a Living in the City

INTRODUCTION

W E have already examined, in Part III, the urban labour market. We have noted that there is no evidence to support the notion that the rapid growth in the urban population of Third World countries has led to a disproportionate growth in unemployment. Urban unemployment, as well as underemployment however defined, is high, but the overwhelming majority of urban dwellers manage to earn a living, even as the numbers of their competitors in the labour market increase. Some hold secure appointments in public administration, others enjoy quite stable jobs with well-established firms. But many have a rather insecure existence. We will now explore the varied ways in which they manage to survive, and analyse the contribution they make to the urban economy.

In 1971 Hart (1973) presented a paper distinguishing an 'informal' from a 'formal' sector in the urban labour market. On the basis of research in a low-income neighbourhood in Accra, Ghana, Hart emphasized the great variety of both legitimate and illegitimate income opportunities available to the urban poor. The response to his plea that a historical, cross-cultural comparison of urban economies in the development process must grant a place to the analysis of 'informal' as well as 'formal' structures was nothing less than overwhelming: a great deal of research, much of it sponsored by the International Labour Office, has been carried out on the 'informal sector' in the 1970s. Moser (1978) provides a review of research and analytical approaches. Research on the 'informal sector' has served to direct attention to a work-force that is typically under-enumerated, commonly characterized unproductive, and all too often dismissed altogether as making little, if any, contribution to the urban economy. On closer inspection, however, the concept of the 'informal sector' appears to cover a great variety of activities. It thus obscures rather than elucidates—with serious implications for both analysis and policy (Bromley 1978; Gilbert and Gugler 1982, 72–6).

Ray Bromley, in Chapter 10, proposes that we conceptualize a continuum of employment relationships ranging from career wage-work to career self-employment. Between these two types of career work he distinguishes four types of casual work that range from short-term wage-work, through disguised wage-work and dependent work, to precarious self-employment. These distinctions serve to make the point that most of those engaged in the least stable and least secure work, while seemingly self-employed, in fact enjoy little autonomy and have rather inflexible working regimes and conditions. 'Disguised wage-work' is paid according to output, like much wage-work—the difference is that it is conducted off-premises. And 'dependent workers' have contractual obligations that substantially reduce their freedom of action: they have to pay rent for premises and/or equipment, to repay credit, and to purchase or sell at disadvantageous prices.

Bromley presents a survey of street occupations in Cali, Colombia, the most comprehensive survey of its kind ever reported. More than 8 per cent of the city's labour force is involved in a great variety of street activities. If some are illegal or illegitimate, most make important contributions to the urban economy: street retailers and small-scale transport provide food and other goods cheaply to consumers, scavengers transform waste into low-cost inputs for artisans and industry. While close to half the street-workers are precariously self-employed, nearly as many are wage-workers in disguise.

Women and children are found disproportionately in the least remunerative and/or lowest status occupations, whether in the streets of Cali or in a squatter settlement in Nairobi. Nici Nelson (Chapter 11) observed women in Mathare to be much more restricted than men in their choice of economic activity. Few of the local business establishments were run by women, and most working women were involved in illegal beer-brewing or prostitution. Women were at a disadvantage because they were less well educated than men, had fewer skills of commercial value, and supported and cared for children. The last point is demonstrated by the fact that a disproportionate number of successful women entrepreneurs were childless, that other women began to expand and consolidate their business only in their late forties when most or all of their children had grown up and perhaps contributed to a joint household income. The handicaps experienced by women may be seen as structural constraints, but they are a function, as Nelson points out, of the cultural context: the education and training thought appropriate for girls, the occupations considered suited for women, the emphasis on women as mothers who bear children, many children, and have the primary responsibility for raising them. For an overview of the issues arising from the position of women within the household on one hand, from discriminatory employment practices on the other, see Jelin (1982).

Lea Jellinek's account, in Chapter 12, of one woman's struggle to support herself and her dependants in Jakarta is unique. Survey research presents a snapshot at one point in time, and even participant observation rarely stretches over more than two years, but Jellinek has regularly visited Bud since 1972. She has thus been able to observe in detail what usually can only be reconstructed from what interviewees or informants are prepared to tell. Indeed, Jellinek's experience illustrates the vagaries of such reconstruction even when close rapport has been established.

The story of Bud demonstrates the importance of a diachronic approach. It illustrates the precarious existence of many urban dwellers in the Third World, but more importantly it shows how their lives are directly affected by economic changes and administrative fiat. As we follow the dramatic ups and downs experienced by a remarkable woman, we begin to understand how the changing fortunes of her commerce relate to changes in the composition of her household and affect the very membership of her family.

References

Bromley, Ray (1978) Introduction, 'The Urban Informal Sector: Why is it Worth Discussing?' *World Development*, 6, 1033–9.

Gilbert, Alan, and Gugler, Josef (1982) *Cities, Poverty, and Development: Urbanization in the Third World* (Oxford and New York: Oxford University Press).

Hart, Keith (1973) 'Informal Income Opportunities and Urban Employment in Ghana', *Journal of Modern African Studies*, 11, 61–89.

Jelin, Elizabeth (1982) 'Women and the Urban Labour Market', in Richard Anker, Mayra Buvinic, and Nadia H. Youssef (eds), *Women's Roles and Population Trends in the Third World* (London and Canberra: Croom Helm), 239–67.

Moser, Caroline, O. N. (1978) 'Informal Sector or Petty Commodity Production: Dualism or Dependence in Urban Development?' *World Development*, 6, 1041–64.

10

Working in the Streets: Survival Strategy, Necessity, or Unavoidable Evil?*

Ray Bromley

THIS essay is concerned with some of the theoretical, moral, and policy issues associated with the low-income service occupations found in the streets, parks, and other public places of most Third World cities.[1] These 'street occupations' range from barrow-pushing to begging, from street-trading to night-watching, and from typing documents to theft. They are often grouped together in occupational classifications, and they are generally held in low esteem. The street occupations are frequently described by academics and civil servants as 'parasitic occupations', 'disguised unemployment', and 'unproductive activities', and they are conventionally included within such categories as 'the traditional sector', 'the bazaar economy', 'the unorganized sector', 'the informal sector', 'the underemployed', and 'sub-proletarian occupations'. It seems as if everyone has an image and a classificatory term for the occupations in question, and yet their low status and apparent lack of developmental significance prevent them from attracting much research or government support.

This discussion of 'street occupations' is based mainly on research in Cali, Colombia's third largest city with about 1.1 million inhabitants in 1977.[2] The research was conducted between mid-1976 and mid-1978, and the present tense is used here to refer to that period. The occupations studied in the streets and other public places of Cali are remarkably diverse, but they can be crudely described under nine major headings:

Retail distribution: the street trading of foodstuffs and manufactured goods, including newspaper distribution.

Small-scale transport: the operation of *motocarros*—three-wheel motor cycles, horse-drawn carts, bicycles, tricycles, and handcarts, as well as porters for cargo transport.

* This is a revised and enlarged version of a paper that appeared first in Alan Gilbert, Jorge Hardoy, and Ronaldo Ramirez (eds), *Urbanization in Contemporary Latin America* (Chichester and New York: John Wiley, 1982). It is published here with the permission of the publishers.

Personal services: shoe-shining, shoe repair, watch repair, the typing of documents, etc.

Security services: night-watchmen, car-parking attendants, etc.

Gambling services: the sale of tickets for lotteries and *chance*, a betting game based on guessing the last three digits of the number winning an official lottery.

Recuperation: door-to-door collection of old newspapers, bottles, etc., 'scavenging' for similar products in dustbins, rubbish heaps, and the municipal tip, and the bulking of recuperated products.

Prostitution: or, to be more precise, soliciting for clients.

Begging.

Property crimes: the illegal appropriation of movable objects with the intention of realizing at least part of their value through sale, barter, or direct use. This appropriation can be by the use of stealth (theft), by the threat or use of violence (robbery), or by deception—('conning').

Of these nine categories, 'retail distribution' is the largest, accounting for about 33 per cent of the work-force in the street occupations, followed by small-scale transport and gambling services, each accounting for about 16 per cent of the total work-force. The six remaining categories each account for 2–10 per cent of the total work-force.

With the exception of small-scale public transport, all of these occupations can be conducted in private locations as well as in public places, and private locations are generally considered more prestigious. Private premises give a business a certain stability and security which is not available to the persons who work in streets, parks, and other areas of public land. Those who work in public places may try to obtain a degree of stability by claiming a fixed pitch, and by building a structure there to give them some protection, but their tenure is almost always precarious, and the overall level of investment in 'premises' is likely to be very limited in these circumstances. Thus, the street occupations are classically viewed as 'marginal occupations', as examples of how the poor 'make out', or as the 'coping responses' of the urban poor to the shortage of alternative work opportunities and the lack of capital necessary to buy or rent suitable premises and to set up business on private property.

Even though they are an integral part of the street environment and interact strongly with the street occupations, four major groups of economic activities are not considered as street occupations *per se*: the off-street private shops, supermarkets, market stalls, etc. which open onto the street; government and company employees who are responsible for building, maintaining, and cleaning the streets; the police and soldiers who are responsible for law and order on the streets; and the operators of larger-scale public transport vehicles such as buses, trucks, and taxis. These occupations are considered in relation to the street occupations, but not as part of the street occupations. All of them have a strong off-street base, most have working regimes and relationships rather different from those prevailing in the street occupations, and many have much higher levels of capital investment in premises, equipment, or merchandise.[3]

Because they are neither practised in conventional (off-street) establishments, nor in the homes of the workers, the street occupations are usually severely under-represented or excluded altogether in statistics based on sample surveys of establishments or households. In spite of their under-representation in most official statistics, however, their highly public location ensures them a prominent place in the urban environment and popular consciousness. Even if he has no direct dealings with those who work in the street occupations, a member of the general public can hardly fail to be aware of their existence.

The streets of the city serve a wide variety of interrelated purposes: as axes for the movement of people, goods, and vehicles; as public areas separating enclosed private spaces and providing the essential spatial frame of reference for the city as a whole; as areas for recreation, social interaction, the diffusion of information, waiting, resting, and, occasionally, for 'down-and-outs and street urchins', sleeping; and as locations for economic activities, particularly the 'street occupations' (Anderson 1978a,b). Within the functional complexity of the street environment, the street occupations are both strongly influenced by changes in other environmental factors, and also contributors to general environmental conditions. Thus, for example, street-traders and small-scale transporters depend upon the direction, density, velocity, and flexibility of potential customers' movements, and are immediately affected by changes in traffic flows and consumer behaviour. At the same time, they influence patterns of movement and overall levels of congestion.

Average incomes per worker in the street occupations in Cali in 1977 are only equivalent to about $US3 per day, and are roughly comparable to the sums paid to unskilled urban wage-workers in casual employment (i.e. without long- or indefinite-term contracts). The actual distribution of incomes in the street occupations is highly skewed, with the majority of workers having incomes below $US3 per day, but with a minority having incomes far in excess of this mean value. On a 'good day', a prosperous street-trader with a large capital stock of merchandise may make $US30 or more in profit, and a skilful street-thief might make much more than that. The success of a few of the most prosperous, skilful, or simply lucky people working in the street occupations, however, should not blind us to the bare economic subsistence of the great majority of the participants in these occupations. Most of them are intensely competitive, and improvements in profit levels and incomes tend to lead to additional workers entering occupations, and/or to increased competition from larger, more capital-intensive enterprises based in off-street locations, forcing average incomes down again to their previous level.

As well as low status and low average incomes, the street occupations are characterized by relatively low inputs of capital in relation to labour, and by low 'formal' educational requirements. Most street enterprises operate with a total working capital equivalent to less than $US100 in terms of equipment and/or merchandise. Indeed, some porters, watchmen, scavengers, beggars, and thieves incur no monetary expenditures in order to be able to work, beyond the costs of their normal clothing, food, and transport to a work-place. Even the most elegant and well-stocked street stall is unlikely to have a value of more than $US1,200, and the most expensive small-scale public transport vehicle, the *motocarro*, has a maximum capital cost of about $US560. Basic literacy and numeracy are generally useful to participants in the street occupations, but even these relatively low educational standards are rarely required. Instead, the occupations are characterized by skills learnt outside the government educational system, such as hard bargaining, quick-wittedness, manual dexterity, good memory, an engaging personality, and physical endurance. On-the-job experience and effective utilization of social networks are particularly important in the street occupations, together with such difficult-to-alter variables as 'an honest face', physical toughness, or beauty.

The most basic need of the urban poor is an income in goods and/or money to provide for food, drink, housing, clothing, and other necessities. An income may come from government or private charitable institutions, from investments, moneylending and renting, from windfall gains, or from work. For the poor, work is the normal way to obtain an income, and for an estimated 29,000 people in Cali—around 8.4 per cent of the city's working population—some or all of their work is in the street occupations (Bromley and Birkbeck 1984, 187–90).[4] Ideally, work should be both enjoyable and rewarding, yielding an income and a sense of personal achievement and satisfaction. Instead, to most people, including those working in the street, work is boring and exhausting, and even dangerous or degrading. Furthermore, work opportunities are usually scarce and insecure, and work is often inadequately remunerated, leading to poverty and deprivation. All of the occupations under consideration here have some of these negative characteristics of work, and together the street occupations form a complex of low-status, poorly remunerated, insecure forms of work. Although many who work on the streets comment that their occupations are less exhausting than heavy manual labour like cutting sugar-cane and carrying bricks on construction sites, the street occupations usually require long hours and uncomfortable conditions. The 'curse of Adam' weighs heavily on the urban poor, and most perceived advantages in street occupations reflect even worse conditions in alternative occupations.

Work and Employment Relationships

Work is defined here as 'any activity where time and effort are expended in the pursuit of monetary gain, or of material gain derived from other persons in exchange for the worker's labour or the products of such labour'. In other words, work is the labour involved in producing goods and services for exchange, and it is 'income-generating'. The category of 'work', thus defined, excludes the equally important category of 'expenditure-reducing' activities which can be described collectively as 'subsistence labour', for example growing food for household consumption, self-help house construction and repair, unremunerated housework and child-minding, voluntary unpaid help given to friends and neighbours, and walking or cycling to places of work or recreation so as to avoid paying transport fares. Any form of work which is regularly performed by a given person may be described as an occupation.

Under these definitions, such classic lumpenproletarian occupations as begging, prostitution, and theft can all be viewed as work, and hence can be analysed together with the remaining street occupations. The presence of these illegal or illegitimate (not illegal, but considered shameful or a public nuisance by most) occupations in the category of 'street occupations' serves to emphasize the high degree of differentiation which exists within this category. Of course, illegality and illegitimacy extend much further than begging, prostitution, and theft, as some traders deal in illicit merchandise, various forms of street gambling are illegal, and substantial numbers of persons working in transport and gambling services and the overwhelming majority of street-traders do not possess the licences and documents required by official regulations.

Before we consign the street occupations to the dustbin of immorality, it is worth considering how much corruption and tax evasion occurs in government and big business, and the extent to which élites make laws to further their own interests. Illegal and illegitimate income flows are received by both rich and poor, and arguably the poor have greater moral justification for breaking the laws and norms of society than the rich. The poor did not make the laws and norms that they are expected to live under, and they can legitimately claim that their own poverty, combined with the presence of wealth and conspicuous consumption around them, lead them to break these laws and norms. The poor are often acutely conscious that their own income must be earned by work, whilst the rich have inherited wealth and access to unearned income through investments, usury, and renting out property. The problem of poverty is not just a shortage of income and wealth, but more importantly a shortage of bona fide income opportunities. Illegal and illegitimate opportunities help to make up some of this deficit.

In this analysis, the term 'work' has a different meaning from the term 'employment'. 'Employment' is used to denote a relationship between two parties, an 'employer' and an 'employee', the former paying the latter to work on the former's behalf for a significant period of time (at least a working day), or for lesser periods on a regular basis. When there is a direct two-tier employer–employee relationship based on some form of contract (an oral or written agreement), there are two main forms of working relationship: 'on-premises working', when the employee works at a site owned, rented, or operated by the employer; and, 'outworking', when the employee works away from the employer, usually in his own home, in the streets, or in some door-to-door operation. An employee may be paid wages per unit of time worked, per unit of 'output', or by some combination of the two. When work is remunerated wholly or partly by the unit of time worked, whether as 'on-premises working' or 'outworking', it is generally recognized as a form of employment. When it is remunerated solely per unit of 'output', however, it is usually only viewed as employment if it is conducted 'on premises'. When conducted off-premises, piece-work is conventionally viewed as a form of self-employment, and this is certainly the conception embodied in Colombian labour legislation and in the perceptions of most middle- and upper-class Colombians. In contrast to such views, and in keeping with our definition of employment, 'outwork on a piece-work basis' is viewed here as a form of employment remunerated by a piece-wage. The fact that it is not officially or widely recognized as such means that it may be viewed as 'disguised wage-working', as distinct from the more widely recognized forms of wage-working which take place 'on-premises' or which involve off-premises work remunerated per unit of time worked.

When a worker is not employed by someone else, two alternative working relationships are commonly found; 'dependent working' and 'self-employment' (Bromley and Gerry 1979, 5–11). Although dependent workers are not employees even on a piece-work basis, and they do not have fixed margins and commissions, they do have obligations which take a contractual form and which substantially reduce their freedom of action. These obligations are associated with the need to rent premises, to rent equipment, or to obtain credit, in order to be able to work. Although the appropriation of part of the product of the worker's labour is not as clear and direct in the dependent-working case as in the disguised wage-working case, there is normally an appropriation process through the payment of rent, the repayment of credit, or purchases and sales at prices which are disadvantageous to the dependent party in the relationship. In contrast, true self-employment has no such relationships; the workers work on behalf of themselves and other persons that they choose to support. Self-employed workers must of course, rely on inputs provided by others, on the receipt of outputs by others, and on a system of payment. However, the bases of their self-employment are that they have a considerable and relatively free choice of suppliers and outlets, and also that they are the owners of their means of production. They are dependent upon general socio-economic conditions and on the supply and demand conditions for their products but they are not dependent upon specific firms for the means to obtain their livelihoods.

The employment relationships described above form a continuum ranging from wage-work to self-employment. Two major variables, 'relative stability and security of work opportunities', and 'relative autonomy and flexibility of working regimes and conditions', can be used to divide up the continuum into six major categories whose relationships and characteristics are set out in Figure 10.1. The two extremes of the continuum, described as 'career wage-work' and 'career self-employment', have relatively high levels of stability and security, and they are not

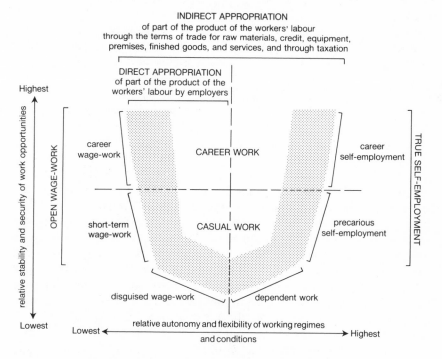

Fig. 10.1 The continuum of employment relationships

encountered among the street occupations of Cali. The four intermediate relationships have relatively low levels of stability and security, can be described collectively as 'casual work', and are found in varying proportions among the street occupations. The six categories of the continuum are not commonly distinguished, and ambiguous cases will arise as in all classifications, but they do provide a much richer and more appropriate focus for studies of contractual relationships, economic linkages, and changes in employment structure than the more conventional 'dualist' distinction between wage-work and self-employment with no intermediate categories.

In the streets of Cali, self-employment is a much less common phenomenon than might at first glance seem apparent, and there are signs that it is diminishing in significance in the face of the expansion of disguised wage-working and dependent working (Birkbeck 1978a; 1979; Bromley 1978; Gerry 1985; Gerry and Birkbeck 1981). Estimates of the proportions of those working in the street occupations falling into the different categories of working relationship suggest that only 40–45 per cent are in precarious self-employment.[5] A further 39–43 per cent of those working in the street occupations are disguised wage-workers, 12–15 per cent are dependent workers, and only about 3 per cent are short-term wage-workers.

Disguised wage-workers and dependent workers have a variety of obligations to employers, contractors, suppliers, property-owners, and usurers, yet do not have the employees' rights specified in government labour legislation. They are 'disenfranchised workers'. Thus, for example, *motocarro* drivers notionally own their vehicles, but a substantial minority are buying them on hire-purchase, or have borrowed money for the full purchase, so that they are really dependent workers. Similarly, most night-watchmen who work on the streets watch over a single property, or a group of neighbouring properties, guarding the property of the same owner(s) night after night. Though they are officially viewed as self-employed, they are paid a fixed sum by the owner(s) every day, week, or month, and their working relationship is effectively one of disguised wage-working. As a further example, those street-traders who have kiosks or fairly sophisticated mobile stalls usually rent the units or buy them on some form of credit (dependent working), or, as in the case of soft drinks company kiosks and ice-cream company carts, they have been lent these units by companies on the condition that they sell company products on a commission basis (disguised wage-working). Even in the street-trading operations requiring least capital and yielding the lowest incomes, such as door-to-door newspaper selling by small boys and small-scale fruit and vegetable selling in the street markets, dependent working relationships are formed when, because of their own poverty, the sellers have to obtain the merchandise on credit from other sellers or wholesalers. The condition of this credit is that they must sell that merchandise and pay for it before they can have another lot of merchandise on credit.

The low status and economic dependence of many of the people working in street occupations, combined with the intensely competitive, individualist mentality associated with most of these occupations, contributes to a lack of political and economic organization. The street occupations are moreover fractionalized by the high socio-economic differentiation within and between occupations, low general levels of education, insecurity, instability, official ignorance, and persecution, and conflicting loyalties to a variety of different 'masters'. Although trade unions and co-operatives exist, these associations are usually small, unstable, and relatively ineffective. Worse still, they are frequently corrupt and/or personalistic, there are often several different rival organizations within a single occupation, and only a small proportion of the total work-force belongs to any organization. Even among lottery-ticket sellers, the most unionized group in the street occupations, only half of the total workers are paid-up members of trade unions or occupational associations, and these are divided between five different organizations. Only about one-eighth of street-traders are in unions and associations, and they are split between six different organizations (Bromley 1978, 1167). Among most of the other street occupations, the proportion of workers in such organizations is even lower, and several occupations have either never had a 'formal' organization, for example theft, begging, and knife-grinding, or have only defunct organizations which no longer hold meetings or collect subscriptions, for example shoe-shining and garbage-scavenging. Viewing the street occupations as a whole, worker solidarity is ephemeral or non-existent, and group interests tend to be subordinated to individual concerns and ambitions.

The Diversity of Street Occupations

The typology of street occupations in Cali shows considerable variety of types of work within the broad category of the street occupations. Further variety is emphasized by the consideration of the nature of work, and the 'rag-bag' character of the general descriptive category of services.

In classifying work and economic activities, it has become customary to identify three major economic sectors (Fisher 1933; 1939): 'extraction' (primary production), 'manufacturing' (secondary production), and 'services' (occasionally described as tertiary production). Of these three sectors, the third is by far the most heterogeneous, and many attempts have been made to subdivide it into smaller categories (e.g. Foote and Hatt 1953; Gottmann 1970). All of the street occupations fit loosely into the service sector, and the diversity of occupations found in the streets reflects some of the diversity in this sector. The term 'services' ranges from the highest levels of government, education, and research, to such low-status workers as watchmen, porters, and scavengers. The category of services also includes work in such diverse institutions as the Church, the fire brigade, the police, and the armed forces, so that the general character of the service sector is really that of the 'everything else sector' after extraction and manufacturing have been separated out from the total range of economic activity. Within the broad category of services, I include begging and property crime, occupations which are usually viewed as 'parasitic' in that the worker serves himself at the expense of the victim (Hirst 1975, 225). We should not forget, however, that the beggar may provide a service to the donor by giving him some moral satisfaction, and that the thief may serve a 'fence' by supplying him with cheap stolen goods.

Most people's attitudes to services tend to be coloured by whether or not they consider services to be 'productive', and as a related issue, how they define 'productive'. The word 'productive' has a good connotation, whilst the word 'unproductive' is little short of insulting. In some circles, productive is used only to describe the extraction or manufacture of goods, and in this sense, services are clearly not productive. In other circles, productive is used to describe any activity which 'creates surplus-value' or even simply 'adds value', and there is no distinction between primary, secondary, and tertiary sectors in terms of their innate productivity or non-productivity.[6] In a sense, this is merely a terminological debate. Whether or not services can be *produced* like goods, some services must be *provided*, if only to sustain the output of the primary and secondary sectors. No economy could function for long without transport or commercial distribution, and some services are clearly 'essential', whether or not they are defined as productive. In contrast, the production of such superfluous goods as electric toothbrushes and plastic beads can hardly be described as satisfying any basic human need, or as making any major contribution to the economy. Production may not only be 'superfluous', but also 'antisocial', in that for some forms of production, the social costs clearly outweigh the private benefits, and social benefits are virtually non-existent. For example, forms of extraction or manufacturing which cause severe environmental damage, or which produce dangerous materials, may be productive in terms of both material output and profit, but still socially very undesirable.

Most street occupations provide some wanted services, as evidenced by the fact that the public is prepared to pay for these services. Of course, we may feel that some of these services, such as prostitution and the sale of gambling opportunities, should not be offered to the public, but we should recognize that they are available off-street as well as on-street, and that even worse social problems may result from their indiscriminate repression. The lessons of the 'Prohibition Era' in the United States are as valid today as they were in the 1920s. Furthermore, of course, the repression of any income opportunity may endanger the livelihoods of its workers, and also the income flows and accumulation patterns associated with it. Thus, for example, a complex set of 'moral dilemmas' faces anyone seeking to formulate policies towards the sale of lottery and *chance* tickets. Not only is it necessary to bear in mind the welfare of approximately 4,750 sellers and their dependants, but also the facts that the lotteries are government-organized to generate revenues to finance part of the social welfare system, and that *chance* is organized by capitalist enterprises paying taxes to local government (Gerry 1985). Further dilemmas arise when one tries to assess the overall effects of gambling on the gambler and his/her household, and the degree to which the 'ethic of the windfall' implicit in gambling may strengthen individualistic and competitive traits amongst the poor, reducing the potential for worker solidarity and class conflict.

For those services which are offered both on-street and off-street, the street occupations play an important role by increasing levels of competition in the economy, and hence reducing the likelihood of the formation of oligopolies working against the interests of the consumer. In two price surveys in Cali, for example, street-vendors were shown to sell most basic foodstuffs in small quantities at prices significantly below those of supermarkets, with the sellers in the street markets generally recording the lowest prices in the whole city.[7] Thus, some street occupations contribute towards lowering the cost of living in the city, and particularly towards holding down the prices of foodstuffs for the urban poor, who buy a greater proportion of their food in the streets than better-off social groups. By holding down the cost of living for the urban poor, these street occupations contribute to holding down the costs of wage-labour (see Oliveira 1985; Williams and Tumusiime-Mutebile 1978). Street-retailing also plays a role in encouraging consumption, both by selling at relatively low prices and by making items available in a wider range of locations and for longer periods of time on each day of the week. Further encouragement to consumption is given by the cheapness of many transport services, including such small-scale transporters as *motocarros* and hand-carts, which enable the consumer to get the goods bought back home at relatively low costs. It should be evident, therefore, that most of the street occupations are eminently functional to the socio-economic system as a whole, rather than simply 'marginal occupations' of no social or economic significance beyond providing subsistence to those who work in them.

Three street occupations stand out from the others as not clearly offering a service to the general public: recuperation, begging, and property crime. Recuperation as a street occupation is distinct from the municipal collection of refuse, and is oriented mainly towards the recovery of paper, metal, glass, bone, cloth, and other products which can be sold to artisans and industrial establishments. As an occupation, it serves manufacturing industry directly by collecting (or, in a sense, extracting) useful materials, reducing industrial costs and national imports. The benefits of recuperation can potentially be passed back to the consumer in terms of cheaper manufactured goods and even increased manufacturing employment. Recuperation is not parasitic, but rather symbiotic, to some extent benefiting all parties involved, though providing the scavenger with only a low-income job and rather degrading and frequently unhealthy working conditions. Begging is a very different case, in that it is genuinely parasitic, but as there is no victim, it cannot be considered seriously antisocial. Property crime is yet again a different case, being parasitic, having a clear victim, and hence, given that almost everyone is at risk, being decidedly antisocial.

Underemployment: Concept and Reality

In Colombia, any discussion with an academic or civil servant on the characteristics of the street occupations is likely to be cut short by a kindly, 'Ah, yes, you mean underemployment.' All work in low-income services, and often also the work of artisans and peasant farmers, is lumped together as underemployment, and the street occupations are considered the classic type-examples. The implicit assumption is that the workers are doing something, but not much, and that they are in some sense 'less employed' than those who work in government, public services, factories, or capitalist agriculture. When one asks what underemployment means, answers are usually tautologous or contradictory, revealing the different meanings attached to the popular usage of the word 'employment' as well as general confusion about underemployment. Four main approaches are taken individually or in combinations to the definition of underemployment: that the workers work less hours than they wish to or ought to according to some élite-defined norm; that the workers have very low productivity in terms of the amount of work completed per unit time; that the workers are inadequately remunerated for their labour; and that the workers do not have a 'normal relationship' with an employer, in other words, they are not on-premises wage-employees. The variables embodied in the four approaches are not necessarily mutually correlated, and none applies exclusively to those occupations which are usually described as underemployment.

Underemployment is a decidedly derogatory term mainly used by the upper and middle classes to describe the work of the poor. Those who work on the streets do not usually describe themselves as underemployed, but rather stress the long hours that they have to work to earn a subsistence income, their lack of capital, and the hard or tedious nature of their work. If underemployment simply means low incomes, clearly the street occupations reflect underemployment. If underemployment means low productivity per worker because of low capital investment and intense competition between workers, then again the street occupations reflect underemployment. However, these two criteria for defining underemployment are simply the criteria for defining poverty: low incomes and little or no capital. If, instead, we look at how much those who work in the streets actually work, there is no evidence that they work less than those in most other occupations. Most people working in the streets work rather long hours and a seven-day week. Although some sellers of goods and services spend substantial periods between sales waiting for the next customer, such slack periods are also found in most government and private offices and in many off-street commercial enterprises. A few years ago, a Communist Party member of Cali's Municipal Council suggested that many municipal offices were being refurnished at public expense 'so that bureaucrats can have larger desks to stretch their newspapers over'!

One remarkable feature of many street occupations is, discounting the cost of labour, the high return they can give on very small amounts of capital invested. One hundred pesos loaned to a retailer at the beginning of a day can yield 150–200 pesos at the end of the day, enabling the retailer to pay back the loan with an exorbitant 5–10 per cent interest, and to keep the balance for his/her own subsistence. Thus, large numbers of workers can make an income around or somewhat above the national minimum wage (which was equivalent to $US1.60 per day for the first part of 1977, and rose to $US2 per day by the end of the year) with an extraordinarily low capital investment. Capital is used very efficiently, though of course this efficiency is conditional on the low cost and high availability of labour, and upon the donor of credit limiting the number and size of loans so as to be able to closely supervise the recipients and avoid frequent non-repayment.

On-premises wage-working governed by written contracts and official labour and social security legislation is only available to a minority of the labour force in Colombia and most other Third World countries. However desirable regulated, protected on-premises wage-employment and its associated bonus payments, holiday pay, redundancy pay, family benefits and social security may be, it is necessary for governments to spread their attention more widely so as to cover short-term wage-workers, disguised wage-workers, dependent workers, and the self-employed. Simply to consign all of these forms of working away as underemployment which will eventually disappear, is to ignore the worst forms of poverty and exploitation associated with work.

The urban poor are loosely tied into a vicious circle of low capital, low training, shortage of remunerative work opportunities, and low incomes. Only a minority with considerable luck, talent, or initiative can break out of this situation, and the success of a minority is often conditional on the relative stagnation of a great majority. The vicious nature of the circle within which the urban poor 'usually work is accentuated by a variety of exploitative contractual relationships, and by the lack of effective organizations and participatory structures. The application of the term 'underemployment' to a wide variety of the occupations of the urban poor can have serious negative effects, both in lowering the status of the occupations concerned, and in tending to throw the blame for low incomes and productivity caused by poverty upon the poor themselves—in effect, blaming the poor for being poor.

Official Intervention in the Street Occupations of Cali

It is not surprising, and in total accordance with the legal system, that clearly illegal activities such as property crimes, trading in contraband goods, and the sale of prohibited gambling opportunities, are persecuted occupations in Cali. Indeed, many complain that these occupations are not persecuted enough. It is also hardly surprising that such occupations as prostitution and begging, viewed as 'disreputable' by most of the population, are officially regulated and suffer from periodic police harassment (see Bromley 1982). In the case of prostitution, however, official attitudes are decidedly ambiguous, and there are many complaints that the organizers of prostitution and the upper-class prostitutes are free from harassment, while the lower-class prostitutes are frequently persecuted.

The intervention of the authorities in occupations which are not clearly criminal, immoral, or antisocial according to conventional, élite-defined standards, is much more complex and diverse. In general, the concern is to regulate activities by introducing checks and controls on prices, standards, and locations, and by limiting entry to the occupations. Government personnel are appointed to enforce these regulations, and penalties are prescribed for offending workers. Thus, for many street occupations, e.g. the operation of a street stall, the commercial use of a *motocarro*, horse-drawn cart or a handcart, and shoe-shining, registration procedures have been introduced, and regulations have been made as to when, where, and how these occupations should be practised. Hundreds of pages of official regulations (MSC 1971; GV 1976) specify the municipal, departmental and national government's regulations on street occupations, and substantial bodies of police and municipal officials are expected to administer these regulations. In reality, however, these regulations are excessively complex, little known, and ineffectively administered, resulting in widespread evasion, confusion, and corruption. Thus, for example, in one of the poorest squatter settlements on the extreme eastern periphery of Cali, the inhabitants were subjected to harassment from the local police for a three-month period in 1978. The inhabitants of the settlement include a number of thieves working on the streets, and the period of harassment began when four of the inhabitants who had been arrested and locked up in the jail at the local police station paid a bribe of 500 pesos to the police on duty so as to secure their own liberty. From then on, until there was a change of personnel at the police station, the police visited the settlement two or three times a week, each time 'detaining' someone and extracting a bribe of 100–200 pesos so as not to make a 'legal arrest'. This harassment was partly directed at known offenders, even though there was no evidence of any specific act which might merit arrest, but also at anyone else against whom the police could find the slightest pretext for detention. In most cases, the detainees were never even taken to the police station, and in others they were released immediately upon arrival (after payment), without the case being recorded.

A more complex example of corruption associated with the regulation of the street occupations is the case of the Calle Trece-Bis street market, close to the main shopping area of central Cali (Bromley 1978, 1170). This street market functioned without official permission and against the wishes of virtually all higher-level municipal officials from 1972 until July 1978, and it has been active again since October 1978. When the market was functioning, the traders made a daily collection so as to gather together the necessary funds to pay off each shift of police patrolling the area, a routine bribe to enable the market to continue its activity. In the period from July to October 1978, the street market was temporarily 'eradicated' in a very determined municipal campaign to control street activities in that section of the city, an area bisected by the inner-city ring-road, and highly visible to passing tourists and rich Caleños. After a change of municipal administration, however, the political will for 'eradication' was sharply reduced, and despite the hardships caused by the recent persecutions, most of the traders returned to their previous locations. In general, those who administer law and order on the streets complain that there are so many people working in the street occupations, and that there is such widespread ignorance and disrespect for the official regulations, that controls must be very selective. The main objective is usually 'containment': to hold down the numbers of people working in the streets of such priority areas as the central business district, the upper-class shopping-centres and residential zones, and the main tourist zones.

Although there are occasional cases of assistance by the authorities to street occupations, as when help was given in improving street stalls and providing uniforms for street-vendors and shoe-shiners at the time of the Pan-American Games in Cali in 1971, official intervention in the street occupations is essentially negative and restrictive, rather than supportive. The basis of government policy is that 'off-street occupations' should be supported in the hope that they will absorb labour from the street occupations. However, this objective has not been achieved because insufficient investment funds have been mobilized, and because investment has tended to be concentrated in areas which generate relatively few work opportunities. In the meantime, the street occupations have tended to be persecuted, and opportunities to improve working conditions in these occupations have generally been neglected.

Street occupations conflict strongly with the prevailing approaches to town planning. Although Cali has a warm, dry, and congenial climate for economic activities and social interaction in the open air, city planners have usually been concerned with reserving the streets for motorized transport and short-distance pedestrian movement, and concentrating economic activities into buildings. In general, no special provision has been made for street occupations, and restrictions on *motocarros*, non-motorized transport, and the sale of goods and services in the streets have been partly intended to reduce the incidence of these occupations. Cali is officially twinned with Miami and has strong links with other North American cities. Urbanistically, Cali is being planned along North American lines, and the street occupations are, from the planners' point of view, an unfortunate embarrassment to such plans.

Women and Children in the Street Occupations

Amongst the urban poor, conventional official definitions of 'labour force' and 'economically active population', based upon the idea that neither children nor housewives earn an income, are simply irrelevant. When personal incomes are low, and when the membership of households is often unstable, there is a strong pressure on all household members to seek work opportunities. Work is a form of personal security as well as a contribution to the household budget, and women and children cannot assume that they will be supported by an adult male breadwinner. Instability and insecurity of work and income opportunities are endemic amongst the urban poor (McGee 1979; Rusque-Alcaino and Bromley 1979), and reliance on only one breadwinner increases the risk of disaster. An adult male breadwinner may be the victim of theft, arbitrary arrest, or the eradication of job opportunities (Cohen 1974), or he may choose to abandon family responsibilities and to spend his money on himself. As poverty may also contribute to family breakdown, or to heavy reliance on tobacco, alcohol, drugs, or gambling as potential escapes from a depressing reality, it is especially important for each member of a poor household to have his/her own potential income opportunities.

About 70 per cent of those working in the street occupations in Cali are males, at least three-quarters of these falling into the 18–55 age-range. Around 15 per cent of those working in the street occupations are aged under 18, about three-quarters of these being boys.[8] Some occupations, for example virtually all work in transport and security services, shoe-shining, and most forms of street theft, are almost exclusively male preserves. In general, therefore, males and adults predominate in the street occupations, though females and children are numerically quite significant. Only prostitution is almost exclusively a female preserve, though women predominate in many forms of retailing, particularly the sale of fruit, vegetables, and cooked foodstuffs. Child workers are mainly concentrated in scavenging, newspaper-selling and other small-scale retailing, shoe-shining, and petty theft.

In general, women and children are especially concentrated in the least remunerative or lowest-status street occupations, and have less access to capital than men. Particular occupations are age- and sex-specific, and although this division of labour may at times be convenient, it mainly acts to reduce the range of work opportunities and the potential income available to women and children. Children working in the streets have great difficulty in obtaining and keeping any significant capital equipment or merchandise for their occupation, and most young women without access to significant capital are aware that prostitution may be potentially their most remunerative form of work.

Conclusions

This rapid summary of the characteristics of street occupations, based on the example of Cali, has emphasized the diversity of these occupations and the impossibility of applying a uniform set of policies to all street occupations. These occupations deserve a greater degree of attention and respect than they have received, and it is necessary to convert the predominance of negative policies to a predominance of more positive measures. Most potential improvements to the working conditions in street occupations are relatively inexpensive, and some would actually save government money by reducing the number of regulations and the costs of enforcing these regulations. There is no reason, therefore, why a 'humanization' of the street occupations should hold up vital investments in agriculture, manufacturing, and public services. An improvement of the working conditions in the street occupations will not greatly increase the number of workers in these occupations and lead to accelerated rural–urban migration if appropriate investments are made in agriculture, manufacturing, and public services, and if investment funds are to some extent diverted away from the largest cities towards smaller cities and rural growth-centres.

For substantial numbers of the urban poor, working in the streets is a survival strategy. The great majority work in legal occupations, though they may contravene minor bureaucratic regulations, and the legal street occupations are often their only alternatives to parasitic or antisocial occupations, or to destitution. It is important to realize that many of the negative features of the street occupations are reflections of much wider social malaises which cannot be resolved simply by regulating and persecuting the street occupations. Gross poverty and social inequality are institutionalized in Colombia, and it is unreasonable to blame the poor for their own situation and to fail to tackle the conditions which underlie their poverty.

There is a growing awareness that our concern to increase production should be tempered by periodic consideration of *what* we are producing, and for *whose* benefit. Thus, for example, the work of slaves or serfs, or even of low-paid wage-workers, to enrich their employers, is not necessarily productive for society as a whole, but only for a privileged minority. The creation of material wealth is a less important objective than welfare, and so-called unproductive work may be desirable if it contributes to the welfare of members of society other than the worker, or even if it simply contributes to the welfare of the worker without prejudicing the welfare of others.

Many of the street occupations, and particularly those concerned with public transport and food retailing, are important to the functioning of the urban socio-economic system. Indeed, in some cases where significant capital investments are required or where official controls are exercised over-zealously, there appears to be a shortage of service provision resulting in increased costs and inconvenience for the consumers of those services. In Cali, for example, there are serious deficits of public transport and food-retailing facilities in some sectors of the city, and these deficits raise living costs and reduce the numbers of work opportunities available. More generally, the removal of street-traders would encourage price speculation in the shops and supermarkets and might reduce the sales of some agricultural and manufacturing enterprises, the removal of street newspaper-sellers would severely damage the sales of the press and would bankrupt some of the newspaper companies, the abolition of street sales of lottery tickets would reduce government revenues and lead to the closure of several social welfare institutions, and the elimination of recuperation would lead to increased imports and higher costs for Colombian manufacturers.

A few street occupations, of course, are decidedly antisocial and/or parasitic, and are obviously undesirable. For the moment, however, hardly anyone would claim to be able to eliminate criminal or immoral activities, and some presence of these occupations is probably an unavoidable evil. Antisocial and/or parasitic occupations can be controlled, and efficient control may reduce their incidence, but there is no evidence that they can be eliminated. Any hope of their disappearance must depend on general improvements in social conditions and the creation and improvement of alternative work opportunities, and not simply on the repression of those involved.

In summary, therefore, working in the streets encompasses a very wide range of service occupations, almost all of which can be described as survival strategies for those who work in them, many of which can be described as necessities for those they serve, and for the efficient functioning of the contemporary national economy, and a few of which can be described as unavoidable evils. In policy terms, the most urgent requirement is to generate more work opportunities both outside the street occupations, and in the more necessary street occupations, so as to divert manpower from antisocial and/or parasitic occupations, and to improve the general range of income opportunities available to the urban poor. It is also necessary to adopt a more positive series of policies towards those street occupations which are not clearly antisocial or parasitic, simplifying rules and regulations and administering them more equitably, providing work-places and sources of credit and training for workers, and encouraging the formation of workers' organizations without co-opting them into the web of governmental paternalism. Unfortunately, of course, there can be no assurance, either in Colombia or in most other Latin American countries, that the sort of government which would adopt such policies will assume power in the near future. In such circumstances, increased worker consciousness and the mobilization of those working in the street occupations in alliance with the more conventional 'proletariat' of unionized wage-labour, may represent the only means of achieving greater bargaining power to enforce more favourable government policies and to break the present dependent links with employers, contractors, suppliers, usurers, and the owners of sites and equipment.

Notes

1. The author is indebted to Chris Birkbeck and Chris Gerry for critical comments on an earlier draft. The research was financed by the UK Ministry of Overseas

Development, and was conducted in association with the Servicio Nacional de Aprendizaje, Regional Cali. The views expressed here, however, are exclusively those of the author and do not imply the agreement of any British or Colombian institution.

2. In the 1973 census, Cali was recorded as having 923,446 inhabitants (DANE 1975), and the projection to 1.1 million inhabitants in 1977 is based on an estimated growth rate of 5 per cent per annum.

3. In public transport, for example, there is a sharp distinction between the vehicles used in small-scale transport, none of which costs more than $US560 new, and those used in larger-scale transport, none of which costs less than $4,000 new, and most of which cost over $20,000 new. This distinction is paralleled by the much heavier involvement of government regulation, subsidies, and credits, and of capitalist companies employing wage-labour in the larger-scale forms of transport (Birkbeck 1978b).

4. Estimates of the numbers working in the street occupations are based on a combination of different sources and research methods: official registers, particularly of transport vehicles and *chance* sellers; street counts at different times of the day, week, and year in all concentrations of people working on the streets, and in sample residential neighbourhoods; estimates given by knowledgeable individuals, particularly the more experienced workers, union leaders, administrators, and policemen; and, calculations derived from estimates of supply, demand, and/or turnover. Estimates were made for each occupation, and then added together to produce a general total for the city.

5. These estimates are based on the same sources as the estimates of numbers of persons working in the street occupations, though they have a lower level of accuracy. Most of those workers who are in true self-employment work on their own, but a few are involved in partnerships and a very small number actually employ others as wage-workers.

6. For a useful review of these issues, see Hirst's (1975, 221–30) discussion of Marx (1969, 155–75, 387–8, and 399–401); see also Birkbeck (1982).

7. The main survey was conducted on Friday, 3 Sept. 1977, and the other survey was held on the following day. A 'shopping basket' of basic foodstuffs was purchased in a wide variety of different sectors of the city and types of establishment, according to a prearranged sampling scheme. All purchases were made by the same person, Carmen Rosario Asprilla, and the author is indebted to her for her assistance with the survey. She is a relatively poor black person who has been resident in Cali for several years. Once the purchases had been made, all products were weighed and details of quality were noted.

8. Estimates of the proportions of males, females, adults, and children in the street occupations are based on the results of the street counts mentioned in n. 4, and on general observations at points of concentration of the street occupations. The main count was made on the afternoon of Tuesday, 13 Sept. 1977, but further sample counts were made at different times of the day and night, on all of the different days of the week, and in all of the months of the year.

References

Anderson, Stanford (ed.) (1978a) 'People in the Physical Environment: the Urban Ecology of Streets', in *On Streets* (Cambridge, Mass.: MIT Press), 1–11.

—— (1978b) 'Studies toward an Ecological Model of the Urban Environment', in *On Streets* (Cambridge, Mass.: MIT Press), 267–307.

Birkbeck, Chris (1978a) 'Self-employed Proletarians in an Informal Factory: The Case of Cali's Garbage Dump', *World Development*, 6, 1173–85.

—— (1978b) 'Small-scale Transport and Urban Growth in Cali, Colombia', in *The Role of Geographical Research in Latin America*, William M. Denevan (ed.), Muncie, Indiana, Conference of Latin Americanist Geographers, Publication 7, 27–40.

—— (1979) 'Garbage, Industry, and the "Vultures" of Cali', in *Casual Work and Poverty in Third World Cities*, Ray Bromley and Chris Gerry (eds) (Chichester: John Wiley), 161–83.

—— (1982) 'Property-Crime and the Poor', in *Crime, Justice and Underdevelopment*, Colin Sumner (ed.) (London: Heinemann), 162–91.

Bromley, Ray (1978) 'Organization, Regulation and Exploitation in the So-called "Urban Informal Sector": the Street Traders of Cali, Colombia', *World Development*, 6, 1161–71.

—— (1982) 'Begging in Cali: Image, Reality and Policy', *New Scholar*, 8, 349–70.

—— and Birkbeck, Chris (1984) 'Researching Street Occupations of Cali: The Rationale and Methods of What Many would call an "Informal Sector Study"', *Regional Development Dialogue*, 5 (2), 184–203.

—— and Gerry, Chris (1979) 'Who are the Casual Poor?', in *Casual Work and Poverty in Third World Cities*, Ray Bromley and Chris Gerry (eds) (Chichester: John Wiley), 3–23.

Cohen, D. J. (1974) 'The People Who get in the Way', *Politics*, 9, 1–9.

DANE (Departamento Administrativo Nacional de Estadística) (1975) *XIV Censo Nacional de Población y III de Vivienda: Resultados Provisionales* (Bogotá: DANE).

Fisher, A. G. B. (1933) 'Capital and the Growth of Knowledge', *Economic Journal*, 43, 379–89.

—— (1939) 'Production, Primary, Secondary and Tertiary', *Economic Record* (June), 24–38.

Foote, N. N., and Hatt, P. K. (1953) 'Social Mobility and Economic Advancement', *American Economic Review*, 43, 364–78.

Gerry, Chris (1985) 'Wagers and Wage-working: Selling Gambling Opportunities in Cali, Colombia', in *Planning for Small Enterprises in Third World Cities*, Ray Bromley (ed.) (Oxford: Pergamon), 155–69.

—— and Birkbeck, Chris (1981) 'The Petty Commodity Producer in Third World Cities: Petit Bourgeois or Disguised Proletarian?', in *The Petite Bourgeoisie: Comparative Studies of the Uneasy Stratum*, Brian Elliott and Frank Bechhofer (eds) (London: Macmillan), 121–54.

Gottmann, Jean (1970) 'Urban Centrality and the Interweaving of Quaternary Activities', *Ekistics* 29 (174), 321–31.

GV (Gobernación del Valle) (1976) *Código de Policía del Valle del Cauca* (Cali: Imprenta Departamental).

Hirst, Paul Q. (1975) 'Marx and Engels on Law, Crime and Morality', in *Critical Criminology*, Ian Taylor, Paul Walton, and Jock Young (eds) (London: Routledge & Kegan Paul), 203–32.

McGee, T. G. (1979) 'The Poverty Syndrome: Making out in the South-east Asian City', in *Casual Work and Poverty in Third World Cities*, Ray Bromley and Chris Gerry (eds) (Chichester: John Wiley), 45–68.

Marx, Karl (1969) *Theories of Surplus Value* (Vol. 1) (Moscow: Foreign Languages Publishing House).

MSC (Municipio de Santiago de Cali) (1971) *Compilación de Disposiciones Legales sobre Urbanismo, Saneamiento y Otras Materias* (Cali: MSC).

Oliveira, Francisco de (1985) 'A Critique of Dualist Reason: The Brazilian Economy since 1930', in *Planning for Small Enterprises in Third World Cities*, Ray Bromley (ed.) (Oxford: Pergamon), 65–95.

Rusque-Alcaino, Juan, and Bromley, Ray (1979) 'The Bottle Buyer: an Occupational Autobiography', in *Casual Work and Poverty in Third World Cities*, Ray Bromley and Chris Gerry (eds) (Chichester: John Wiley), 185–215.

Williams, Gavin, and Tumusiime-Mutebile, Emmanuel (1978) 'Capitalist and Petty Commodity Production in Nigeria: a Note', *World Development*, 6, 1103–4.

11

How Women and Men Get By: The Sexual Division of Labour in the Informal Sector of a Nairobi Squatter Settlement*

Nici Nelson

IN this paper I examine the sexual division of labour in the informal sector activity of a Nairobi squatter neighbourhood, Mathare Valley.[1] I describe briefly the range of activities, with data on the capital investment, costs, and revenues where available, and discuss the relative involvement of men and women in each. I close with a number of observations, some descriptive and some theoretical, on the sexual division of labour in Mathare in particular and the position of women in the informal sector in general.

The Setting

Kenya has a population of eleven million people. Though only 8–10 per cent of the people live in urban areas, these areas are growing at a rate of 6 per cent per annum. The high differential in income to be found between rural and urban areas, the rapid population growth in Kenya as a whole (3½ per cent per annum, one of the highest in the world), combined with a shortage of arable land, has made migration to the city an attractive option for many people. Until recently it was mainly men that migrated; in the last census (1969) men still substantially outnumbered women in the urban areas. Recently, however, there has been an increase in the number of women migrants, especially single women.

Kenya is a relatively successful mixed capitalist economy. A number of major earners of foreign exchange, such as tea, coffee, pyrethrum, and tourism are completely or partially in the hands of national parastatal bodies, but there is much private enterprise at all levels, including a great deal of investment by multinational companies. The government also has a

* Reprinted, in a revised form, from Ray Bromley and Chris Gerry (eds), *Casual Work and Poverty in Third World Cities* (Chichester, New York, Brisbane, and Toronto: John Wiley & Sons, 1979), by permission of the publishers.

controlling interest in a number of strategic industries such as banks and oil companies. Leys has described the contradiction in the government's economic nationalism and its dependence on foreign capital (Leys 1975). Though there has been a respectable growth in the GNP recently, this growth has not resulted in a significant increase in jobs in the non-farm sector, and there is growing unemployment among school-leavers and even university graduates. It is, however, impossible to measure this unemployment with any exactitude due to inadequate data and the existence of widespread underemployment (ILO 1972). One result of this lack of jobs in the formal sector has been the growing number of urban migrants who support themselves in the urban areas by petty commodity production in self-built squatter areas.

Mathare is one of these squatter areas, which may house as many as one-fifth of Nairobi's population. It is a young community, whose present building stock was begun after the end of the Mau Mau emergency in 1962. Between 60,000 and 70,000 people occupy mud-and-wattle houses with cardboard roofs, or wooden blocks with tin roofs built on the slopes of the Mathare River Valley, 15 minutes' drive from the Nairobi Hilton. Approximately one-third of the inhabitants are owners of housing units; the rest are renters. The inhabitants are largely poorly educated Kikuyu (or Kikuyu-related groups such as Meru or Embu) migrants to Nairobi; 75 per cent are without land, and 80 per cent support themselves in the informal sector (Ross 1973, ch. 7).

There is some disagreement over how many women live in Mathare, and how many of them are independent heads of households. To summarize briefly, the various sources (Etherton 1973; my own sample survey; data obtained from the National Christian Council of Kenya Survey in Mathare), there are probably more women than men in Mathare, and anything from 60–80 per cent of them are heads of households. Less than 10 per cent of them are employed in the formal sector, as opposed to 20 per cent of the men.

The Informal Sector as a Conceptual Tool

The concept of the informal sector arose in the early 1970s, mainly as an attempt to deal conceptually with the so-called 'underemployment' and 'disguised unemployment' that was continuing to grow in Third World cities. It forms part of a two-sector dualist model in the tradition of the traditional–modern or capitalist–peasant models. The formal sector is seen as the enumerated, large-scale, capital-intensive, wage-earning sector. The informal sector, on the other hand, is seen as a sector composed of the unenumerated self-employed, mainly providing a livelihood for new entrants into the city (Hart 1973). The International Labour Office Report on Kenya (ILO 1972, 6) defines the informal sector by its ease of entry, reliance on indigenous resources, family ownership of enterprises, small scale of operations, labour-intensive and adapted technology, skill acquired outside the formal school system, and unregulated and competitive markets.

I realize that of late the concept of the informal sector has been disputed and many people have proposed substituting the concept of petty commodity production (see Moser 1978, for a clear discussion of the pros and cons of these two concepts). The informal sector suffers from all the problems that beset any bipolar model; life is infinitely more complex than any two extremes. For example, the formal–informal dichotomy does not clearly set out the articulations between the two sectors. It cannot deal satisfactorily with such anomalies as small-scale owner-operated establishments that supply on contract to large formal sector firms. For the purposes of this paper, however, and despite its deficiencies, I will continue to use this concept.

Informal Sector Activity in Mathare Valley

It could be said that Mathare Valley exists by virtue of the informal sector. Its housing is 'informal' and self-built; according to Ross (1973) 80 per cent of the total population (and by my calculations 90 per cent of the women) make their living in various types of informal sector activities (some visible and some not so visible) within the Valley. I shall discuss briefly the organization, the costs and revenues (where I have figures), and the sexual division of labour of informal sector activity in Mathare. I have divided them into the following categories: (1) the entertainment industry; (2) self-built housing for rental; (3) shops; (4) other small businesses; and (5) hawking.

The Entertainment Industry

As I have pointed out elsewhere (Nelson 1977), Mathare is ideally located to provide the services which I have grouped together in the entertainment industry. The Valley is surrounded by areas which have large populations of men living alone, either unmarried or with wives and families who have stayed behind in the rural areas. The Mathare Mental Hospital has large staff quarters; similarly the Police Lines, the Army Camp, and the Kenya Airforce. The neighbours to the west and south are dormitory areas for men working in Nairobi, who rent rooms with their workmates. The 'entertainment industry' (a phrase coined by Ross 1973) consists of the provision of alcoholic drinks and commercial sex. In the evenings, on their days off, or at the weekend, men from the surrounding neighbourhoods and elsewhere in Nairobi flood into the Valley looking for beer and companionship. They stay till late at night roaming the alleyways looking for prettier faces, fresher beer, more lively talk or excitement, or a woman willing to let them stay the night.

Buzaa (maize beer) brewing is so prevalent in Mathare it could almost be called a local industry. It is the product for which the area is known throughout Nairobi. Eighty per cent of the women in both my samples brewed beer; this constituted all the independent heads of households without formal wage-work as well as many of the married women. This is primarily a women's business in Mathare; very few men actually engage in either brewing or retailing *buzaa*. Women brew and sell individually, working out of the rooms they rent to live in. Thus each woman's room is used for domestic purposes, and for manufacturing beer while also serving as an informal bar. I have described the process of brewing and selling *buzaa* elsewhere (Nelson 1978, 1979). Suffice it to say that it is an uncertain and dangerous process principally because it is illegal; women run the risk of being arrested or having their beer, raw materials, and/or equipment destroyed in a raid or of being forced to bribe the police. Though women work as individual entrepreneurs, they have evolved complex networks of co-operation between patron-clients, neighbours in local residential clusters, and friends, to enable them to cope with the difficulties engendered by police action and the uncertainties arising from having to sell on credit to men over whom they have no sanctions.

Despite the uncertainties, *buzaa* brewing is a rational choice for many women in Mathare. The only wage-labour job options easily open to women in Mathare are those of house-servant and barmaid. Both of these jobs involve long hours, often ten hours a day, six days a week and in many instances the wages are only Ksh 100 or less a month.[2] To provide a basis for comparison, the Kenyan minimum wage in 1974 was Ksh 200 monthly. Women brewing beer in Mathare could in 1974 earn anything from Ksh 250 to 400. In addition, the work is done in their own homes, which enables them to care for their children while they are brewing and selling. By contrast, a woman working as a barmaid or house-servant cannot care for her children at all during the day and only rarely during the evening. Thus she finds her already small wage eaten into by the need to hire someone else to care for her children, which may cost Ksh 30 a month. *Buzaa* brewing is a convenient option which needs little training or skills, gives a decent return, and requires relatively little initial capital. Thirty Ksh provides the capital for a brew and the loan of the simple equipment necessary, the purchase of a few clean oil tins in which to serve the beer, and a room from which to operate.

Though *buzaa* brewing and selling is primarily women's work in Mathare, there are subsidiary activities which are conducted by men. There are a number of men who sell firewood which they bring from the rural areas; this is used by beer brewers to fry the fermented maize flour because a wood fire produces a quick, hot flame. People normally cook with the slower-burning charcoal in Nairobi. There are also men who hire themselves out to women who have a large batch of *buzaa* to strain, squeezing it through a burlap sack. These latter are usually alcoholics who cannot sustain any type of consistent work, and are often glad to accept a few free drinks of the *buzaa* as payment.

Some gin (called locally *changaa* or Nubian gin) is distilled and sold in Mathare Valley. The production of Nubian gin is a male task in Mathare for several good reasons. It too is illegal, but it is a more risky proposition than *buzaa* brewing because it must be distilled in the open and heavy equipment must be set up at the river's edge. The police are even more diligent in seeking out Nubian gin manufacturers than they are in apprehending *buzaa* brewers. Additionally, the penalties are much greater; in 1974 the fine for gin-distilling was over Ksh 1,000 as compared to Ksh 100–200 for *buzaa* brewing. Some distillers prefer to seek out a river that is more secluded and isolated than the Mathare River, which is completely open to police observation. However, this requires a vehicle. The only woman I encountered who distilled Nubian gin had a boy-friend with a car who provided transport and 'muscle' to deal with a police raid. Distilling gin is very lucrative; for this reason it is often rumoured that important people in Nairobi invest in it as a sideline. Most of the Nubian gin from Mathare is distilled by relatively few rich men in an area of the Valley where I did little work. It is largely distributed by 'runners', young boys who carry it to individual women *buzaa* sellers in plastic bleach bottles. These boys are well capable of avoiding any attempts by the police to arrest them. While most of the gin is produced by men on a relatively large scale, it is sold much in the same manner as maize beer, namely through the informal bars run by individual women retailers. A couple of men who distill Nubian gin also run illegal 'bars' where the gin as well as bottled beer is sold. These bars are modelled on the Western-style bars found elsewhere in Nairobi, with tables, benches, and portable record-players blaring out the latest hit tunes.

The third aspect of the entertainment industry in Mathare is commercial sex. It is not within the province of this paper to go into this subject in any detail here (see Nelson 1977, for a longer exposition on sexual relations in Mathare). Most of Mathare's independent women make the greater part of their income through retailing *buzaa* and Nubian gin; they supplement their income through sexual liaisons with some of the men whom they meet as customers. Married women have been known to do the same with varying degrees of success, depending on how predictable are their husband's sojourns at work. There are a variety of forms of commercial sex. There is 'Quick Service' (a name taken from a local bus company which is self-explanatory), which in 1974 was Ksh 5 for a short 'bed ride'. Staying all night cost anywhere from Ksh 5 to 20, depending on how close was the man's pay-day and the woman's relative need for money. These are the more casual forms of commercial sex and the extent to which they are practised by Mathare women is obviously impossible to ascertain with any degree of certainty. It is my impression, gleaned from two years of observation, that many women only form these casual relationships when they are in great need of money. There are however a few women in the Mathare area where I worked who depend on casual sex as their only source of income. These women are generally regarded by their female peers as stupid and too lazy to earn money in any other way, mainly because this is a dangerous activity and women who constantly entertain large numbers of strangers run the risk of being badly beaten or robbed by unscrupulous, vicious, or drunken customers. Most women prefer to limit their practice of this form of commercial sex to times when they need money badly and favour longer-term relationships with lovers or with *town bwanas* (town husbands). Lovers are men who do not live with their women, but merely visit and spend an occasional night with them. *Town bwanas* live with their women, or 'town wives', and act as real husbands. However both parties recognize that the relationship is of limited duration, that the woman who rents the room retains the right to it after the man has gone, and that the *town bwana* has no responsibilities for, or rights over, the children born of the union. Both these relationships have a strong monetary component, so I have included them in commercial sexual arrangements. Women themselves view these relationships in explicitly money terms, saying such things as: '*Bwanas* should help with the rent and give money for food. Else what is their use?' 'Any *bwana* who does not help me with money when I want it would have to leave.' There is no direct payment for sexual services as there is with Quick Service; the economic exchange is couched in terms of gifts, help, or loans to capitalize brewings (loans that rarely get repaid). Though women recognize and value the aspects of sexual attraction, affection, and companionship in these relationships when sitting around gossiping with their friends, most of them would assess the relative merits of their men on the basis of the latters' generosity with money. Women manipulate the various kinds of sexual relationships to maximize their economic security in the urban area.

Self-built Housing for Renting

Mathare Valley was repopulated after the lifting of the emergency prior to

Independence in 1962; many Kikuyu migrants from nearby rural areas or from other parts of Nairobi flocked into the valley to build mud-and-wattle housing for occupation and renting. In 1968, a number of the wealthier local inhabitants in collaboration with some investors from outside the Valley legally incorporated themselves as companies to buy land from the Indian owners, and then to build large blocks of concrete-floored, tin-roofed wooden houses. In 1969, after a serious confrontation with Nairobi's City Council in which the fate of Mathare literally hung in the balance, a moratorium was declared on further house-building. This ruling is currently enforced by the District Officer and his *askari* (police). Therefore it must be clear that being a landlord in the Mathare Valley is the province of those whom I have called the Old Timers. No one who arrived after 1969 could build a house. Though it is still just possible to purchase one that is already built, very few such cases came to my attention.

House-building was a very lucrative financial proposition. In the early period most people built mud-and-wattle housing with cardboard roofs. The mud housing units cost, by my calculations, approximately Ksh 400 to 500 each, which at the monthly rents prevailing before 1969 (Ksh 30; Ksh 45 by 1973) meant that the investment could be paid off in less than a year and a half. After the original investment is recouped, rents on houses are pure profit. Owners pay no rates, no taxes, no water or electricity bills, and repairs are limited to remudding after a severe rainy season or patching the roof with a few more flattened cardboard boxes bought cheaply from hawkers. The demand for rooms in Nairobi is high, and Mathare rooms are rarely empty for more than a day or so, regardless of their condition. Investing in housing has yielded a quick and steady return.

Investment through a registered company was a more costly and complicated procedure. Builders of the early mud houses merely had to apply for permission to build from the local village leader. Mathare is divided into ten named, spatially and administratively distinct villages, each with a core of self-appointed leaders, usually older men and women who arrived early in the settlement period. However, to join a company, the individual had to pay a membership fee that was perhaps as high as Ksh 1,000. After that he or she was permitted to buy shares in the company; the number of shares determine the number of rooms listed under the member's name. Contractors built these housing units, which adhere more closely to Nairobi City Council building standards, though they lack water, electricity and drainage. Each share cost approximately Ksh 1,000; but the rents were higher in this more 'staff' (meaning high class or elegant) housing area. At an average rent of Ksh 80 per room, the period of recouping one's investment was much the same as it was for the mud-and-wattle housing.

The numbers of men and women owners of housing is a fascinating puzzle to work out. In the period after the struggle for survival with the City Council, KANU (the political party in Kenya) conducted a survey of all individually owned housing in the whole valley. The KANU committee of each village of Mathare kept its own register and these were to be used as the basis for distribution of units from the new housing and site-and-service schemes being built to rehouse owners of the mud houses. In 1973, 800 such units were allocated, a drop in the ocean when one considers that there are probably 10,000 owners of mud housing in the Valley. I was only able to examine the register for Village II. There 66 per cent of registered owners were women, though they owned only 50 per cent of the units registered. The Secretary of Village I confirmed that the proportion of women owners was much the same in his village; but the Secretary of Village III (a village settled after I and II) said that he thought, without actually adding up, that the numbers were fewer in his village. It was my impression that the proportion of women is lower in the areas of the Valley settled after 1963–4. Just after the emergency there were undoubtedly a large number of independent, older Kikuyu women (many of whom had husbands who were killed or detained during the emergency), living and working in the Nairobi area. It was somewhat easier for Kikuyu women to stay in Nairobi at this period than it was for Kikuyu men. The surge of building in Mathare may have come at just the right time to release the small capital savings of these women. The investment was not high; especially if a woman could get together a work party of friends to build the houses with her rather than hire a builder. In this case she could bring the costs down to as little as Ksh 250 a room. The companies tend to have a smaller female membership, usually only about one-third. These companies involved a higher capital investment and attracted many prominent businessmen and politicians from outside the Valley; it was inevitable that there would be fewer women capable of raising the necessary Ksh 2,000 subscription.

Shops

The results of a survey of 204 shops and other businesses located in two of the 'villages' of Mathare, with ownership broken down by sex, are presented in Table 11.1.

In the category of shops I have included *duka* (which are small general stores), *hoteli* (restaurants), butcheries, charcoal shops, vegetable sellers, and maize and yeast wholesalers. All these retail a product and require some sort of business establishment, even if it is only a rudimentary roof. Needless to say all these businesses are unlicensed and therefore illegal.

Dukas are usually converted rooms with an enlarged front window and a counter area. There are a great many of these small establishments in

TABLE 11.1   Survey of 204 Small Businesses in Villages II and III in Mathare Valley, 1974

A. Shops

	General stores	*Hoteli* (small restaurants)	Butchers	Charcoal sellers	Vegetable sellers	*Kimera* and maize flour sellers
Male	40	11	19	24	8	16
Female	4	1	1	5	15	9
Total	44	12	20	29	23	25

B. Other small businesses

	Tailors	Carpenters	Shoemakers	*Dobis* (washermen)	Water sellers (Village II only)
Male	12	7	6	6	4
Female	3	0	0	0	0
Total	15	7	6	6	4

Mathare. In Villages II and III, there are 44 to serve a population of approximately 7,000 (see Table 11.1). Most of these are small and badly stocked with a limited range of domestic goods and foodstuffs, usually non-perishables (with the exceptions of milk, eggs, and potatoes). A well-stocked *duka* might contain such articles as beans, rice, sugar, tea, flour, cocoa, cooking fat, salt, chilli and curry powder, jam, maize meal, milk, eggs, potatoes, onions, matches, kerosene, cigarettes, needles, thread, safety-pins, aspirin, rubber baby-bottle teats, soap, wicks for lamps, and wirewool. People in Mathare buy their food daily in small quantities, so these *duka*, run as they are by family labour, make up for a low volume of trade by staying open 12–14 hours a day. They give credit only to customers who live close by, as some measure of insurance against a creditor moving on without paying the debt.

Few if any of the owners of these *dukas* keep accounts or have any precise idea of cash flow or costs of stocking. One woman claimed that when she had stocked a *duka* for her son to set up in business, she had spent Ksh 1,500 for a moderate amount of stock. 'Money just comes in and goes out!' was a typical response to my questions about gross or net profits. Money and food from the shop are used for the daily running of the family as needed, further complicating any attempts to estimate income. One store-owner guessed that he made a gross profit of Ksh 1,000 monthly, out of which he paid Ksh 100 in rent for house and shop. He was more businesslike than many others and his was a particularly large and well-stocked store. However, even he was vague as to how much he paid out monthly to restock, saying that it depended on whether he had any cash when the wholesalers came around in their vans. The vans that stock the legal stores in the neighbouring areas of Nairobi often go to Mathare at irregular intervals to sell to the larger *dukas* there. Smaller shopkeepers merely walk into Eastleigh, a nearby 'formal' suburb, and buy from shops there. One woman would buy a dozen pints of milk a day, paying 75 cents each and charging her customers 80 cents in her small *duka*.

Dukas are almost invariably owned by men. Forty out of 44 *dukas* are owned by men, though many of them have wives to help them run the shop. The same is true of *hotelis* (restaurants); only one woman owned a *hoteli* in Villages II and III. Men who run *hotelis* also have wives to do much of the daily work of the business. If owners do not have wives to do this they hire women to cook. Failing this they hire very young boys or very old men. Perhaps the fact that cooking is considered a task that is not manly has something to do with this division of labour; only the very young or the very old of the male sex are willing to consider doing it for a living. The standard fare at such a place is limited to tea, white bread, eggs, a slightly fermented porridge, and *irio* (a mixture of maize kernals, beans, and potatoes boiled for long periods and then partially mashed together). The prices are reasonable (50 cents for a large and very filling plate of *irio*) but the overheads are also low. Foodstuffs are bought each day at the Municipal Market. I could obtain no figures on relative profit margins, but one owner thought he could make Ksh 1,000 in a good month. Janet Bujra, who was simultaneously working in and studying Pumwani, estimated that some of the larger *hoteli* in that area (very close to Mathare in both senses of the word) make Ksh 3,000 gross profit.

Butcheries are very much in evidence in Mathare; there are 20 in Villages II and III. This is not because the local population normally eat a great deal of meat; on the contrary meat is a luxury for most residents. However, drinking beer and eating roasted meat is a fine old custom in Kenya. Therefore the butcheries cater for the many men that come to Mathare to drink. They do a roaring trade in the early part of the month, after everyone has been paid. In order to cope with the leaner periods after the middle of the month, butchers cut down on the stocks of meat they order. In one butchery, the owner claimed he could sell 100 kilos of meat in the first week of the month, and only 30 kilos in the last. He sold on average 240 kilos of meat per month; meat cost Ksh 2 per kilo at the Kenya Meat Commission, and he retailed it at the standard government controlled price of Ksh 6 a kilo; therefore he made, before deducting costs, a gross profit of Ksh 960. His rental and the salary of his cook came to Ksh 200.

Butcheries not only sell raw meat, but also have a charcoal fire to roast meat on the spot. Usually a couple of shaded benches are provided for clients to wait while the fatty meat roasts appetizingly over the coals. Some butchers also provide a soup made of bones, blood, and entrails.

All butcheries in Mathare, except one, are owned and operated by men. One woman owned a butchery, but she hired a man to operate it. This may have been due to a traditional belief that women should have nothing to do with the slaughtering of cattle or the portioning out of the meat.

Charcoal shops are usually the most rudimentary of establishments, often consisting only of an awning to protect the charcoal from the rain. They are supplied by lorries that come into Nairobi from the rural areas where the charcoal is prepared. Most charcoal sellers have an arrangement with a rural charcoal-maker, often a relative or friend from home. The charcoal is delivered in large burlap sacks at regular intervals, and is sold by the measure, a large tin bowl or *karai*. Profit margins are high. A bag of charcoal from the rural wholesaler costs Ksh 7.50 to 11.00. One bag divides into approximately 36 *karais*, which are sold at Ksh 1 each. Thus the profit on one bag of charcoal is Ksh 25–28.50. Most charcoal retailers seem to be able to retail a bag a day. Thus the average charcoal seller can earn Ksh 680 a month, and a larger operator more. Rental costs, if any, are low. Most operators merely store the coal in their rooms and set up a plastic awning each day on a piece of waste ground near by, often near a well-frequented road.

It is surprising that so few women sell charcoal: the proportion is nevertheless higher than for other shops, five out of 29, or nearly 20 per cent. Capital investment and skill content are low, and I could isolate no traditional Kikuyu beliefs that might have contributed to a reluctance of women to participate in such an enterprise or of customers not to patronize a woman so employed. On the contrary, in Kenya, women have a strong traditional link with fuel and its provision for the household. This may in fact account for why as many as 20 per cent actually are involved in the trade.

The traditional image and role of Kikuyu women may also explain why more than 65 per cent of the vegetable sellers are women. This is the only informal sector activity (outside of *buzaa* brewing and commercial sex) which has more female than male entrepreneurs. The low capital investment necessary may also contribute. Vegetable sellers, like charcoal sellers, usually have the most basic of business establishments. A seller usually has a plastic sheet spread on the ground. The more substantial establishments have an awning, or even tables to display the goods. This is not a very lucrative form of self-employment. The seller goes very early to the Municipal Market and hangs around the periphery trying to purchase small amounts of vegetables in season from the licensed dealers. This is illegal and the Mathare sellers are often pushed around by the police for their pains. Periodically the City Council orders a clean-up campaign and then these luckless individuals may be arrested and fined for obstruction. Most days they obtain a few vegetables, often at a cost higher than that the licensed vegetable dealers pay. The profit margins on these perishable goods are low, and the seller may find that at the end of a day towards the end of the month, he or she is reduced to selling at cost in order to clear his stock of items that will not be saleable the following day. One woman claimed that she barely made enough to feed her four children by this daily activity. Certainly vegetable sellers, unlike other shops and businesses discussed above, are the least permanent form of business to be found in Mathare. Women frequently stop brewing beer when the police raids or drunken customers become too much for them, and some of them take up vegetable selling. Perhaps because it needs no premises, little capital, and no network of contacts (as the charcoal sellers must cultivate, though in other respects charcoal selling is very similar to vegetable retailing), it is very easy to drop and pick up at will. Women who take up vegetable selling to escape the rigours of *buzaa* brewing, soon start complaining about the roughness of the City Council's *askari*, the long hours spent sitting in the sun, and the low financial returns. Eventually they return to making *buzaa*.

The numerous brewers involved in *buzaa* manufacture mean that there exists a strong demand for the ingredients: maize flour and *kimera* (a form of sorghum flour). Most women did not have the time or transport facilities to buy these ingredients themselves in neighbouring areas. A single brewing uses 45 lb. of maize flour and 10 lb. of *kimera*. In addition, a woman carrying a 45-lb. bag of flour on her back would be particularly conspicuous and vulnerable to police harassment. For these reasons brewers prefer to buy from wholesalers who transport the materials by truck into the Valley for them.

This form of selling is fairly lucrative. Two women who ran such an enterprise alternated going into the country where they would buy seven bags of *kimera* weighing 180 lb. each and transport these back to Nairobi, hoping that the police would not apprehend them: a licence is required to transport food in bulk in Kenya. After deducting the costs of their journey, they estimated that they made a net profit of Ksh 400 per trip. They averaged three such trips a month.

More than one-third of the *kimera* and maize flour dealers in Mathare are women. This type of selling thus ranks above charcoal selling and under vegetable retailing in the numbers of women entrepreneurs involved. Most of these are older women who have already made a great deal of money brewing *buzaa*. They may wish to shift their business activities to a less arduous and risky area, one which takes less energy and mental wear and tear. It was perhaps a natural transition for them to begin providing materials for the brewing process. Many of these women have been brewers on a scale larger than the average (brewing perhaps three or four times a week instead of the more normal once or twice a week). As a result they become known as reliable sources of this *buzaa* if women want to buy wholesale a large tin to retail to their own customers when they themselves have not had the time or capital to brew. This means that these larger brewers have formed networks of female clients who buy *buzaa* regularly from them. When the brewers set themselves up in the *buzaa*-ingredient wholesaling business, their former *buzaa* clients make a point of buying the raw materials from them instead. In the case of the two women cited above, I witnessed the inauguration of their business. These women had excellent reputations among the younger, smaller-scale brewers for fair dealing and the reliable quality of their *buzaa*. When they started selling *kimera*, many of their former *buzaa* customers switched their trade to them, saying that they knew these two women would only buy the best quality ingredients. The quality of *buzaa* can vary enormously with that of the *kimera* used, as the inferior types produce a sour, unpalatable brew.

Other Small Businesses

In this category, I have included those entrepreneurs who produce a product (tailors, carpenters, shoemakers) and those who produce a service (*dobis* or washermen, water-carriers, traditional doctors, female circumcisors, and hawkers). I have less complete data on many of the activities in this category.

Tailoring involves the use of a sewing-machine, usually treadle-operated. There are 15 tailors in the two villages surveyed, 12 of whom are men. Two of the factors preventing women entering this business may be: (1) the high cost of the machine (more than Ksh 1,500 second-hand); and (2) restricted access to informal or formal 'on the job' training; a woman tailor claimed to have spent Ksh 20 a week for nearly two years' part-time instruction at a local Nairobi school for tailoring.

There are seven shoemakers and six carpenters in Villages II and III. All of them are men. Many of the carpenters have learned their skills as casual labourers on building sites or as apprentices to other small carpenters. Most shoemakers undoubtedly learn as apprentices to other shoemakers, though I encountered one who had learned his skills at a mission trade centre. There is no clear pattern for apprenticing to a trade as far as I know. Most of the apprentices I came across seemed to work for several years doing increasingly responsible tasks for little or no pay. Only the car repairers can demand payment from their apprentices, since this is a more lucrative and prestigious trade that many young men are eager to enter.

Dobis are not very numerous in Mathare, mainly because having one's clothes washed is a real luxury for most of the inhabitants. They wash the clothes of the single male employees of offices in town who have to present a smart appearance by wearing well-creased trousers and crisp shirts. The six Mathare *dobis* are all male, which is interesting because clothes washing is specifically a woman's task, and few men wash their own clothes if they have a wife or girlfriend to do it for them. It is perhaps a colonial 'residue' inasmuch as most white Europeans hired male house servants and taught them the skills of washing and ironing. All the *dobis* in these villages were single old men who lived very poorly and always complained about how poor business was.

Water selling is another male occupation. It is arduous and does not bring in much profit. In Village II, the four water carriers are young single men. All they need is a cart costing Ksh 30, a number of metal drums, and a strong back. One such vendor made only Ksh 90 to 120 a month. When brewers brew *buzaa* more frequently (as for instance during an important public holiday that will attract many customers) they are more likely to need to buy water from a water carrier.

In 1974, there was one traditional doctor (a man) and one female circumcisor (a woman) in Village II (not included in Table 11.1). Both of these practitioners were reputed to be extremely well off. I was told that most traditional doctors are male, and I never heard mention of a female traditional doctor. As well as treating the sick these 'doctors' make love potions and good luck charms, and reveal the identity and whereabouts of thieves.

Female circumcisors are always women, as the whole female circumcision ceremony is an exclusively female gathering. The Kikuyu form of female circumcision entails the removal of the clitoris. The female circumcisor that I interviewed boasted of being very 'modern' in her techniques, and claimed to have a certificate in circumcision from a hospital, though I doubted its existence. She eschewed the traditional 'dirty, dull knives' using instead a double-edged razor. She dressed the wound with a sanitary towel and administered aspirin afterwards 'like a real doctor'. It is interesting that this shrewd and calculating woman has attempted to update her very traditional skills.

Hawkers were not included in the business survey of Mathare because of difficulties in locating and counting them. Hawkers are numerous in Mathare, especially on the first days of each month, when wage-workers have recently been paid. They hawk vegetables, used as well as new clothes, trinkets, and cleaned half-litre oil tins which women use to serve *buzaa*. Some vegetable hawkers are women, but in general, hawkers are predominantly male.

Conclusions: How Women and Men get by in Different Ways

I can now conclude with a number of observations on the sexual division of labour found in informal sector activity in Mathare and some rather more general statements about the woman's role as well as the comparative position of women and men in the informal sector.

Division of Labour in Mathare's Informal Sector

Men in Mathare are much less restricted than women in their choice of economic activity. I cannot fully support this assertion with concrete data; I concentrated on women in Mathare, and in writing this paper I recognize a gap in my own data-base. At the end of my study I conducted a survey of a 10 per cent sample of the adult females in one village. From this survey I can say with certainty that most women working in the informal sector brewed *buzaa* or practiced commercial sex within the Valley itself. In my survey of informal business establishments in Villages II and III, I noted that there were only 38 women entrepreneurs as opposed to 153 men, and that these were effectively concentrated in three out of eleven categories, namely those of charcoal, vegetable and *kimera*/maize flour selling. On the other hand I did not conduct a similar survey of adult males, so I cannot state definitely the range and location of their occupations. From observation it is my impression that men who are employed or self-employed in the informal sector are involved in a much wider range of activities, and most of them work outside of Mathare. Men can be motor mechanics, small-scale building contractors, house repairers, glaziers, plumbers, electricians, casual wage-workers on building sites or in industry, drivers of illegal taxis, fare-boys on buses, delivery boys, thieves, fences for stolen goods, street thieves and 'muggers', barbers, hawkers of anything portable, street sweepers, dustbin scavengers, bicycle repairers, makers of shoes from old tyres, carvers of wooden utensils or tourist curios, makers of *sufurias* (aluminium pots) and charcoal stoves and a variety of other activities too numerous to mention.

There are a number of reasons for this difference, some of them structural and others cultural. Women are not only less well educated than men in Kenya, they have fewer skills with commercial value. This obviously limits the choices open to any individual woman when she has to choose an economic activity by which to support herself. This is a structural constraint arising out of a cultural constraint, namely the Kenyan view of what is feminine, and what economic roles women should and should not play. Finally women are limited by the presence of young children who need care. In Mathare and other areas like it, peopled with recent migrants to the city, there are no real and/or acceptable substitutes for mother care—neither the older relatives of the more traditional village setting nor any kind of alternative provision, such as crêches. This again is a constraint with a cultural component. Women are defined as primarily mothers, and women are expected to have a closer commitment to their children.

In Mathare, if we look at the list of activities practised by men and women, we reach the following conclusions about the sexual division of labour. First, men do the work which involves 'new' skills (new to traditional East African village society) learned in the educational system or on the fringes of the modern industrialized economy: tailoring, carpentry, shoemaking, and other activities not present in Mathare Valley Villages II and III, such as repairing cars and driving taxis. Women have not been permitted to learn these skills. Their skills have remained those of the traditional subsistence economy or the domestic sector. This has far-reaching consequences as we shall see in a moment.

Secondly, men, in the main, carry out all the activities that require a relatively high capital investment, such as running *dukas*, tailoring, or owning a taxi or private bus. The only exception to this is the *kimera* and maize flour wholesaler. Men who are in the informal sector in Mathare do not have to be the sole support of their dependents. Such men do one of two things. They may keep their family on a small farm where the wife grows food and perhaps does casual agricultural labour, such as coffee- or tea-picking, to supplement the family income. Or, if the family is landless, the man expects the women to brew beer or carry out some other business to add to the family income. Only women who are married to well-paid men in the formal economy or whose husbands have made a success in informal sector self-employment, do not work for money outside the home. A man struggling to survive in a highly competitive economic world has the advantage of being able to channel part of his income into capital investment. An independent woman in the informal economy is at a disadvantage if she has small children. I found there is a correlation between successful women entrepreneurs and barrenness; a surprising number of big women in Mathare's informal economy are barren. Either that, or women begin to expand and consolidate in the informal economy only in their late forties when most or all of their children have grown up and perhaps help to contribute to a joint household income (see Nelson 1979). Most women with children spend more of their meagre earnings on essentials and school fees than men in the same economic bracket. While men see their financial future in building up a business and accumulating capital, women see it in terms of educating their children. This is not to say that men do not invest in their children's education, but the men in the informal economy can expect financial contributions from their wives. I am reminded of an interview of a successful American woman executive, who expressed the opinion that more women did not succeed in the world of big business because they lacked one essential ingredient for success—women executives do not have wives!

I would also like to make a brief point about perceptions. I found that there are two different ways of viewing the informal economic activities pursued by Mathare residents. This is determined by a combination of sex and age. Young men in Mathare view their particular activity in the informal economy as a stop-gap measure. They always talk about getting away and finding a 'real job'. This is a combination of both rationalizing and salving one's pride for doing a job considered to be demeaning (both by oneself and by others) by virtue of one's high educational status, and a reflection of a very real hope that soon one's luck will change. All women (regardless of age) and older men tend to think of themselves as more or less permanently in the informal economy, perhaps because as groups they both have lower educational standards than young men. Because women and older men have fewer options, they view informal sector activity as a way of life. Whether or not this leads to greater commitment (in skills and investment) can only be found out by a specific study of this aspect of the informal sector. Young men often spend their profits on fashionable clothes and 'high status' consumer goods, but it is impossible to tell whether this is a manifestation of their youth or their high educational status which leads them to see a higher standard of living as theirs by right. Perhaps all youth hope for something better and eventually come to terms with a harsher reality only with the passing of years.

Women in the Informal Sector

My data on the sexual division of labour in Mathare lead me to consider the specific position of women and whether or not it is possible to generalize from the specific (Mathare) to a wider context, that of the informal economy in Third World cities, to make some observations that pertain to women's position in that wider context. One of the first problems is the lack of rigorous, detailed primary data of any kind. If men are badly enumerated in the informal sector, then women are even more so. This is no doubt partly due to an assumption made by census takers and economists that women in urban areas who are not employed in the formal economy must therefore be dependants of men. Only middle class women with high qualifications are listed in national employment statistics; though often men without any qualifications are listed in national employment statistics as a matter of course. Secondly, researchers who have done micro-studies in cities concentrating on the informal economy have shown little interest in the numbers of women working in it, or the types of work that they do. It is not uncommon to find a study listing people's occupations with no breakdown by sex. Ross (1973), Jocano (1975), and Hake (1977) are three such examples, Jocano gives a listing of the occupations of 567 residents of a Philippine slum but does not say which are women. Some of the categories (such as waitress) are self-evident. Others, like storekeeper, are not; and it becomes clear in the text that there are women in these undisaggregated categories without stipulating how many. Hake, who gives fascinating micro-data on the informal economies of several neighbourhoods of Nairobi, makes the same error. Further research in Third World cities must be done by researchers willing to consider the involvement of both sexes in the formal as well as the informal economies.

Only one study (Arizpe 1977) has come to my attention which attempts to deal theoretically with women in the informal sector, but even this deals with the subject in a somewhat superficial manner. There are a few micro-studies that concentrate on women in the informal sector (e.g. Bujra 1975; Hellman 1948; Longmore 1959; and Obbo 1980), all of which deal with low-income African women. One thing is very clear from all of these; *women sell in the informal market-place skills they normally practice in the home.* In any given cultural situation women will be found clustered in certain types of activities, and most of these relate to women's roles as providers of food, beer, child care, and sex-companionship. This is not only true of the informal sector in Third World cities, but also of Western industrialized economies where women are primarily found in service jobs, nursing and teaching. The boundary between women's unpaid domestic labour and the paid labour they perform in the informal (and even the formal) sector is often a tenuous one. Both in the domestic realm and in the market-place, women provide food, child care, domestic services, 'charm', companionship and sex.

Commercialized sex deserves a special word, since it is a very problematical area which relates almost entirely to women (there are very few males who make a living selling sexual services, except perhaps in areas which attract many Western tourists) and which evokes a great deal of emotion. Very little accurate, in-depth information exists on the incidence and organization of prostitution in Third World cities. There are no models to deal with it as a form of informal sector activity. It is an issue that raises a great deal of hostility either because it is seen as an aberration from the 'normal' emotionally intimate act between man and woman (an attitude fraught with moral connotations) or because it is seen as exploiting women. Is prostitution a moral issue? Is it an exploitation of women by men in which women are degraded victims? Or is it merely a service, different in degree but not in kind from other services normally rendered without pay in the domestic sphere (e.g. cooking food)? Mathare women at their frankest often made explicit the parallel between marital sex and commercial sex. As one woman said to me: 'The only difference between me, a married woman, and her, a prostitute, is that I do it with one man for food and rent and she does it with many men for direct payment.' When commercial sexual activities take place in a social situation without pimps or madames (as they do in Mathare), I think it is possible to view commercial sex as merely an economic activity that brings in fairly good returns for relatively little work, though perhaps some physical risk either short- or long-term. It is also an economic activity in which the owner-operator cannot have the means of production taken away from her. Janet Bujra comments for women in another neighbourhood of Nairobi, 'prostitution allowed them (the women) the independence and freedom from exploitation that would not have been possible to them had they chosen any of the other socioeconomic roles open to them' (1975, 215). More work must be done to devise models to deal with commercialized sexual relationships which avoid both pseudo-scientific moralizing and a simple assumption of social deviance. I feel that an approach which relates commercial sex to women's domestic labour theoretically could well be a fruitful one.

Finally, I would like to point out that the informal sector may provide many women in the Third World with the only viable alternative to marriage. Men do not have to choose between economic independence and marriage. Women often have to, or are forced to. In Africa, at any rate, the city may be the only option available for women who are widowed or divorced without hope of remarriage, or who have children out of wedlock. Informal economies provide openings for female migrants from rural areas both in Africa and in Latin America (Arizpe 1977, 36). Women with few 'modern' skills and little education can market their domestic skills in a socioeconomic setting which allows them to care for their children. Without question the women of this sector are exploited, poor and overworked. Yet, faced with considerable structural and cultural constraints, unable to either secure employment at the lower end of the formal sector job hierarchy, or marry someone with a relatively well-paid formal occupation, it is undoubtedly rational for women to seek a more lucrative and less exploited life.

Notes

1. I am an anthropologist by training. The data used in this paper was collected between Feb. 1972 and June 1974 in Mathare Valley in preparation for a Ph.D. degree. The 'ethnographic present' of this article is 1974.

   The methods used in this research were participant observation, interviews, and surveys. Participant observation was carried out in three of Mathare Valley's ten named 'villages', but I visited the other villages as frequently as possible in order to ascertain whether what I observed in Villages I, II and III was similar to what seemed to be happening elsewhere.

This data was supplemented by a number of long, structured, but open-ended interviews with 89 selected respondents. Due to the tense situation in Mathare *vis-à-vis* city and national authorities, it was difficult to interview people at random. I tried to stratify the sample on the basis of age, education, and status as owner or tenant. At the end of my period of research I conducted a random sample survey collecting basic demographic data. I used the results of this survey to compare my intensive interview sample with the total universe of women in Mathare. My sample was younger and better educated than the norm, possibly because young, educated women were able to understand and sympathize with what I was trying to do. In addition, there was a larger number of house owners in my sample than was the norm. It is possible that better-off women had more time to devote to this curious stranger with her interminable questions.

2. In mid-1974 there were Ksh 7.14 to the $US, which was equivalent to £0.43.

References

Arizpe, Lourdes (1977) 'Women in the Informal Sector: The Case of Mexico City', *Signs: Journal of Women in Culture and Society*, 3, 25–37.

Bujra, Janet (1975) 'Women Entrepreneurs', *Canadian Journal of African Studies*, 9, 213–31.

Etherton, David (1973) *Mathare Valley* (Nairobi: University of Nairobi, Housing Research and Development Unit).

Hake, Andrew (1977) *African Metropolis: Nairobi's Self-help City* (Brighton: Sussex University Press).

Hart, Keith (1973) 'Informal Income Opportunities and Urban Employment in Ghana', *Journal of Modern African Studies*, 11, 61–89.

Hellman, E. (1948) *Rooiyard*, Rhodes Livingstone Papers no. 13 (Manchester: Manchester University Press).

ILO (1972) *Employment Incomes and Equality: A Strategy for increasing Productive Employment in Kenya* (Geneva: ILO).

Jocano, Landa (1975) *Slum as a Way of Life* (Quezon City: University of the Philippines Press).

Leys, Colin (1975) *Underdevelopment in Kenya* (London: Heinemann).

Longmore, Laura (1959) *The Dispossessed: A Study of Sex Life of Bantu Women in and around Johannesburg* (London: Corgi Books).

Moser, Caroline (1978) 'Informal Sector or Petty Commodity Production: Dualism or Dependence in Urban Development?' *World Development*, 6, 1041–64.

Nelson, Nici (1977) 'Dependence and Independence: Female Household Heads in Mathare Valley, a Squatter Community in Nairobi, Kenya', Ph.D. thesis, University of London, School of Oriental and African Studies.

—— (1978) 'Female-centred Families: Changing Patterns of Marriage and Family among *buzaa* Brewers', *African Urban Notes*, NS, Special Issue on East Africa, Summer.

—— (1979) 'Women must help Each Other: The Operation of Personal Networks among *buzaa* Beer Brewers in Mathare Valley, Kenya', in *Women United: Women Divided*, Janet Bujra and P. Caplan (eds) (London: Tavistock).

Obbo, Christine (1980). *African Women: Their Struggle for Economic Independence* (London: Zed Press).

Ross, Marc (1973) *The Political Integration of Urban Squatters* (Evanston, Ill.: Northwestern University Press).

12

The Changing Fortunes of a Jakarta Street-Trader*

Lea Jellinek

A PASSION for hot lemon-juice and fried prawncrackers brought me by chance to Bud's stall. I loved sitting there deep into the night, watching the pedestrians go by. She parked her stall in the heart of the city, beside a water fountain opposite Merdeka (Freedom) Square. In the mid-1960s and early-1970s there was nothing unusual about street stalls and traders in the centre of Jakarta. They were patronized by most of the city's population, rich and poor. The streets were alive with people passing to and fro. Most went on foot. Many travelled by bicycle or *becak* (trishaw). Only government employees and the very rich went by car.

Bud started her day's work at 7 a.m. with a walk to a nearby market. She bought the foods she needed for her stall from the same few traders every day. Her survival as a trader depended on her ability to manage her modest finances very carefully. Her prices had to be competitive and profit margins were small. Most of what she earned from one day's trading had to be used to buy raw materials for the next day. The more she spent on herself, the less she could buy for her stall. There was little room for error. Bud watched the market traders keenly. She could accurately assess the price of her purchases before they were weighed. She began by offering a price she knew to be too low. Traders responded with prices they knew to be too high. On most occasions a compromise was soon reached and both would seem happy with the deal. Sometimes Bud would hold out for a price which a trader refused to accept, or he would demand a price which she thought excessive. Bud would move on to the next market trader, who would have witnessed the unsuccessful negotiations, and start bargaining all over again. No one harboured ill feelings, though it was all in deadly earnest. Like Bud, the market traders also depended on the daily earnings for their next day's survival.

Bud returned from the market in a *becak* laden down with fresh fruit, live chickens, and the other goods she had bought. The household was soon busily engaged in food preparation. They chopped and peeled and

* This account grew out of my relationship with Bud, whom I first met in 1972. She introduced me to her neighbours and my research of her community evolved gradually.

fried for the next few hours. Bud's tiny house was taken over by delicious odours and baskets of banana fritters and fried chicken. The cramped and primitive kitchen did not seem to limit the output.

Cooking was completed by 3 or 4 o'clock in the afternoon. The food was carefully stacked into Bud's cart which, loaded to the brim with the day's cooking, was then pushed to the site of Bud's stall. Bud operated her stall from 5 o'clock in the evening until the early hours of the morning, serving her prepared food or cooking afresh as required. After 1 a.m. Merdeka Square was reduced to the glow of kerosene lamps from the few night-traders, the bell-sounds of passing *becaks*, and an occasional pedestrian. Before dawn Bud's tarpaulin was taken down and neatly folded, and any left-over food was carefully put aside for the next night's trade. With the help of her husband and another assistant, the cart was packed up and pushed back home for storage near Bud's house. By 4 a.m. she was asleep. Three hours later she would have to awaken to start the day's work. Without her two-hour siesta in the early afternoon while Nanti, the second wife, did the cooking, Bud would never have been able to endure these long hours of work.

Like many of her neighbours, Bud was a migrant to Jakarta. When she first arrived in the city in 1948, the population was about 800,000. Twenty years later the population had increased fivefold to 4 million. Migrants poured into the city at a rate of 100,000 per year.[1] Bud came from Tegal, a small town on the north coast of Central Java. Her father was a clerk in the railways, and her mother was the daughter of a local headman.[2] Bud was lucky to be able to attend school for six years and learned to read and write. Her introduction to trading came during the tumultuous years of the Japanese occupation (1942–45).[3] Bud's father had lost his job, and her mother set up a stall in their home to earn money to feed the family. Bud married in Tegal, and then with her husband, Santo, migrated to Jakarta, where relatives had assured them that there were greater opportunities. From his wartime contacts with the Japanese army and the Allied forces Santo had acquired many new skills. In Jakarta he tried his hand at *becak* repair and, when that did not do well, he turned to repairing radios. In between he tried a variety of jobs, including casual labouring and trade. At the time they were living with relatives, and Bud once again assisted in the running of the household's food stall. By the early 1950s they were able to rent a room of their own. Bud left the family stall and took work as a ward assistant in a hospital. After a few months she found a better-paid position weighing out ingredients for prescriptions in a privately run pharmacy.

After moving several times in the early 1960s, Bud and Santo finally obtained a home of their own. It was a small bamboo hut with two rooms and a thatched roof. During the dry season it had an earthen floor, but that soon turned to mud in the monsoon rains. Wire-mesh windows helped to keep out the hands of passing strangers, but not the noise and cooking smells from neighbouring households.[4]

The amenities of the kampung[5] were primitive. The narrow winding paths which separated the tightly packed shanties from one another turned to quagmires in the wet season. Rubbish accumulated in the doorways. The only toilet facility was the nearby Cideng canal, which flooded several times in each monsoon with subsequent outbreaks of disease. Two wells originally provided drinking-water, later they became polluted and drinking-water had to be brought into the kampung. Vendors squeezed along the narrow pathways carrying cans on long shoulder-poles.

A few years after moving to the new house, Santo failed to come home every night, explaining his absences in one way or another. Bud eventually discovered that he had taken a second wife. She was devastated, and began to feel insecure. To become self-supporting, she decided to start trading by herself. Her business prospered over the next year or two and she was in need of additional manpower. By then Santo was womanizing again. Nanti, the second wife, came to Bud in sorrow and anger, not knowing what to do. Bud invited Nanti to live with her and help her trade. Feeling desolate and betrayed, Nanti accepted gratefully, and a seemingly improbable partnership ensued.[6]

Their enterprise thrived. In 1968, Santo, who temporarily had been brought to heel, built them a push-cart. This enabled Bud and Nanti to offer not only the cooked foods they were now selling, but a variety of new items including fresh bananas, cigarettes, beer and, of course, the hot lemon-juice that attracted me. The business went so well that by 1974 Bud was able to rebuild her house. She was one of the first in her neighbourhood to do so. Bud's renovations transformed the house into a two-storey structure with solid brick walls, a tiled floor and roof, glazed windows and a large, carved wooden door. Her old house had always been open, but Bud's new door was frequently slammed shut.[7] A water-pump was installed, so that the members of Bud's household no longer had to queue with the neighbours at the communal well. As Bud prospered, the members of her household began to withdraw from the surrounding community.

By 1975 the house was well stocked with consumer goods. The front room housed a battery-powered television set and a large radio. Four plastic chairs and a small coffee-table seemed to take up most of the remaining space. Along one wall there was a new sideboard, decorated with plastic flowers and plastic fruit, and full of crockery. Toys were scattered everywhere. Bud's kitchen contained a variety of recently acquired cooking utensils. There were pots and pans, and a new kerosene stove. An imported Chinese-made pressurized kerosene lamp replaced Bud's old locally made kerosene one. Electricity had entered the kampung and by paying a small fee, Bud was able to obtain an illegal connection from a neighbour. This enabled her to install a fluorescent light in the bedroom. There was an ice-chest, a thermos, a cupboard, and two wardrobes full of valuable batik cloths. Upstairs, overlooking the roofs of neighbouring households, family members now slept on beds with mattresses, instead of the bamboo mats they formerly rolled out on the floor.

As Bud prospered, new members were progressively added to the household. Santo surreptitiously took a third wife, Ade, who lived in a village 30 kilometres from Jakarta. Bud, whose only two children had died when still young,[8] agreed to the marriage only when she was promised a child of that union. A year later, Ade duly handed Bud her first child. Bud called the baby Rahmini and brought her up as her own child, although the adoption was never legally formalized. Bud's mother came to live in the house to look after Rahmini, whilst Bud and Nanti continued to trade. The two of them were now often working up to 20 hours a day, so Bud invited people from the third wife's village to assist in preparing food for the stall.

Where once only three people (Bud, Nanti, and Santo) had lived in the house and run the stall, there were now from six to eight. Santo, who was frequently away in the third wife's village, increasingly felt that pushing the cooked-food cart was below his dignity. He persuaded Bud to search for a neighbour to do the job. Bud recruited a young man called Achmad. Santo seldom helped in Bud's enterprise after Achmad was hired. The fact that he spent an increasing amount of his time in the third wife's village annoyed Bud greatly. As the senior wife, Bud felt that he should spend more time with her. Santo nevertheless continued to draw an income from the stall, even in his absence.

Ade's fellow villagers viewed Bud as a rich townswoman and she felt compelled to live up to that image. She held a large festivity in the village as Rahmini's birth approached and another after she was born. The celebrations cost her at least two months's earnings, but they were not the only occasions on which she financed festivities in the village. Once, for example, she hired a film and movie projector. She marked another celebration by contributing a live goat and festive meal.

The growing demands of kinsfolk, family, and friends, both in the village and in Jakarta, imposed a considerable financial burden on Bud. During each Lebaran (the major Islamic celebration after the fasting month of Ramadhan) she bought members of her household new clothes and cooked elaborate meals. During Idul Adha (a festival which emphasizes communion with the poor) she donated substantial sums of money to the local mosque and contributed to the cost of a goat that was slaughtered in the festivities. Rahmini's snacks in a day cost as much as the entire household spent on a meal. Previously Bud's household had modestly eaten the left-over food from her stall, but now Bud bought extra delicacies such as bread, fresh fruit, cakes, and ice-cream. New luxuries such as soap, toothbrushes, toothpaste, and shampoo appeared beside the water-pump. Dishes were no longer cleaned with the ash used by kampung dwellers, but with detergent which cost much more. Bud and Nanti had always sewn their own clothes, but now Bud hired a woman from the kampung to make clothes for them. Each festive season she paid a neighbour to repaint her house.

As Bud's aspirations and living expenses began to soar, the future of her stall, and thus of her income, became increasingly uncertain. Signs that her trade was doomed became more evident. The central city area was rapidly being transformed by new, wide, and ever-busier roads and the construction of numerous multi-storey offices, banks, and hotels. Bud's community prospered from this building boom. Many labourers in the neighbourhood worked on construction sites, whilst other members of the community sold food or provided transport to workers engaged on the projects. But the community was gradually engulfed by the city it had helped to create. In 1974 Bud was directly affected, for the footbridge across the canal to Jalan Thamrin, the main highway running north–south through the city, was dismantled and a high steel fence erected in its place. Kampung dwellers now had to make a long detour to get to the city centre.[9] Bud and her assistants had to find back routes away from the traffic to get to their trade location. On several occasions Bud's cart was struck by speeding cars. The vacant plot of land beside Jalan Thamrin where Bud and many other traders had stored their carts was fenced off for the construction of a multi-storey office block. As available land at the city centre became scarce, it became increasingly difficult and expensive for Bud to find a place to store her cart.

Step by step, Bud's stall was forced out of the central business district. She had started her trade in front of Sarinah department store less than five minutes from her home. After the first anti-trader campaigns in 1970[10] she had moved north along Jalan Thamrin to Merdeka Fountain, where I first met her. To avoid encountering the police clearance teams (Team Penertiban) she became a night-trader. Her main clients were the soldiers and policemen who guarded the nearby offices, government buildings, and banks. Some of her customers were involved in the trader-clearance campaigns, and they warned Bud when it would be unsafe to set up her stall at Merdeka Fountain. As she provided a good service at a time and place where it was needed, her clients protected her against harassment. She in turn gave them generous helpings of food. While other traders went bankrupt, Ibu Bud was able to survive and even thrive.

This happy state of affairs could not last. By 1975 the anti-trader campaign was being so firmly enforced that not even Bud dared venture out to her normal location near the fountain. She had to be content with a more secluded spot along a back street between a number of banks and a military depot. As a result of her move she lost many of her valued customers. However, the move was not the only factor undermining her trade. The period from 1968 to 1972 had been one of great economic stability, especially for a country that in the mid-1960s had been experiencing hyperinflation. By 1973 Indonesia was once again subject to steady inflation, although it was by no means as spectacular as it had been in the 1960s. Bud nevertheless began to find herself trapped in a cost–price squeeze. Her rising transport costs are a good illustration of this dilemma.

Bud had always used a *becak* to ferry her from the market-place to her home when she was loaded down with her early morning shopping. Later she also used a *becak* to ferry her to her stall in style. Bud used the same *becak*-driver each day. He was contented to be paid with a meal from her stall. But just as the government was pursuing its anti-trader campaign, it was also implementing a ban on *becaks* in the streets of central Jakarta.[11] As these policies came into effect, Bud's driver could no longer operate in safety. Instead Bud had to transport her goods from the market by motorized *becak* (*bajaj*) which were twice as expensive as *becaks*. Moreover, their drivers would not accept payment in kind.

Throughout Jakarta the old markets of crowded muddy passages and makeshift stalls were being replaced by multi-storey concrete buildings. The stall-holders in the old markets had paid only nominal site rates, usually illegal taxes pocketed by local officials. They could hold their prices down because their overheads were minimal. As the markets were organized and rebuilt, additional fees were levied for the rent of kiosks, rubbish collection, the maintenance of law and order, and market administration.[12] These additional charges were passed on to buyers like Bud, who in turn should have raised their prices. As the anti-trader campaigns progressively eliminated traders from large parts of Jakarta, however, those who were displaced were forced to congregate in areas where they were still tolerated. Competition became intense. Despite rising costs, traders undercut one another in a desperate bid to attract customers.

By 1976 Bud, and her fellow street-traders, were also competing with an increasing number of office canteens and small cafeterias which could buy in bulk and undercut their prices. Bud's customers were increasingly reluctant to pay for a rickety seat at her roadside stall when, for a little more, they could dine elsewhere in greater comfort. Most of the offices which lined Jalan Thamrin set up their own canteens, so that employees no longer had to seek cheap food on the street. The government even launched a campaign in the newspapers and over television and radio to advise people against eating from street stalls, which were said to be unhygienic. They did not conform with the city's 'modern' appearance and were accused of cluttering up the streets and hampering the flow of traffic. The military men who still patronized Bud's stall were forbidden to eat there after a canteen was set up in their depot. Bud believed that this directive was given to ensure the success of the army canteen, which was run by the wife of one of the senior officers.

Competition from off-street cooked-food outlets was reinforced by the changing tastes of a growing middle class, which benefited from the economic prosperity of the late 1960s and early 1970s. Soldiers and policemen who had once been content to eat at stalls like Bud's became reluctant to do so. To be seen eating on the street was an embarrassment. Kentucky Fried Chicken, McDonald's, and more modest Indonesian-style cafés were increasingly preferred to roadside stalls.

In an almost self-defeating effort to maintain her trade, Bud held her prices constant but lowered the quality and quantity of the food. She offered more rice, less meat, and fewer vegetables. She no longer used the best quality 'Saigon rice'. Her customers noted the decline in quality of the food and complained. But there was little Bud could do. Between 1972 and 1976 the number of her customers dropped sharply. In 1972 she had served over 100 military men, police, drivers, and lower-level civil servants each day. By 1976 she served no more than 50.

In the face of ever-increasing difficulties, Bud somehow continued to trade. In late 1976, however, Nanti, the second wife, ran away. It was a major blow. The quality of the food at Bud's stall was almost entirely due to Nanti's skills as a cook. Yet Nanti was not just a skilful cook. She was also by far Bud's most industrious worker. She was responsible for most of the food preparation. After spending some hours in Bud's hot, steamy kitchen, she helped Achmad to push the loaded cart to its trading-site, while Bud travelled to the site by *becak*. Nanti then crouched on the ground behind the stall preparing food or washing dishes, whilst Bud chatted to customers and handled the money.

Stalls like Bud's rely on cheap family labour. Nanti had received food, clothes and accommodation, but hardly any pay. As Bud's stall prospered, Nanti felt embittered by her lack of a share in the profits. When Bud hired Achmad to help Nanti push the cart to its trade location, the two of them conspired to run away and set up a stall of their own. Before their plans could be brought to fruition, however, Achmad was expelled from the household for flirting with Nanti.[13] Less than a year later, Nanti fled. Bud thereby lost the backbone of her enterprise. Bud's aged and arthritic mother took over Nanti's role as cook and Bud looked for another man to help push the cart. She hired a young man who was recommended by a neighbour. A month later he absconded, taking Bud's radio and kerosene lamp with him. Bud could not push her cart to her trading-site unaided.

Without an assistant, and preferably a male, she could not work.

After searching for many days, Bud recruited another young man, Narto, whom she found sleeping on the streets. He had come to Jakarta from a village near Tegal, Bud's birthplace, but he had no work or place to stay. He was delighted with the prospect of work and more than satisfied with the accommodation provided. Narto proved an able and willing worker, and Bud started to trade again, full of hope that this time her luck would hold. She must have known, however, that her problems were far from over.

Almost a year earlier, some of Bud's military protectors from the nearby army depot had warned her that their base was to be moved to the outskirts of Jakarta to make way for a highway. They suggested Bud move to their new location, for they were sure she would fall victim to anti-trader raids if she stayed in the central city area. Bud toyed with the idea of moving, but decided to struggle on with the difficulties she knew, rather than face the risks of the unknown. Within the year the soldiers moved to their new base. Bud realized almost immediately how vulnerable she had become, for she was ordered to move a few days after the soldiers left. When she tried to defy the order, she was threatened with a fine and the loss of her cups and plates. Within a week of restarting her trade, she was again ordered to leave and realized that this time she would have to go.

The next few days were the worst in her life. She and Narto pushed the food-stall along endless roads in a vain search for a new trading-site. All the sites they came to were either occupied or subject to frequent anti-trader raids and too dangerous to use. Their search seemed hopeless. Bud sat down and wept. They moved on till finally they found a vacant site. Bud dared not leave her cart for fear that it would fall prey to an anti-trader raid. The two of them worked all day and slept on the roadside by their cart at night. Narto was used to life in the open, but Bud found it a strain. They had no toilet or washing-facilities. But they worked on. One day Bud's mother and Rahmini came to ask for some housekeeping money. Bud burst into tears. In the past 24 hours she had only earned the equivalent of 8 US cents.[14] That was as much as Rahmini had spent on a single snack in their better days. When I first met Bud six years earlier she had earned one hundred times that amount in a day.

Bud's managerial skills seemed to have deserted her.[15] The food she sold catered to nobody's tastes. She was serving an odd mixture of coffee, eggs, and rice. Eggs and coffee were too expensive for the poor, who could only afford rice, vegetables, and tea. Rice and eggs were too modest for the better-off, who wanted meat and fruit. Potential customers passing by her stall were not impressed by the range of food and simply moved on to another vendor.

As her trade and livelihood disintegrated, Bud seemed to lose her hold on life. Whilst her stall went well, she thought nothing of working 20 or more hours a day. Now, as her plight became increasingly desperate, she lost the will to work at all. Everything seemed futile and hopeless. Whatever she did went wrong. She forgot to pay the monthly interest on some batik she had deposited with a pawnbroker and forfeited $US11 worth of cloth.

Bud's neighbours, whom she had shunned in her better days, rallied to her cause. Some, who had weathered the collapse of their own stalls, advised Bud to dispose of her cart and become a mobile trader. Carts were hard to store, difficult to push, and almost impossible to move out of the way of a police raid. With the small sales Bud was now making, it seemed pointless retaining her cart and incurring a charge for storage. Regretfully Bud sold her cart to a neighbour for $US40 and obtained more than enough capital to establish herself as a mobile trader.

With lots of advice and help from her neighbours, Bud set about cooking food that would find a ready demand and that she and Narto could carry. On her first day she slipped and fell, and her basket of freshly cooked food was upturned in mud. It was, to put it mildly, an inauspicious beginning. But Bud and Narto persisted. Rahmini realized that something was seriously wrong, even though she was only six years old at the time. She began to feel insecure and insisted on accompanying Bud as she worked. One day as Bud set down her load on the edge of a little alleyway near Hotel Indonesia, her plates slipped and rolled down an embankment into a murky canal. Bud clambered down to retrieve her plates. Rahmini began to cry hysterically, apparently fearing that Bud would drown.

Bud had been unable to pay Narto for some weeks. Narto realized that it was becoming increasingly hard for Bud even to feed him. Rahmini could carry the pot of tea or extra baskets of food that Bud needed. Sadly he said goodbye and promised he would return to his parents' village near Tegal. A year later Bud heard he had gone back to sleeping on the streets. He was too embarrassed to return home destitute. Bud never learnt if he had found another job, returned home, or perhaps, died destitute in Jakarta. Achmad, her former assistant, was found dead under one of Jakarta's bridges. Bud was told he had become ill and desperately poor, though no one was sure why he died.

Despite her set-backs Bud continued to trade. Without her cart she was certainly more mobile. She went from street to street seeking a place where she could stand with her child and sell her few wares. But wherever she went, other traders told her to move on. Every location, it seemed, was already taken. Many of the other traders were sympathetic to her plight. Some had known Bud in years gone by as 'the successful trader by the fountain'. They suggested other places to try, sending her from one street corner to another. There was said to be a potential market of labourers here, or drivers there. Eventually Bud found a location on a small street to the north of Hotel Indonesia. She found out what to cook and when to sell by asking the mini-car drivers and labourers who worked near by. They warned her that her predecessors had been cleared from the site Bud was proposing to use. She realized she would have to exercise great care. Bud put considerable effort into her cooking and her trade started to develop. She could not afford to relax, however, for the threat of an anti-trader raid was ever present.

Bud did not know any of the police at her new site. She had lost the network of informers who could warn her of an impending raid. The local police still tried to extract bribes, often demanding a packet of the most expensive cigarettes she stocked. There was now nothing more costly in her wares that they could demand. Even the loss of a packet of cigarettes was more than she could afford. She realized that, if she gave in, they would return again and again. Their friendship did not seem worth cultivating, even if she could have afforded it, for she felt the police would not help her when she was in need. Instead, she had to rely on her own cunning. She soon learned that the area around Hotel Indonesia was cleared whenever there was an important conference. All the drivers and traders in the area took a keen interest in any forthcoming conference and speedily passed on any information they acquired. Bud and her fellow traders stayed away whenever an important meeting was scheduled, so they would not offend any foreign dignitary who happened to speed past in his chauffeur-driven Volvo or Mercedes-Benz. If she did encounter the odd policeman, she soon learned to exploit her daughter's presence quite shamelessly. She would explain that she was only trying to support her little girl. Her little daughter, in turn, soon became quite adept at bursting into tears as soon as Bud gave her the appropriate cue. They became an accomplished duo.

Bud earned a greatly reduced income (47 US dollars gross a month)[16] from her food selling and learned to live frugally again. She battled to make some headway and counted every rupiah she spent. It became increasingly and painfully apparent that Santo and his third wife, Ade, were an expense she could no longer sustain. From the beginning Bud was critical of Ade and on several occasions, demanded that she leave Jakarta and return to her village.[17] But life in Ade's village had become impossible. It had been a bad year and there was not enough to eat. There was no money in the village to pay for such luxuries as Santo's radio repairs or Ade's washing of clothes. To make matters worse, Santo was clearly ill. He was coughing up blood and had lost weight. He was in all probability dying of tuberculosis, though of course he had not seen a doctor. Even if he had found the money for a consultation, he would not have been able to pay for any medicines that might be prescribed. Ade had given birth to a second child, but she realized she could not afford to keep it. Santo and Ade begged Bud to adopt their child, but Bud sadly explained that she was in no position to do so. The baby died, presumably of malnutrition, less than 12 months later.

Bud felt trapped. She understood why Santo and Ade were spending more and more time with her. When she had earned well, she had welcomed Santo's visits. Now it was a question of her survival. But how could she convince her husband and his third wife that she could no longer support them? Bud decided that her only course was to cease trading and dispose of all her possessions. Santo and Ade would then realize that Bud's position was as desperate as theirs.

In June 1979 Bud stopped trading. She progressively sold off her possessions, and her household lived off the proceeds. By word of mouth it became known through kampung brokers that she had items to sell. Offers were made through intermediaries for her wardrobes, sideboard, cooking utensils and, eventually, even her television set, with which she was most reluctant to part. Santo, of course, demanded a share. Behind her back he tried to sell some of her possessions and pocket the money.

Bud's household economies were far-reaching. She rented out half her house, severed her electricity supply, and reverted to the use of kerosene. To save on fuel she rarely cooked and used only one lamp, so the room was dimly lit. Her house fell into disrepair. The well gave forth just a trickle of murky water. Its pipes had rusted, but Bud could not afford to replace them. Bud no longer asked the vendor or a servant to bring drinking- and cooking-water to her house. She collected it herself from a kilometre away where, if she queued, she could get it free of charge. All these measures enabled Bud to survive without working for about six months. The household became spartan by any standards, let alone those of her so recent past. Gradually Bud's strategy began to work, and Santo and Ade's visits became a rarity.

Less than four years earlier Bud had been one of the richest people in her community; now she was one of the poorest. Her neighbours felt pity for her. One regularly brought Bud's mother and daughter porridge from the hospital where she worked. Another provided them with the flavoured iced-water Bud so loved to drink. A third delivered rice and tit-bits of food left over from her cooked-food trade. Another invited Bud, Rahmini, and the grandmother to watch television. A spiced-vegetable trader paid Bud's mother a small sum for baby-sitting, whilst she went out to trade. When she had money, Bud had alienated many of her neighbours by her individualism, consumerism, and snobbery. But now, in her difficulties, she sought comfort and support from those around her and they responded to her needs. But Bud found it hard to admit to others just how far her fortunes had declined. Once, as she was struggling to carry her household's daily water supply, a kampung dweller, who had known Bud in her better days, asked what had become of her servant. Bud replied she had temporarily given her leave to return to her family in the village.

Like Bud's neighbours, I found it difficult to be critical of her when her fortunes were down. When she was doing well, she was sometimes strong, arrogant, and mean. She made use of the people around her, for her own benefit. Even if her personality basically remained the same, intense suffering had mellowed her and brought out the softer, gentler sides to her character. She was more introspective, more sensitive to other people's hardships and sorrows. As she was no longer busy trading 20 hours a day, she had more time to sit and talk to her neighbours and find out about their problems.

Her months away from trading and the constant fear of a raid started Bud thinking. If she took up trading again she could, at most, look forward to a modest income, and that only if everything went well. She could earn as much if she had a regular job, but she would then be spared the uncertainties and insecurity of her past existence. Bud asked me to help her find a job.

Even though she was not a good cook and had never worked in a Western household, I was able to find her a job amongst my expatriate friends. At first she was overwhelmed by the prosperity of the household where she worked. She could not get over the spaciousness of the house or the size of its garden. Her entire kampung neighbourhood of several hundred people would have fitted inside it. Yet only six people lived there. Bud was amazed by the enormous refrigerator and the pantry stocked with tins of imported food for months in advance. Like other kampung dwellers Bud had always bought just enough food to last for the day. Despite her urban background, she had never used a telephone or seen a shower, a Western-style toilet, or a swimming-pool. The unlimited supply of hot and cold running water in the three bathrooms and kitchen was a novelty. The automatically opening car-port door was a source of astonishment. Her employer's noisy teenage children gulped down bottles of milk and gorged themselves on uncounted mouthfuls of chocolate cake. Yet, she observed, when a little sugar or spice was missing from the pantry there was inevitably an uproar. More than anything else, Bud was amazed to see the family's children shout at their parents. The family were wealthy and secure, but Bud noted that did not seem to make them any happier or more contented with their lot.

Though Bud had at last found a measure of security, at least for as long as her employer remained in Jakarta, she felt imprisoned behind the high walls of the expatriate compound. She missed the ceaseless activity of her kampung's crowded lanes, the familiar faces, the gossip, the smell of kampung food and, above all, her family. Bud worried about leaving her aged mother and young daughter to fend for themselves. Her job was a residential one. She was given one day off a week and that was her only opportunity to see her family. Bud's mother was in her late sixties and had become accident-prone. She had fallen down the steep rickety ladder in their two-storey home and broken an arm. On another occasion she had spilt boiling cooking-oil over her leg. Rahmini, now eight years old, was left to care for her grandmother. When Bud had first left home to work as a cook, Rahmini had cried bitterly. Without telling anyone, she fled alone to the third wife's village to find her father. Then she played truant from school and was rumoured to be roaming the streets. Her grandmother could not control her. Bud was concerned that Rahmini could not yet read or write. Once a week when Bud returned home, she tried to tutor her daughter but her efforts were in vain. Bud worried about Rahmini's future. She feared it would be no better than her own. Whilst her trade had prospered, she had dreamed of her daughter becoming a doctor. Now her aspirations were much more modest: 'If only Rahmini could learn to type or sew'. Bud felt sadly that Rahmini was unlikely to get as good an education as she herself had had 40 years earlier under the colonial administration. Bud would not be able to rely on Rahmini to support her when she was old.

As a domestic servant in 1979, Bud was paid in cash a third of what she had earned as a food-trader in 1972.[18] It was just enough to feed herself, her daughter, and her mother. She did obtain medical care, clothing, accommodation, and some food for herself, and occasional presents for her daughter and mother, but she was cut off from her family and her roots, and forced to live in an alien and lonely world. She longed for the interaction with people, her freedom and the independence of petty trade. The job as a cook had its attractions, however. It was less arduous than trading. Bud was nearly 50 years old and she had to think of her future. Her new position offered her a sense of security. She would have a job so long as she did her work properly. She was no longer at the mercy of soldiers and the police, of rising prices, and a diminishing clientele. Despite the drawbacks of her new job, the lack of freedom, and the realization that it offered little scope for material improvement, life as a domestic gave her economic security. For Bud, that was worth a lot.

Bud's peace of mind did not last long. In 1981 the government announced that her neighbourhood was to be demolished. The government was proposing to replace the houses with flats, which would be made available to the inhabitants of the kampung. Was this the same government that had banned their *becaks* and outlawed their petty trade? The kampung was full of households like Bud's who had lost their livelihoods as a direct result of government policy. Was the government now going to help them? There were families in the kampung who had been the subject of earlier government demolition programmes. Bud herself had once been forced to move as a result. Everyone knew that the government forcibly evicted kampung dwellers and either claimed that the kampung was illegally sited—and thus not entitled to compensation—or made such paltry payments that they scarcely covered the costs of transporting one's belongings to a new site, let alone buying another home. Fear and rumour swept the kampung. For Bud, now isolated from her community, the news was especially alarming. Once a week, on her day off, she made her way to her home, wondering each time what she would find. Would her house still be standing? How could her infirm mother or her nine-year-old daughter stand up for her when officials came? But the officials seldom came, and the neighbourhood had little firm information to go on. Each time Bud returned there was new speculation and new fears. The government held a public meeting at which it outlined its proposals. But the various officials contradicted one another. The brochures they distributed did not seem to agree with what they had said. The confusion and the apprehension increased.

The government proposals were in fact confused. The government announced its intentions before it had finalized its plans or properly defined its aims. The government planned to rehouse Bud's neighbourhood in subsidized flats. The earliest announcements, however, mentioned prices for the proposed flats that were beyond the financial capacity of all but a handful of the inhabitants. Moreover, many kampung dwellers used their houses to store and prepare materials for their livelihood, just as Bud had prepared and cooked food in her home whilst she ran her stall. But the government was going to prohibit its flats being used for commercial purposes. To its credit, the government did substantially modify its proposals to try to accommodate the objections of the kampung dwellers. It offered generous compensation and alternatives for those who did not want to accept a flat. Unfortunately, it never overcame the suspicion of the community. People observed that the proposals continued to change no matter what was announced and believed that the junior officials with whom they had to negotiate were enriching themselves at the kampung dwellers' expense.

I encouraged Bud to take a government flat. It seemed too good an opportunity to miss. The flats would be located in the central city area where Bud's house now stood. Those who lived in the neighbourhood could buy them at a subsidized rate. Repayments on the units could be by monthly instalments at very low interest rates over a period of 20 years. Each unit would have running water, electricity, and gas. When Bud's employer departed for overseas, she would readily find domestic work within walking distance of her home in the élite suburbs of Menteng or Kebayoran. If she carefully placed the compensation she received for her home in a bank, she could pay off the flat with the interest she earned.

Bud did not see things my way. She had never entered a bank, let alone used one. They and the officials working within them were impersonal and intimidating. Most importantly, interest payments were against the teachings of Islam. According to Bud, the usurer would be strangled by snakes in hell. Bud longed to use the compensation for her house to buy the consumer items—radio, television, and sideboard—she had been forced to sell. Rahmini wanted a bicycle like the one promised to the children next door upon their parents' receipt of compensation. Bud's mother want a gold ring like the one Bud had borrowed and pawned when her trade was going badly. Santo insisted on his rightful share, saying that he planned to better equip his radio repair business in the village. Bud had long-standing debts which needed repayment. For example she still owed $US76 on the renovations made to her home in 1974.

Bud agonized over her options. How would she pay for the monthly charges for water, gas, and electricity? Previously she had paid for kerosene and water daily as the need arose and she had money to spare. When times were bad, she simply cut down on the quantities she used, or accepted these items on credit until she could pay. But the government was not offering such flexibility. Failure to pay on time would entail a fine. Minimum charges would be levied even if no water, electricity, or gas were used. She wondered how her flat instalments would be paid if she became too old or too ill. Would she be thrown out of her unit? Would the government carry out repairs free of charge as it promised or would Bud be forced to pay for them? If Bud and her household were assigned units on the fourth floor how would her rheumatic mother, or even Bud when she was older, clamber up so many flights of stairs?

But Bud had made her choice. She overcame her Islamic scruples about interest payments by making a donation to an orphanage. With my help and encouragement she banked most of her compensation payment. The interest on this was almost enough to cover her repayments on the flat. Some of her neighbours were not so wise and embarked on an orgy of consumption. They bought television sets, motor cycles, and clothes. Vast fortunes were spent in a few days. In other instances, men ran off with the money, leaving their wives and children penniless.

After a long and difficult transition period in appalling temporary accommodation, in April 1984 Bud, her mother, and Rahmini moved into a government flat. Although Bud had asked for a ground or first-floor unit, she was allocated a flat on the fourth floor. Yet the household was so relieved to move out of the temporary accommodation that not even Bud's elderly mother complained of the four flights of stairs to the flat. As Bud had feared, the bills for electricity, gas, and water soon began to prove a burden. Even though the flats were subsidized, the mortgage repayments, and the charges for services, represented a threefold increase over Bud's expenses prior to demolition. They consumed half of her income. On her modest and fixed wage it was going to be hard enough meeting her monthly commitments. What would happen if she had an unexpected medical bill or if Rahmini needed a new school uniform? Bud could only pray that she would manage somehow and that she, Rahmini, and Bud's mother would not have to move again.

Santo never saw the new flat. Bud received word of his death just as her old home was demolished. It took her two years to save enough money for a trip back to the village to see Santo's grave and a further two years before she could pay for his funeral rites.

Bud's mother died less than a year after the family moved to the flat. The loss affected Bud badly. She felt intensely sad and isolated. Now she and Rahmini were alone in the world. Her mother had stood by her when Bud was in great difficulty. She could have gone to stay with her other daughters who were all comfortably off but she chose to stay by Bud who, she felt, needed her moral support through marital difficulties and financial hardship. She had remained when the household moved to temporary accommodation whilst they waited for over two years for their flat. The three of them had to squash together in a cubicle eight metres square. The overcrowding, lack of amenities, and dust must have been a nightmare.

After her mother's death, Bud left her job to look after Rahmini. She felt she could not leave Rahmini unsupervised throughout the week whilst she lived and worked elsewhere. Rahmini was now twelve years old and more frequently on the streets than at school. She was terrified of staying in the flat alone and kept running away to neighbouring houses.

Instead of earning an income from domestic work, Bud felt the two of them could live off some of the money in the bank. But Bud gradually became more and more depressed. She was now spending all day in her flat. Most of her neighbours had managed to cover the bare concrete walls and floors of their new accommodation with colourful tiles and paint. They had bought new furnishings, curtains, and blinds. Bud felt embarrassed. Only she and Rahmini lived in what looked very much like a gloomy, empty prison cell.

Bud decided to withdraw all her compensation money and renovate her house. She had the walls plastered and tastefully painted in creamy white. The floors in the kitchen, bathroom, and guest-room were tiled. I was aghast, wondering how she would survive. Bud was no longer working, yet she had spent all the money I had persuaded her to bank on renovating her flat. How would she pay for her gas, electricity, water, and the flat instalments? How would she feed herself and her daughter? I deeply regretted that I had ever meddled in her life. How could I, a secure middle-class Australian, hope to understand the values, the dreams, the pressures, and especially the unpredictability of the life of a Jakartan kampung dweller?

But as my doubts grew, Bud became more convinced that she had made the right choice. She was delighted with her renovated flat and, from the beginning, had appreciated the convenience of running water, drainage, and electricity. She explained she was now able to cover all her overhead costs by subletting a room. She argued that she would never have been able to attract a sufficiently high rent without renovating. Subletting was illegal but Bud—and her many neighbours who were making ends meet in the same way—simply registered her tenant as a member of her household.

Bud may have moved to better housing but she had paid a big price. She had owned her former house and had only a relatively small amount to pay back for the renovations she had undertaken. Moreover she could, and did, delay her repayments whenever she encountered difficulties. Her house had been comfortable and convenient, even if illegal. She had not been subject to any rules and regulations. She could take in tenants as she pleased.

Life in the flats was altogether different. Although the flats were legal, bureaucrats, whom the flat-dwellers never saw, made rules which governed every aspect of their lives. Flat-dwellers could never be sure when inspectors would come around checking that they were not trading from their units, did not have tenants, nor had decorated their flats in the wrong way. They could never be sure that the flat instalments or other costs would not be raised. They could never be sure they would not be thrown out of their units if they failed to pay because of loss of work or illness.

Like most other kampung dwellers, Bud's life had had its share of drama. Like them, she had made her way to the city in the 1950s in the hope of a better future. Like them, she had tried numerous jobs but found petty trade most lucrative. In the mid-1970s her material conditions improved beyond her wildest dreams. Then, equally suddenly, things began to change, and the work, renovated home, and friends she had acquired were lost. Many of her neighbours suffered the same fate as they struggled to escape from the hazards of working on the streets and tried to obtain salaried jobs and legal accommodation.

Bud was perhaps luckier than most. She had an expatriate friend who could help her find alternative employment, advise her on government policies, and occasionally give her money. She would not have been able to find such a well-paid domestic job had I not introduced her to a wealthy expatriate household. She would have fled from the government flat-building programme had I not prevented her from doing so. She would not have put the large amount of compensation she received from the government for her kampung house in the bank, had I not shown her how.

But by intervening in Bud's affairs, I may have distorted her priorities.

She knew, better than I, how to survive in Jakarta. She had survived all those turbulent years before I came along. Her ability to adapt to rapidly changing circumstances had enabled her to tackle each new difficulty as it came along. She was not naïve to the unpredictabilities of kampung life and hazards of entrusting one's fate to government.

By following my advice, she had gained far better accommodation (despite the lack of variety and greenery) than she could ever have hoped for. But she had also acquired commitments which diminished her ability to respond to her changing fortunes. She could no longer reduce her costs to meet her earnings. She had to ensure her income covered her expenses not just for the present but for 20 years in advance.

Bud's life therefore remains precarious. Will she be able to remain in her flat? What if she is forbidden to have a tenant or if the instalment costs rise? Will she, like the other kampung dwellers who accepted flats, be forced to sell her unit surreptitiously to the wealthier middle-class on the black market? Or will she be forced to sell it back to the government at well below the market price? Would she have been wiser to move out of the project area like the majority in her kampung? Or would all of the kampung dwellers have been better off without the project that was meant to improve their lives?

Only the future can tell. Bud will continue to live from one crisis to the next. I will continue to anguish over my intervention in her life; to help her whenever I can and feel guilty that I do so little; to admire the way she accepts and struggles against the difficulties that confront her and to be unendingly grateful that her life is not mine.

Notes

1. See Jakarta City Government, *Gita Jaya* (Jakarta, 1977), 97; Planned Community Development Ltd, *Kampung Improvement Programme Jakarta, Indonesia* (Washington: IBRD and IDA, 1973), Annex 1, 5; J. Heeren, 'The Urbanization of Djakarta', in *Ekonomi Dan Keuangan Indonesia* (Jakarta: University of Indonesia 1955), Vol. 8 (11), 700.
2. Some details of Bud's life changed as I came to know her better. Earlier I had understood her father was a policeman. See L. Jellinek, 'The Life of a Jakarta Street Trader', in J. Abu Lughod and R. Hay jun. (eds.), *Third World Urbanization* (Chicago: Maaroufa Press, 1977), 252. Later, she told me he was a clerk in the railways. Bud's mother had a number of husbands and it is possible that one worked as a policeman, another as a clerk.
3. There is a discrepancy in my data with regard to Bud's first trading experiences. For some years I thought her initial trading experiences were in Jakarta (*Third World Urbanization*, 252), but later I learned that she had helped her mother trade in Tegal during the Japanese occupation.

4. For more details on Bud's house and furnishings and how I came to be invited to her home see *Third World Urbanization*, 244–7.
5. Kampung here means 'urban village'.
6. For Nanti's background and role, and Bud's treatment of her see *Third World Urbanization*, 253–4.
7. For a more detailed account of the old and new house see L. Jellinek 'The Life of a Jakarta Street Trader—Two Years later', Working Paper 13 (Centre of South-east Asian Studies, Monash University, 1976), 2–3.
8. Initially, I thought Bud only had one baby ('Life of a Jakarta Street Trader', 4), but later she told me she had had two, both of whom died.
9. For evidence of other kampungs stifled by the expanding modern metropolis see G. H. Krausse 'The Kampungs of Jakarta, Indonesia: A Study of Spatial Patterns in Urban Poverty', Ph.D. dissertation, (University of Pittsburgh, 1975), 57–8.
10. See G. J. Hugo, 'Population Mobility in West Java, Indonesia', Ph.D. dissertation (Australian National University, 1975), 521–2; G. F. Papanek 'The Poor of Jakarta', in *Economic Development and Cultural Change*, 24 (1) (1975), 9–12.
11. See Hugo, 'Population Mobility', 521; Papanek 'The Poor of Jakarta', 9–12; R. Critchfield 'Desperation grows in a Jakarta Slum', in *Christian Science Monitor*, 12 Sept. 1973, 9.
12. See T. G. McGee, and Y. M. Yeung, *Hawkers in South-east Asian Cities: Planning for the Bazaar Economy* (Ottawa: IDRC, 1977), 50–52; *Laporan Hasil Survey Profile Pedagang Kaki Lima Di DKI Jaya* (Jakarta: University of Indonesia, 1976), 28; *Jakarta Post*, 21 Dec. 1983, 2; *Jakarta Post*, Jan. 1984, 2; Praginanto 'Pak Parto Pengusaha Warung Tegal' *Galang*, Lembaga Studi Pembangunan, Jakarta, 1 (1), (1983) 44–7.
13. For a more detailed description of the relationship between Nanti and Achmad see '*Life of a Jakarta Street Trader*', 7.
14. US dollars are used throughout even though all transactions amongst kampung dwellers in Jakarta are performed in Indonesian rupiah. Though the US dollar also suffered inflation, this was much less than the Indonesian rupiah and makes comparison of prices and incomes over time easier. Furthermore, most readers will be more familiar with the rate of inflation in US dollars than Indonesian rupiah. The exchange rate was:

1971	Rp 420 = $US1.00
Dec. 1978	Rp 632 = $US1.00
Dec. 1981	Rp 655 = $US1.00
Jan. 1983	Rp 701 = $US1.00
Dec. 1983	Rp 998 = $US1.00

15. For Bud's remarkable managerial skills see *Third World Urbanization*, 249–51, 255–6 and 'Life of a Jakarta Street Trader', 11–12.
16. It is extremely difficult to calculate the earnings of food-traders. They seldom distinguish between their business and their domestic budgets. Their sales generate an income which, as mentioned previously, has to provide for their purchases for the next day's sales. But it must also cover any costs that arise

from their trade (such as a tax or the repair of a cart) as well as any household expenses (such as children's schooling or the repayment of a loan). Traders appreciate that the more they spend on themselves, the less stock they will be able to buy and the less they will earn. The household's food is usually obtained from the stall, frequently from its left-overs, or by barter from other food-traders.

17. For more details on Ade's relationship to Bud see Jellinek 'Life of a Jakarta Street Trader', 4–5.

18. I estimate that Bud was earning Rp120,000 or $US285 a month net at the peak of her trade in 1972. Seven to eight years later Bud was earning Rp40,000 or $US63.

VI

Social Organization in the City

INTRODUCTION

THE city presents problems of integration and social control. The individual or family recently arrived from a village, from another town, or from another part of the city may be surrounded by unfamiliar faces in the neighbourhood and at work. In a crisis there may be nobody to turn to for help. At the same time anonymity or superficial acquaintance limit the control that can be exerted collectively as well as in transactions among individuals. Even parents have little control over what their older children do much of the time.

Actually, most urban dwellers are well integrated. For some rural–urban migrants, the village continues to provide social integration as well as economic security, as we have seen in Part III. Most migrants move to a location where they expect to be received by relatives or friends. Some find housing in the same neighbourhood, or a job with the same employer. But even when not so clustered, people of common origin frequently maintain close ties. Most urban dwellers, whether migrant or urban-born, are quite well integrated, whether in long-established neighbourhoods, at the work-place, in religious groups, or in far-flung networks of kin and friends.

Mary Hollnsteiner-Racelis, in Chapter 13, critically addresses the view that the city is characterized by social disorganization, a view that has a long tradition and continues to enjoy considerable popularity. The poor Manila neighbourhood she describes is characterized by close interaction and mutual aid, community-wide activities, and effective social control. Many inhabitants have lived in the city for generations, but in neighbour-hoods such as this they have established 'islands in the city'. Hollnsteiner suggests that middle-class households, in contrast, are little involved in their neighbourhoods, that their members are integrated instead in cross-city networks of relatives, office-mates, or co-members of clubs.

Mutal aid among the urban poor is the focus of Larissa Lomnitz's study of Cerrada del Condor (Chapter 14) reported more fully in her monograph (Lomnitz 1975). She argues that the urban poor in Latin America find their ultimate source of livelihood in market exchange, but cannot survive individually: the market fails to provide any security and the poor are not in a position to accumulate savings. They survive by complementing market exchange with a system based on resources of kinship and friendship, which follows the rules of reciprocity, a mode of exchange among equals, embedded in a fabric of continuing social relationships. In Cerrada del Condor, the patterns of individualism and mistrust prevalent in rural Mexico are superseded by powerful tendencies toward integration, mutual assistance, and co-operation. Recent arrivals are housed, sheltered, and fed by their relatives in the shanty town; the men are taught a trade and oriented towards available urban jobs, in direct competition with their city kin. The migrants thus become integrated into local networks of reciprocity.

Crime is generally seen as the curse of the city. And, indeed, there is good evidence, whatever the shortcomings of crime statistics, that property as well as victimless crime is more common in urban than in rural areas. There is no evidence, however, to support the assumption that homicide is more prevalent in urban than in rural areas in the Third World. The most comprehensive data available, comparing the national homicide rate and the rate for a major city in 11 countries, fail to indicate any consistent pattern (Archer and Gartner 1984, 105–7).

Various explanations can be advanced to explain high urban crime rates. Against ecological determinism it can be argued that deviants subject to the sanctions of the rural community are likely to be disproportionately represented among rural–urban migrants. While rural dwellers are usually assured of subsistence, some urban destitute have no alternative but to steal or rob for survival. The glaring contrast between rich and poor in the city exacerbates relative deprivation. Theft and burglary are facilitated in anonymous urban settings where the stranger goes unnoticed. The city's underworld is sufficiently large for a division of labour among professionals to be established, indeed for organized crime. The greater number of customers brings forth a wide variety of illegal services to cater to their wishes, whether they want sex, drugs, or gambling. Finally, social control is less effective in the city. While most urban dwellers are well integrated, some, especially young adults, are quite footloose, accountable to none, ready to try their hand at theft or burglary, and prepared to cash in on what has been outlawed as vice.

Martin King Whyte, in Chapter 15, describes an all-pervasive system of social control in urban China. It is based on formal control mechanisms at both the neighbourhood level and the work-place and reinforced by a formidable rehabilitation system. Notable are several features of Chinese cities which enhance control: neighbours and co-workers know one another well since there is little turnover; rural–urban migrants are selected through migration control; the regime has effectively dealt with the destitution that characterized China in the past; and the differences in life-style between the élite and the masses are much less in evidence than elsewhere in the Third World.

The all-pervasive system of social control is a distinctive feature of the Chinese city. A Chinese model of urbanism, especially as it was developed during the Cultural Revolution, can be constructed that includes a number of additional features. Whyte and Parish (1984, 358) list the following:

strict migration controls and minimal urbanization in spite of considerable economic development (as we have seen in Blecher's account in Chapter 7); a highly developed bureaucratic allocation system; an emphasis on production rather than consumption; a rejection of schools as the basic mechanism for sorting talent; much stress on citizen involvement in public health, social control, and other realms; and rigid taboos on all forms of dress, expression, ritual life, and communication that do not conform to the official ideology. They report that these structural changes in urban institutions contributed to a number of social consequences that also seem quite distinctive: high stability in jobs and residences; involvement and familiarity with neighbours and workmates; minimal differentiation of consumption patterns and life-styles; low divorce; high female work-participation; and rapid changes in fertility and religious customs. Whyte and Parish (1984, 368) observe how difficult it proved to be to change some urban characteristics, such as bureaucratism and the urban prestige hierarchy. They further emphasize that when aspects of social organization were effectively changed, the results were not always as expected: the pursuit of equality through class struggle and an emphasis on class labels alienated many and failed to eliminate awareness of occupational rank and privilege; the pursuit of comradely relations produced more interpersonal knowledge and concern but also a certain interpersonal wariness and weak attachment to middle-level community and work organizations. And several reforms produced contradictory results: the pursuit of equality through class struggle, for example, destroyed much of the former unity of purpose of the goal of national strength and growth. Since the death of Mao Zedong in 1976 there have been major reversals on policies affecting urban social organization, changes·expressly directed to deal with some of these unexpected consequences. The outcome of the policies of the Cultural Revolution, of the subsequent policy reversals, and of future policy changes, provide opportunities to analyse separately the effects of three variables affecting urban social organization: the demographic characteristics of the city—to use Wirth's (1938, 1) classic definition—the fact that the city is a relatively large, dense, and permanent settlement of heterogeneous individuals; the cultural patterns of a society; and a nation's political economy and level of economic development.

References

Archer, Dane, and Gartner, Rosemary (1984) *Violence and Crime in Cross-National Perspective* (New Haven and London: Yale University Press).
Lomnitz, Larissa Adler de (1975) *Como sobreviven los marginados* (Mexico City, Madrid and Buenos Aires: Siglo Veintiuno Editores). Rev. Eng. version,

Networks and Marginality: Life in a Mexican Shantytown (New York and London: Academic Press, 1977).

Whyte, Martin King, and Parish, William L. (1984) *Urban Life in Contemporary China* (Chicago and London: University of Chicago Press).

Wirth, Louis (1938) 'Urbanism as a Way of Life', *American Journal of Sociology*, 44, 1–24.

13

Becoming an Urbanite: The Neighbourhood as a Learning Environment*

Mary Hollnsteiner-Racelis

GENERATIONS of anti-urban writers have portrayed the city as impersonal, by nature evil, and naturally conducive to social disorganization. Its indictment follows directly, they say, from the high frequency of serious crimes and the proliferation of narcotics rings and prostitution dens. They blame the city for the mental strains allegedly induced by the extremes the individual must face. The urbanite is characterized as lonely in the midst of crowds, a man forced to accept both the rapid pace of city life and the slow rate at which urban institutions respond to his needs. In contrast, the countryside retains the romantic glow of a lost Eden.

Convincingly though these prophets of gloom have argued their case, apparently millions of city dwellers have either not heard or not cared about their pronouncements. Migrants flock to cities never to return to the countryside. In the process they commit their offspring to permanent urban dweller status.

The stranger wandering into a low-income neighbourhood for the first time will be likely to conclude that here are the folk in the city. A group of men drink together at a corner store, while some women squat at the public faucet, gossiping as they scrub the last pile of laundry for the day. Children shriek and play in the streets, setting stray dogs, cats, and chickens bolting in all directions. How like the village, the observer thinks, until he discovers that many of his 'villagers' have in fact lived in the city for generations. Further investigation will show him that while external manifestations of social relationships bear strong rural qualities, they in

* Reprinted, in a revised form, from D. J. Dwyer (ed.), *The City as a Centre of Change in Asia* (Hong Kong: Hong Kong University Press, 1972), by permission of the publishers. The research on which this article was based was done mainly in 1965 through a grant from Procter and Gamble Philippine Manufacturing Company to the Institute of Philippine Culture, Ateneo de Manila University. The data were derived from participant-observation in the Velasquez Street section of Tondo, Manila, and from structured interviews of 52 key informants. Twenty-four of them resided in the neighbourhood and represented a number of statuses, for example, those of storekeeper, factory-worker, policeman, and political or religious block leader. The other 28 were district-level leaders in charge of government, business, and civic organizations involved in Tondo life.

fact combine with less noticeable aspects of the city milieu to form a distinctive urban, lower-class life-style.

This study attempts to describe a lower-class neighbourhood along Velasquez Street in Tondo, Manila. In many ways it resembles its counterparts all over the world; in others, it reveals a distinctive cultural stamp. However, an attempt to seperate the two aspects, the universal and the particular, would be a fruitless exercise. For the Tondo people living that life have integrated the various components into a meaningful total life-stream.

Life in Off-Velasquez, Tondo

Manila's Tondo lies immediately north of the downtown core and, in the mind of the average Filipino, denotes slums, squatters, gang fights, and crime. The resident, however, will usually defend his particular Tondo neighbourhood, blaming some other part of the district for Tondo's unsavoury reputation. Some other objective measures may give further clues to the area's characteristics.

Tondo in 1960 had 351,949 people, or 30.9 per cent of the City of Manila's population (as distinct from the population of the greater area comprising metropolitan Manila) living in households averaging 6.5 persons. Like the nation as a whole, its residents are young, the median age being in the 15–19 year-old range. With 89 per cent literacy for those over ten, it is hardly surprising that this most populous of Manila's administrative districts boasted an educational attainment ranging from high school to college graduate for 25 per cent of the population of ten years and over. The comparable figures for the City of Manila and the Philippines as a whole are 36 and 10 per cent, respectively.

Tondo's intense involvement in Manila life was already indicated in sixteenth-century accounts of Legaspi's conquest. His men soon organized the Tondo parish in 1574 on the site of Raja Matanda's former trading settlement. This ecclesiastical unit, according to the current pastor, has the dubious distinction of being the most heavily populated Roman Catholic parish in the world, with only three harried priests attempting to minister regularly to its many-faceted needs. Serving as the debarkation point of inter-island ships, the railroad line, and almost all provincial buses, Tondo has absorbed large numbers of Manila-bound migrants. In the 1948–60 intercensal period, its population increased by 24.2 per cent, reaching the highest density level in the Philippines, estimated at 41,000 persons per square kilometre: the national average is 101, while Metropolitan Manila's comes to some 6,260 per square kilometre.[1]

Although residents recognize their status as Tondo dwellers, their sense of spatial identity corresponds with the particular street in which they live. It is in this neighbourhood context that a *tagatundo* becomes part of his lower-class surroundings, finding security as well as occasional annoyance at the constant press of close neighbourhood ties. At the same time he does learn to cope with the larger urban setting. A select minority even master this urban environment, utilizing the neighbourhood as a springboard for achieving the upward social mobility coveted by so many.

The Sense of Community

The community notion within neighbourhoods leading into Velasquez Street stems somewhat from the physical layout of the average block, but even more from the social characteristics that encourage a consciousness of kind in that setting. Every side-street harbours a multitude of people living in high-density conditions. The small, crowded dwellings permit little circulation of air, forcing their residents to seek relief in shady spots outside. Through traffic goes along Velasquez but only infrequently into its cross-streets. These side-streets thus double as play and meeting areas. Although an occasional resident possesses a Jeep, pedicab, or *calesa* (usually because it is his source of income), the general reliance on public transportation automatically provides daily opportunities for interaction with neighbours riding the same jeepney to and from work. Children meet their neighbourhood playmates at the free public school, and frequently provide their parents with the opportunity of getting acquainted. Corner stores supplemented by mobile pedlars of meat, fish, vegetables, and sweets cater to the small-scale, frequent buying that fosters further local contact. The native pool table, the cheap *turo-turo* restaurant, or counter-cafeteria, the refreshment stand, and the public water faucet all attract a small, constantly changing clientele of street dwellers throughout the day and night.

Dramatizing the community mode is the common life-style of off-Velasquez residents. It has had many decades to develop into a heritage unobtrusively passed on to newcomers by long-time residents. The newcomer rarely finds himself a stranger in the area, for he probably moved there at the insistence of a nearby relative or at the suggestion of a friend. Being poor forces a closeness beyond mere sociability, for crises arise frequently enough to encourage strong patterns of neighbouring. Mutual aid consists largely of contributions of food, money, or service upon the death of a household member or the happier celebration of a baptism or marriage. It surfaces again in the borrowing and lending of household items and money, maintaining surveillance over a neighbour's house or children while the mother runs an errand, notifying one another of job openings, and (particularly for adolescents) support in the event of a gang fight with rivals from other blocks.

Poverty also encourages family members to remain in the neighbourhood during their leisure hours, for there is rarely money enough to see a movie or journey to the few free recreational spots the city provides. As a consequence, large numbers of residents, but especially men and children, hang around the streets, finding their entertainment in small drinking and play groups. The Happy Birthday Club, for example, meets every evening at the corner store to celebrate a member's birthday—actual or contrived. The honoured man is greeted with shouts of congratulations and then has to treat his well-wishers to a round of drinks. Any out-of-the-ordinary occurrence elicits intense interest in the street, and scores of onlookers collect almost instantly. They crowd around the stage when a local association sponsors a dance complete with stage entertainment, and are attentive at a respectful or fearful distance when a gang fight takes over the street.

Community Events

The ecology of poverty thus enforces frequent contact and a mutual identification among neighbours, fostered by a common life-style. The day-to-day informal interaction of local residents receives further support in community-wide events, the most traditional among them being the annual fiesta. Held on the second or third Sunday of January the Santo Niño celebration draws thousands of metropolitan visitors to *handaan*, or specially prepared foods, waiting for them in virtually every household.[2] In order not to discriminate against either the northern, mainly residential sector of the parish, or the heavily commercial portion in the south, the procession devotes one evening to each section. It wends its way through main streets enlivened by fireworks displays and flanked by rows of spectators holding flickering candles. Elaborate tissue-paper bells strung up over the procession's route shower rose petals down upon the image of the Infant Jesus, the Virgin Mary, and favourite saints, as their *carrozas* roll by. The procession touches only the southern end of Velasquez, but comes close enough to the area to link it to the parish by a somewhat abstract bond.

On the various off-Velasquez side-streets, energetic youth groups have placed arches over the entry points, painting on them the names of their members and patrons. In the morning the children participate in the group games managed by the youth clubs, from traditional ones like climbing a greased pole to claim the cash prize at the top, to American-style dunking for apples in a water-filled wash-tub. These activities require weeks of preparation, especially for the adolescent clubs that decorate the street with paper streamers. Partly financing these activities are the 20-centavo weekly payments turned over during the year to the fiesta fund collector. Since many households default in their contributions, and because stage entertainment programmed for fiesta week costs a good deal, donations from outside sources are also sought. The quality of the performers hired reflects the residents' fund-raising skills, for the larger the contributions solicited the more prominent the movie stars or starlets who appear. Their attendance also testifies to the organizers' influence on artists in the entertainment world.

Another community-wide activity that elicits intense local interest emerges in the biennial municipal and national elections. While ward leaders never really stop their year-round campaigning, the tempo accelerates and neighbourhood excitement over prospective winners intensifies in the few months before elections. Being on the victorious party's side is crucial to many, for the politician represents almost their only access to power. Choosing the right candidate enhances the likelihood of one's receiving return favours later on, from getting a wayward son out of jail to acquiring a new set of basketball uniforms for the neighbourhood team. Unlike squatter neighbourhoods, where a solid vote in support of a sympathetic politician guarantees extended illegal occupancy, lower-income neighbourhoods can afford to split their vote in terms of party loyalties. Their positions are less vulnerable than those of the squatter. In any case, whichever party wins, the process of campaigning fosters neighbourhood interaction.

A third community event is the summer excursion by chartered bus to favourite vacation spots in nearby provinces, like Antipolo, Tagaytay or Los Baños. Excited neighbourhood families able to afford the sponsoring club's tickets board the gaily decorated buses, banners proclaiming their destination and the club's name and location. Their few hours in the countryside end all too soon, but not before the links of friendship have been further enhanced.

Social Control

The physical arrangements and social relationships in off-Velasquez streets give rise to strong patterns of personalized interaction. These in turn encourage a degree of social control virtually unknown in more affluent neighbourhoods. What the neighbours think matters in a very personal sort of way; the deviant from local norms soon finds himself an outcast, the subject of gossip or the butt of jokes and group mischief.

One resident explained why she would move away to suburban Quezon City, given the opportunity:

There people go their own way and don't meddle in their neighbourhood affairs. They don't bother you by coming to ask favours when someone in their family is sick. There are fewer community activities that one has to get involved in, such as being forced by the block rosary to prepare refreshments for the third-day procession. People in Quezon City can be independent and isolate themselves from their neighbours.

Another suburbanite related the rigours of Tondo living:

Tondo is noisy, with adults and children talking aloud or shouting, and using indecent language. The mothers do not bother to correct their children when they are wrong, or clean them up. The environment is unsanitary. Garbage is not dumped into the empty can but around it. Even if you try to keep your home and immediate surroundings clean, bugs continue to crawl under the door. Put up a nice garbage can and next thing you know it is gone. Sooner or later a person just gives up trying to live differently from his neighbours—he simply has to move out. Now that I reside in Cavite I look forward to going home in the afternoon. Before when I lived in Tondo, I hated the thought of going home at the end of the day. We would turn on the TV and people would climb the wall, destroying my plants and house just to watch the programme.

He did not add, as other disgruntled residents have, that the television-set owner living in a ground-floor apartment is expected in the name of neighbourliness to position it so that it faces the window onto the street, and even to tune in the programmes that the street viewers favour.

Still other residents judge Tondo the best neighbourhood for them:

I was born here, and I intend to live here, raise my children here, and eventually die here. People are very helpful and get along well with one another. I can leave the house unlocked with no one guarding it, and nothing gets lost. The neighbours will keep an eye on it for you. Furthermore, when a family member dies everyone helps with contributions and other services. Even the rich don't have this advantage. And besides in Tondo one has no time to be lonely. There are always people in the street. . . .

The community consensus on behaviour and its control fills in the gaps which among higher-class sectors fall into the control domain of formal institutions and associations. Off-Velasquez residents certainly recognize the existence of association rules, municipal ordinances and national laws. None the less, should any one of them cause him some personal difficulty, he appeals to the cultural code of mercy for the weak and repayment for past favours from power figures, even if this means breaking the rules. Formal voluntary associations exist among the better-educated adolescent groups, older household heads in a block association of very limited activity, and a few housewives spurred on by clubwomen organizers from national headquarters in Manila. This pattern of active involvement has not, however, gained popularity among the adult married set. Young parents are expected to avoid these 'frivolous' activities and concentrate on their serious adult role of raising their children properly. The energetic young family man who does take an interest in community improvement automatically earns the reputation of being a political aspirant. Experience has taught residents that few persons help the community at large for purely altruistic reasons. The crusader is seen as bidding either for the neighbourhood's vote or the attention of an established politician willing to launch a political career by gambling on the newcomer's vote-getting ability.

Metropolitan Influences

Also impinging on local life are the evidences of urban institutions like the police and fire departments, the welfare worker, city health department personnel, and the public and private schools. They prevent the neighbourhood resident from adopting an insular stance, for sooner or later he seeks out their services. Employment in factories and other business establishments likewise familiarizes families with the more cosmpolitan norms of minimum wage, fringe benefits, vacation and sick leave, and the discipline of the time clock. Residents can formulate a rating scheme of the best and worst kinds of work sites. They give the best judgements to those companies offering the most generous and regular material benefits on the basis of performance rather than personal connections. Reward for achievement on the basis of merit is important to people with few powerful contacts. They know that others, not they, will be likely to reap the benefits of favouritism in a bureaucracy where they have no personal acquaintances.

This dual orientation of the off-Velasquez dweller towards the neighbourhood, on the one hand, and the metropolis, on the other, greatly distinguishes him from the ruralites he is supposed to resemble. While the Filipino of higher class may not readily see the difference, he has only to watch these 'urban villagers' operate in the village once more. The clothing styles, so *gauche* by fashionable Manila standards, stand out as avant-garde in the village. The returned lower-class urbanite knows the popular songs and dances and is enthusiastically called upon to display them before the hometown folk. The supposedly tradition-oriented little man of the city metamorphoses into a modern pace-setter in the province. Should he remain there, the training he received in Tondo in the uses of technology, a modified mastery over his environment, organization, and a certain amount of planning ahead may well introduce a radical modernizing element into the really traditional rural community. The judgement, therefore, of just how urbane is a city resident, depends very much on the relative class statuses of rater and rated. By city standards, the off-Velasquez resident seems provincial or folk-like; by provincial standards he is citified.

Off-Velasquez and the Surrounding City

The more important question, however, involves not so much the city–village comparison as it does the neighbourhood dwellers' perspectives on the city. Two aspects merit specific mention, namely, the reaction to outsiders and local expectations of city institutions.

The aura of 'our neighbourhood' or 'being from here' (*tagarito*) pervades off-Velasquez streets. Residents radiate the security of belonging, and self-appointed guardians defend their land against unwanted strangers. Yet the right of bona-fide outsiders to enter the area is firmly established, especially along the wider commercial routes. Workers *en route* to their jobs, truck-drivers bound for the city slaughterhouse, welfare workers, pedlars, and the like enter the area in the pursuit of their livelihood. They remain generally safe, although they may have to pay a small 'tong', or protection money, euphemistically called cigarette money, for safe conduct. The outsider who moves into the neighbourhood will find ready acceptance as one who belongs if he exhibits the ideal characteristics of the good neighbour, namely involvement in the street life, with the people in the vicinity, and the activities they cherish.

In contrast, the non-resident outsider without a clear purpose, especially the male, runs the risk of being challenged, threatened, or actively molested by local guardians of the area. The propensity for trouble increases as night falls, for by then drinking bouts have emboldened street-corner vigilantes. Residents rationalize these fights as justified because 'they' started trouble by invading and challenging 'us' in 'our' place. Since the local boys initiate their own trouble-making activities in other neighbourhoods, their elders can conveniently ignore the same streak of violence among their youth. The image thus persists of one's own community as peace-loving while others near by deliberately court trouble.

Perspectives of the City

Off-Velasquez residents have clear though varying ideas regarding the potential contribution of city institutions to their lives and the kinds of leadership appropriate to their needs. Ability to get along well with all kinds of people ranks first. Next is the strong, unwavering leader who remains approachable and manages to steer his following subtly and ostensibly of their own free will to his way of thinking. This emphasis on the personal qualities and interpersonal skills of leaders is reiterated by a sample group composed of Tondo leaders at the district level. The latter add the criteria of higher moral standards, greater interest in serving the people, administrative competence, and concrete evidence of an ability to improve the lives of the people whose standard-bearers they are. Whether or not they personally measure up to their own combination of particularistic and universalistic standards, district leaders reveal a higher degree of sophistication in assessing leadership qualities than average off-Velasquez residents.

Asked to comment on local problems, the district leaders show far greater concern over the social problems they perceive in Tondo's neighbourhoods than do the block residents themselves. The latter mention the inadequate public facilities more frequently, especially the lack of water in local faucets. The crime, delinquency, and street fights stressed by the district sample rank a poor second among off-Velasquez dwellers. Apparently, each group gives greater weight to the immediate difficulties arising out of its own contact with the off-Velasquez blocks.

Only rarely do off-Velasquez respondents suggest that residents themselves can contribute anything substantial to the solution of their problems. Either 'the government should do something', or 'people must have better discipline'. They then add specific suggestions that presuppose the initiative of the administrators involved, as in proposals to enlarge the water-pipes, close the liquor distillery, or relocate the slaughterhouse. Perhaps this narrow orientation makes sense in that the problems they perceive do not in fact lend themselves to individual solutions but rather to large-scale, institutional planning.

In discussing ways by which the city and the nation can be improved, off-Velasquez people do not place the burden for progress completely on the government or on society's institutions. They also emphasize the need for changes in people's values, attitudes, and behaviour even more than do district leaders, ranking it higher than institutional change as a strategy. Despite their poverty, they have not adopted a fatalistic stance. All agree that the Philippines can improve; one group adding 'if needed steps are taken', and the other a more negative 'because the country is so badly off now it couldn't get much worse!'

Our off-Velasquez neighbourhoods may look like 'islands in the city' but their view of the larger world around them from the vantage-point of poverty points not to a displaced rural group but rather to true urbanites.

Contrasts in City Neighbourhoods

While Asian cities harbour so many people that one can almost claim lower-class neighbourhoods as the typical arrangement, the realities of political power simply do not permit this conception. Proportionately small though the middle and upper classes may be, by and large metropolitan policy moves in the directions they favour. More and more the economic differences between them and the urban poor receive ecological support in the proliferation of new middle-class and upper-class subdivisions.

The development of Manila's suburbs as residential sites for the better-off has been accelerating since the late nineteenth century.[3] Its post-World War II expression appears in the exclusive 'villages' ringed in by fences and gun-carrying security guards at selected entry points. As in the lower-class neighbourhood, those at the other extreme exhibit a spatial sense of community, but they institutionalize it through voluntary associations rather than informal face-to-face contacts. Their goals include the maintenance of a beautiful, safe, and orderly residential setting worthy of the nation's élite. These enclaves proclaim to the world at large that the Philippine upper crust can compete favourably with their peers anywhere in the world.

The predominantly middle-class neighbourhoods of greater Manila exhibit another community pattern, namely individual households acting as independent self-contained units. Lodged in single, detached houses, residents build walls, sometimes lined with broken glass on top and supplemented by a watch-dog below, to keep strangers out and residents in. Their children attend private schools scattered all over the city, finding their personal friends there rather than strictly in the home territory. Their parents similarly operate in comparable cross-city networks of relatives, office-mates, or co-members of clubs. The possession of a telephone and a car maximizes the potential for activating these networks.

Middle-class householders may have a nodding acquaintance with neighbours, and even shift the particularly compatible ones to the status of close friend. The choice is theirs to make. Symbolic walls and limited visual contact allow a mere acquaintance to remain that way if one prefers. The occasional extrovert attempting to organize a neighbourhood association must be prepared to take on the burden of work. Once the acquaintance party at his expense is over, his middle-class neighbours settle back once more into their apathy regarding local matters. They simply do not need one another as lower-class neighbours do.

Conclusion

We have attempted to show through research conducted in Tondo, Manila, that:

1. The city as a whole should not be identified with impersonal social relationships; lower-class neighbourhoods, in particular, exhibit a high frequency of personal ties fostered by a physical and social environment conducive to their formation.
2. While lower-class neighbourhoods exhibit many characteristics associated with folk or rural qualities, these are in fact attributes of an urban lower-class life-style; hence the model of 'the folk in the city' or the 'urban villagers' provides a romantic but inappropriate analytical framework for this group.
3. Urban lower-class neighbourhoods provide the poor with the security of personal ties necessary in recurring crises and mediate their interaction with the larger metropolitan environment; they do not keep the migrant rural but allow him to turn into an urbanite at a rapid rate.
4. The social relationships in middle-class and upper-class residential neighbourhoods vary greatly from those in lower-class neighbourhoods.

The implications of these conclusions are clear. To say that heterogeneity ranks high as an urban characteristic does not by any means imply, as Wirth seemed to think, 'a necessary association with impersonality'.[4] Moreover, the notion of the impersonal city reveals a particular ecological bias in assuming that when primary group neighbourhoods liké Tondo no longer predominate in a city, personal alienation and disorganization follow. Technological improvements in transportation and communication have made possible primary ties and socially close links across a city in contexts not based on the neighbourhood unit but personalized none the less.

One final note of caution: to speak of a lower-class life-style (or middle or upper) as though there were only a single uniform type grossly over-simplifies the matter. Further research will undoubtedly reveal distinctions between squatter and non-squatter lower-class subcultures, for example, or between old, fairly stable neighbourhoods and new ones. The off-Velasquez description typifies only one sector of the vast urban poor. It represents only one of the variations in life-style that contributes to the vibrant, ever-changing city.

Notes

1. Bureau of Census and Statistics, *Population and Housing*, Vol. I (Manila, 1963),

Bureau of Census and Statistics, *Facts and Figures about the Philippines 1963* (Manila, 1965).

2. For a more detailed description of the Santo Niño fiesta see Frank Lynch, 'Organised Religion: Catholicism', in F. Eggan *et al.*, *Area Handbook on the Philippines* (Chicago, 1956) 471–4.

3. See Mary R. Hollnsteiner, 'The Urbanization of Metropolitan Manila', in W. F. Bello and A. de Guzman (eds), *Modernization: Its Impact in the Philippines IV*, IPC Papers, no. 7 (Quezon City: Ateneo de Manila University Press, 1969) 147–74.

4. See Louis Wirth, 'Urbanism as a Way of Life', *American Journal of Sociology* (1938), 1–24.

14

The Social and Economic Organization of a Mexican Shanty Town*

Larissa Lomnitz

Introduction

A COMMON prejudice found in the sociological literature on poverty consists in portraying the urban poor as people bedevilled by a wide range of social pathologies, amounting to a supposed incapacity to respond adequately to social and economic incentives. More social scientists have directed their attention toward the material and cultural deprivation that meets the eye than towards the socio-cultural defence mechanisms which the urban poor have devised. My work in a Mexican shanty town, as summarized in the present chapter, deals with a basic question: How do millions of Latin Americans manage to survive in shanty towns, without savings or saleable skills, largely disowned by organized systems of social security?

The fact that such a large population can subsist and grow under conditions of extreme deprivation in Latin American cities has important theoretical implications. Obviously, the members of such a group can hardly be described as 'unfit' for urban life in any meaningful sense. On the contrary, the proliferation of shanty towns throughout Latin America indicates that these forms of urban settlement are successful and respond to some sort of objective social need (Mangin 1967; Turner and Mangin 1968). My own work in Mexico City tends to support this view, by providing evidence that shanty towns are actually breeding-grounds for a new form of social organization which is adaptive to the socio-economic requisites of survival in the city. In this paper, I show that the networks of reciprocal exchange among shanty-town dwellers constitute an effective stand-by mechanism, whose purpose is to provide a minimum of economic security under conditions of chronic underemployment.

This study is the result of two years of field work in a shanty town. I have

* Reprinted, in a revised form, from Wayne A. Cornelius and Felicity M. Trueblood (eds), *Anthropological Perspectives on Latin American Urbanization*, Latin American Urban Research 4 (Beverly Hills and London: Sage Publications, 1974). Copyright Larissa Lomnitz.

used participant observation and unstructured interviews in the anthropo-logical tradition. However, I have also made a number of quantitative surveys covering the totality of households in the shanty town, in order to substantiate the major conclusions. This eclectic approach works better in the urban environment, where the homogeneity of the unit of study cannot be taken for granted. Statistical results are not presented for their own sake; rather, they are used to fill out the conceptual model based on direct observation.

The Research Site

The shanty town of Cerrada del Cóndor sprawls over a ravine in the southern part of Mexico City, facing a cemetery on the opposite slope. The ravine represents the natural boundary between two residential middle-class neighbourhoods of fairly recent development. This area makes up the hilly outskirts of the ancient township of Mixcoac, a part of urban Mexico City since about 1940. Prior to that time, a few small entrepreneurs raised flowers and tree seedlings on the hills and worked the sand-pits in the ravine.

The earliest settler bought a tract of barren land about 1930, at the location of the present shanty town. There he settled with his family and began to manufacture adobe bricks. He was later joined by a caretaker of the sand-pits. Within ten years, there were about a dozen families living in Cerrada del Cóndor, all of them workers in the adobe industry. Around that time the owner decided to sell fifteen small lots to new settlers, who immediately started to build their own homes.

By the end of the 1940s, the whole surrounding area was being urbanized. The southward growth of the city had begun to swallow up the towns of Mixcoac and San Angel. The shanty town, however, was unfavourably located on the slopes of the ravine and was bypassed by developers. During the 1950s, 30 new families arrived. These included workers in the sand-pits, the adobe works, and in the housing projects of the neighbouring hills, particularly the suburb of Las Aguilas. A considerable number of relatives of the original settlers also came directly from rural areas. After 1960, the shanty town began to grow very rapidly: 111 families arrived during this period, plus around 25 families who left during the period of the study (1969–71) and are therefore not included in the survey.

As both the adobe factory and the sand-pits closed down, their owners became the slumlords of the shanty town; yet they did not move away. The settlers pay rent for their houses or for plots of land on which they erect houses of their own. At present the shanty town of Cerrada del Cóndor includes about 176 households, most of which were surveyed during the study.

Origins of the Settlers

A full 70 per cent of the heads of families and their spouses (hereafter referred as the 'settlers') are of rural origin, having migrated to the Mexico City metropolitan area from localities of less than 5,000 inhabitants. The remaining 30 per cent were born in the Federal District, either as sons or daughters of rural migrants, or as inhabitants of the small towns which are now part of the southern residential area of the city.

Eighty-six per cent of the rural migrants moved directly to Mexico City, without intermediate stops. This high proportion applies to all age-groups. In fact, about 70 per cent of all migrants moved in family groups, and only 30 per cent were single. The rural migrants came from the most impoverished sectors of the Mexican peasantry. Eighteen states are represented in the shanty town, but the states of Guanajuato, México, and San Luis Potosí account for 56.6 per cent of all migrants. Veracruz, Zacatecas, and Hidalgo come next with about 6 to 7 per cent each. In discussing their reasons for migrating, nearly all migrants declared that they had been landless field-workers, or that their landholdings had been too poor for subsistence.

Thirty-five per cent of the migrant heads of families and their spouses were illiterate; another 9 per cent had never been to school but knew the rudiments of reading and writing. Another 33 per cent had had one to three years of schooling. It is probably fair to say that more than half the settlers of rural origin were functionally illiterate at the time when they reached the Federal District. They had neither savings nor skills of any value in the urban labour market.

Among those born in the Federal District, the illiteracy rate was significantly lower. Only 17 per cent had never been to school, and nearly half of those had taught themselves the rudiments of reading and writing. We shall see later that there is a significant correlation between schooling and economic status, as measured by income and material possessions.

When migrants reach the city, they normally move in with relatives. The presence of a relative in the city is perhaps the most consistent element within the migration process. The role of this relative determines the circumstances of the migrant family's new life in the city, including place of settlement within the metropolitan area, initial economic status, and type of work. There is no escaping the economic imperative of living near some set of relatives: the initial term of stay with a given kinship set may be variable, but subsequent moves tend to be made with reference to pre-existing groups of relatives elsewhere. Unattached nuclear families soon manage to attract other relatives to the neighbourhood.

The Villela Group: An Example of Kin-mediated Migration

Among the 30-odd households from the state of San Luis Potosí, 25 came from the *hacienda-ejido* Villela near Santa María del Río. These families are related through consanguinal and marriage ties. The experience of the Villela group will serve as an example of the process of kin-mediated migration as observed in Cerrada del Cóndor.

The settlement of migrants from Villela goes back to the early 1950s, when two young men from the village decided to try their luck in Mexico City. They found work in the adobe factory and settled in Cerrada del Cóndor. One year later, one of them brought a sister and two nieces with their offspring to Mexico City. These nieces later brought their mother and brothers, and other relatives in successive waves. Two other Villela families also migrated to Cerrada del Cóndor and became related to the first family group by marriage or *compadrazgo* (fictive kinship).

After working at various trades, one of the migrants was fortunate to find work as a carpet-layer. Later migrants were lodged, fed, and counselled by those among their kin already in residence, with the result that practically all men now work in the carpeting trade. This pattern can be observed quite generally among family networks and is by no means unique to the Villela network. Thus, all the men in one network polish tombstones; in another, they work as bakers; still others are members of bricklayer crews, and so on.

Villela settlers in Cerrada del Cóndor maintain a closely knit community within the shanty town. They founded the oldest functioning local association, the Villela football club, with three teams in constant training which participate in league tournaments in Mexico City. Social contact among Villela families is intense, and there is a great deal of mutual assistance among them. All migrants express satisfaction at the positive results of their move to Mexico City, and not even the grandmothers show any nostalgia for Villela, where, they say, 'we were starving'.

Moves within the City

It has sometimes been assumed that migrants to Mexico City initially tend to gravitate towards the crowded tenements in the old downtown area (Turner and Mangin 1968). Our research in Cerrada del Cóndor does not confirm this hypothesis. Instead, the place of initial residence is determined by the residence of pre-extant cores of relatives in the city. In general, the migrants continue to move within the city, but always in the same general sector. Thus, the settlers of Cerrada del Cóndor were born in or migrated initially to the southern part of the metropolitan area, and few of them have more than a very superficial acquaintance with other parts of the city, including the downtown area. Few of the men venture farther into the city than their jobs require. Women and children barely know anything of the city beyond a church, a market, or the home of some relative.

The mechanism of moves within the urban area was studied in some detail. In general, a family moved once every five years on the average for the first ten or 15 years of married life; some families never seemed to settle down. Moves seemed to be caused largely by displacement due to the southward growth of the city, coupled with a desire to seek better work opportunities and more congenial kin. There is an important turnover in Cerrada del Cóndor: during the period of my study, about 25–30 families moved away, and some 40 new families moved in. As previously noted, Cerrada del Cóndor has been bypassed by developers and represents an area of refuge for those displaced by urban growth.

Most of the new settlers in Cerrada del Cóndor, however, merely follow the pull of relatives who already live in the shanty town. These relatives have told them about cheap available housing and have offered the exchange of mutual help, without which life in a shanty town is extremely difficult. Thus, kinship is also the determining factor in the process of residential mobility within the city. When job opportunities seem sufficiently bright for a nuclear family to move into a new area, they soon bring in other relatives from nearby areas or directly from the countryside. New migrants may be recruited during trips to the village, as migrant families keep in regular contact with their place of origin. Visits occur normally on festive occasions, such as holidays and celebrations.

In conclusion, it may be said that each migrant helps several new migrants to settle in the city, or to move from another part of the city to this shanty town. This he does by providing temporary or permanent lodging, food, information, assistance in job-hunting, moral support, and the basis for a more permanent form of exchange to be discussed later.

Economics of Shanty-town Life

The general economic setting of Cerrada del Cóndor is one of extreme poverty. A typical dwelling consists of a single room measuring 10 by 12 ft., containing one or two beds shared by members of the family. There may also be a table, a chair, a gas or petroleum stove, and sometimes a television set—33 per cent of all households own one. There are three public water faucets in the shanty-town, which are used by most of the population (a few clusters of dwellings have a faucet of their own). There is little public sanitation and drainage; more than four-fifths of the population use the bottom of the gully for a latrine. Sanitary conditions are made worse by the presence of a large public garbage dump next to the shanty town. There is no regular electric service; power is obtained by illegal hook-ups to the power lines. There are no paved streets, only alleys and gutters left between residential units.

Families residing in Cerrada del Cóndor can be classified according to the following levels of living:

Level A. Three or more rooms, running water, bathroom or privy, brick construction, cement or tile floor; dining-room furniture, living-room, electric appliances such as sewing-machine, washer or refrigerator; gas stove.

Level B. Two rooms, cement floor, no running water, some furniture such as a trunk or closet, table and several chairs, some electrical appliance(s), gas stove.

Level C. The same as level B but no electrical appliances (except for radio or television), lower-quality furniture, petroleum-burner for cooking.

Level D. One room with or without small lean-to for cooking; no furniture (except beds and an occasional table or chair); clothing kept in boxes or under the bed; no appliances (except for radio or television); petroleum cooking only.

Analysis by means of contingency tables showed that the four criteria used (housing, furniture, type of cooking, and electrical appliances) were highly intercorrelated.

The distribution of levels of living within Cerrada del Cóndor was found to be as follows:

Level A	7.8%
Level B	8.9%
Level C	23.8%
Level D	59.5%

The level of living was also found to be highly correlated with other economic indicators, particularly the occupational status of the breadwinner. Table 14.1 shows that the total of unskilled labourers, journeymen, servants, and petty traders corresponds closely to the total of families classified in levels C and D.

Unskilled labourers or apprentices include hod-carriers and other construction workers (foremen excepted), house painters, sand-pit workers, brickmakers, bankers' helpers, truckers' helpers, carpet-layers, electricians, gardeners, and other unskilled labourers paid by the day who earn the minimum legal wage or less. Semi-skilled or skilled journeymen or

TABLE 14.1   Occupations of Heads of Households in Cerrada del Cóndor

	Men		Women	
	n	%	n	%
Unskilled labourers or apprentices	51	32.9	1	4.5
Semi-skilled or skilled journeymen or craftsmen	48	31.0	–	–
Industrial workers	16	10.3	–	–
Service workers	5	3.2	12	54.5
Traders	7	4.5	4	18.2
Employees	8	5.2	1	4.5
Landlords	5	3.2	1	4.5
Unemployed	15	9.7	–	–
Housewives	–	–	3	13.6
Total	155	100.0	22	100.0

craftsmen include independent or freelance workers such as bakers, carpet-layer foremen, construction foremen, electrician foremen, truck-drivers, tombstone-polishers, carpenters, cobblers, blacksmiths, potters, and so on. These may earn higher wages, but their job security is usually as low as that of the unskilled workers. Some of them have developed a steady clientele and work with their own assistants, usually relatives. Industrial workers are those who work in an industrial plant, usually with the lowest wages and qualifications: watchmen, car-washers, janitors, and unskilled labourers. The service workers include waiters, water-carriers, watchmen, icemen, and domestic servants. Traders include all kinds of street vendors. None of these has a steady income or social security. The employees are unskilled workers who earn fixed salaries: municipal workers (street-sweepers, garbagemen), and a few similarly employed with private corporations. These have a relatively high job security and other benefits. Finally, there are six households whose income is mainly derived from rentals of property in the shanty town.

About 10 per cent of the household heads were out of work at the time of the survey. However, more than 60 per cent of those who said they were working consider intermittent joblessness for variable periods of time to be normal. Thus, the majority of the working population in the shanty town are underemployed ('eventuales'), and have no job security, no social security, and no fixed income. They exist from day to day, as urban 'hunters and gatherers'. Members of level-of-living D belong to this group. More than half the settlers in level D were illiterate. None owned both his home and the lot it was built on; nearly two-thirds paid rent for both. All lived in a housing unit consisting of a single room. The average number of people per room was 5.4, if the cooking was done inside, and 6.2 if there was a lean-to for cooking. Of all acknowledged cases of problem drinkers, more than 75 per cent belonged to level D.

In contrast, members of level A were practically all owners of their homes and lots. Most of them were either born in the Federal District or have lived in town for many years. There was practically no illiteracy, and most settlers had completed third grade. They tended to belong to the upper types of occupations: landlords, employees, traders, and industrial workers; their most distinctive trait was job security. More than half the households in this group included two or more breadwinners. The men were either abstemious or moderate drinkers.

Levels B and C are intermediate, but they can be sharply distinguished. Level B is urban in type of dwelling, furniture, and life-style, while Level C is still rural in most of these respects. Households of either type can usually be recognized by a glance at their belongings. The transition from C to B is not determined so much by gross income as by the degree of cultural assimilation to urban life: hence, time of residence in the city is a significant factor. Even the highest income in Cerrada del Cóndor could easily be used up by a single heavy drinker. Wives receive a weekly allowance and have no direct knowledge of their husbands' income. Working wives contribute their total income to household expenses; likewise, sons and daughters hand their earnings to the mother. The husband's contribution to raising the economic level of the household is largely limited to major appliances which are purchased on the instalment plan. However, if the husband enjoys a steady income and has a tolerant view of a working wife, the economic improvement of the family may be rapid by shanty-town standards. Nevertheless, the transition from level C to level B is rarely accomplished before a household has completed ten years of residence in the city.

Social Organization

The pattern of social organization which prevails in the shanty town can be described as follows. Most nuclear families initially lodge with kin, either in the same residential unit (47 per cent), or in a compound arrangement (27 per cent). Compounds are groups of neighbouring residential units which share a common outdoor area for washing, cooking, playing of children, and so on. Each nuclear family in such a cluster forms a separate economic unit. Families in the compound are related through either consanguinity or marriage ties; each compound contains at least two nuclear families.

Extended families, e.g. two brothers with their wives and children, may share the same residential unit temporarily; in the case of newly married couples with the parents of either husband or wife, the arrangement may be more permanent. Any room or group of rooms having a single private entrance is defined as a residential unit: this excludes tenements of the '*vecindad*' type, consisting of a series of rooms opening on an alley with a public entrance gate, which may contain several independent family groups. Extended families contain at least two nuclear families; these share the rental expenses or own the property in common. Sometimes they also share living-expenses.

Extended households are less stable than compounds. Nuclear families in an extended household tend to move into a nearby room of their own, or to join a different set of relatives elsewhere. However, those who move away in search of independence and privacy eventually return for security or assistance. In the case of compounds the rate of desertion is much lower. Of 44 nuclear families who joined a compound arrangement since the beginning of married life, only seven moved away in an attempt to form an independent household.

Thirteen couples began their married lives as independent households within the Federal District; these represent the major exception to the pattern of social organization described above. The heads of these households had all been born in Mexico City or had lived there for many years. Yet even these households do not remain independent for long, since they tend to attract other kin who join them in an extended or compound arrangement. The complete data for household types is summarized in Table 14.2.

TABLE 14.2   Types of Households in Cerrada del Cóndor

Extended families	29
Nuclear families in a compound-type arrangement	68
Independent nuclear families:	
(a) without kin	30
(b) with kin in Cerrada del Cóndor	28
Other unknown	7
Total households in survey	162

Independent nuclear families are in the minority; those who live within walking distance of relatives are usually waiting for a vacancy to move into a compound-type arrangement. In this case, there is much visiting, mutual assistance, and other types of interaction even though the related nuclear families are not yet fully integrated into a compound. The term 'nuclear family' is used in a broad sense here, as each nuclear family may include one or more individually attached kin. Most of them are older persons or young children of unmarried recent arrivals from the country. Nuclear families may also include the offspring of a previous union of the mother; in two such cases, there was no offspring from the present union.

Thus, the social organization of the shanty town may be described as a collection of family networks which assemble and disband through a dynamic process. There is no official community structure; there are no local authorities or mechanisms of internal control. Co-operation within the family networks if the basic pattern of social interaction. There is a pattern of movement from the extended family towards the compound type, as illustrated by a survey tracing the moves of each household over the past years. The results show that there is an increase of 29 per cent in compound arrangements, against a decrease of 46.7 per cent in extended family arrangements, as compared with the initial state of residence.

This pattern can be viewed as the outcome of a dynamic process, which depends on economic circumstances, the stage in the life cycle, the availability of housing vacancies, personal relationships with relatives, etc. The initial choice of moving in with the family of either spouse is usually an economic one. Since young husbands or wives often do not get along with their in-laws and conditions in an extended family may be very crowded, the couple tends to move out. However, new circumstances, such as the arrival of children, desertion of the husband, loss of employment, and so on frequently compel the family to return to the shelter of relatives. The preferred arrangement is the compound which combines proximity of kin with an adequate amount of independence and privacy.

Kinship Relationships Outside the Shanty Town

A pattern of residential moves such as we have described, sometimes over large distances, implies a substantial amount of contact among kin extending beyond the physical boundaries of the shanty town. The existence of such contact was confirmed initially through personal observation and later by means of a kinship census covering all households in the shanty town.

Contact with relatives within the Federal District depends on kinship distance and on physical distance. Informants tend to list first their nuclear family of orientation, then other relatives by order of spatial proximity: first those who live in the shanty town, then those who live in a nearby shanty town such as Puente Colorado, and so on. If a relative is not particularly close and lives as much as two hours away by bus from Cerrada del Cóndor the contact is unlikely to be significant, and may be lost after a generation. The mother is often the only nexus between such relatives, and contact vanishes after her death. Of course, if there exists a true closeness of relationship each set of relatives will exert a great deal of 'pull' on the other, in order to encourage them to move into as close a neighbourhood as possible.

Changes in socio-economic status become a factor which influences the intensity of contact between kin. A female informant commented that she hardly ever saw her sisters, who were married to skilled industrial workers: 'To tell the truth, I don't like to go and see them because they can dress very nicely and I can't afford to and so . . . I feel ashamed.' The informant is the daughter of a skilled worker, who married a man who 'never finished grade school and is worse off' than her sisters' husbands. Her parents were opposed to the match because her husband had no skills, 'not even that of a truck-driver, a barber, or a carpenter', but she was in love and they went to live with his parents. At first they lived in the same residential unit in Cerrada del Cóndor; now they have a room near by, because 'each on his own is better'.

Contact between migrant families and their relatives in the countryside usually takes the form of visits to the village on festive occasions, such as Mother's Day, All Saints' Day, and the festival of the patron saint of the village. Unmarried migrants often return to the city accompanied by a smaller brother, sister, or cousin whom they help out until they find a job. Married migrants frequently maintain a share in a small plot of land which they own jointly with a brother, and they time their occasional visits to coincide with harvest time, and so on. Most migrants send money home to their parents or close relatives. Through word of mouth or correspondence, they keep up with village gossip; eventually they are instrumental in promoting the migration of their close kin from the village. For years there is a steady stream of relatives from the country, who are lodged and fed for indeterminate periods of time depending on resources and needs.

Contact with the village gradually wanes over the years. About one-fourth of all informants said they had relatives in the country but had lost all touch with them: 'I haven't seen my folks since I got married and moved to the city eighteen years ago' . . . 'I never went back home since my mother died—my father married again and I don't get along with my stepmother' . . . 'My brothers have moved to the city; I used to visit them and bring them money; now I don't go any more' . . . 'I never went back since my grandparents and parents died.' Other migrants, however, said they maintained significant contact through visits to and from the village; through economic interests (land· owned in common); through correspondence; through remittances of money; through sentimental pilgrimages (visiting Mother's grave on All Saints' Day), and so on.

Local Groups

The shanty town is not organized around central institutions of any kind. Instead, there are several types of groupings, of unequal importance: (a) the family network; (b) football teams; (c) the medical centre; (d) temporary associations.

The family networks will be discussed in greater detail below, as it is our thesis that they represent the effective community for the individual in the shanty town. They are composed of members of an extended family, or a compound, but may include neighbours who are assimilated through fictive kinship. We shall see how these networks have developed into systems of reciprocal exchange of assistance, which provides an important explanation for the fact of survival of large numbers of people under the severe economic handicaps of shanty-town life.

Other forms of organization at the community level are relatively rudimentary. There are four football teams in Cerrada del Cóndor. Three of these teams belong in effect to a single large family network, the Villela network described above. The fourth is a more recent team whose membership is recruited among young people of the shanty town irrespective of family origins. Football teams represent one of the few vehicles of social contact between men of Cerrada del Cóndor and men who live in other parts of town. After a game, there are drinking sessions which reinforce the team spirit and friendship among members of the team.

The shanty town's medical centre was organized and financed by a group of middle-class ladies from the neighbouring residential district, with some assistance from a nearby church. Later, the national Children's Hospital agreed to staff the centre, but this help has recently been withdrawn. In spite of the modest assistance offered, the centre has become an important part of shanty-town life. It is a place where children are welcome during most hours of the day, and where many girls and women receive guidance from an understanding social worker.

There is no local organization for solving the common problems of shanty-town life. Groups of neighbours may band together for specific issues; this has happened three or four times in the existence of Cerrada del Cóndor. The first time was to request the installation of a public water outlet. Another time, a group of women jointly requested an audience with the First Lady, in order to lodge a complaint about spillage of oil from a refinery that was causing brush fires in the ravine. These exceptional instances of co-operation merely serve to highlight the absence of any organized effort to solve community problems.

The residents of Cerrada del Cóndor have little contact with city-wide or national organizations. Articulation with Mexican urban culture occurs mainly through work and through mass media such as radio and television. School is, of course, very important for the children. Adult reading is limited to sports sheets, comics, and photo-romance magazines. Only about one-tenth of the men belong to the social security system. About 5 per cent are union members. In general, extremely few people belong to any organized group on a national level, such as political parties, religious organizations, and so on.

Social Networks

According to Barnes (1954), a network is a social field made up of relations between people. These relations are defined by criteria underlying the field. While Barnes saw a network as essentially unbounded Mayer (1962) showed how certain types of migrants encapsulate themselves in a bounded network of personal relationships. In the case of Cerrada del Cóndor, we find networks defined by criteria of neighbourhood social distance and exchange of goods and services.

Each network is constituted of nuclear families, not individuals. Initially we shall use an operational definition of networks as clusters of neighbouring nuclear families who practise continuous reciprocal exchange of goods and services. A total of 45 such networks were identified in Cerrada del Cóndor. The social relationships which form the basis of these networks are as follows: 30 are networks based on consanguineal and marriage ties, seven are based on kinship but included also one or more families not related by kinship, and eight are formed by families not related by kinship. The number of nuclear families per network is shown in Table 14.3. The average network contains four nuclear families. Of course, the number of families in a network is not static but changes with time. Initially, the network may be composed of two or three families; it may grow until a part of the network is split off because of lack of room or facilities. Table 14.3 does not include unattached nuclear families (estimated at less than ten).

It is possible for several networks to be interrelated through kinship. Thus, the Villela macro-network includes about 25 nuclear families grouped into five networks. Each of these networks internally displays a high degree of reciprocal exchange of goods and services on a day-to-day basis. The resources of the macro-network are used more on ritual occasions, in important matters such as job placement, in the expression of kin solidarity (football teams), and drinking. Reciprocal exchange does occur between families belonging to different networks within a macro-network system, but the recurrence of such exchange is less frequent,

TABLE 14.3   Number of Nuclear Families per Network

Nuclear families	Cases
2	9
3	13
4	10
5	6
6	5
uncertain	2
Total	45

because no single nuclear family in the shanty town has enough resources to maintain a generalized day-to-day exchange with such a large group of families.

All nuclear families in a network practise reciprocal exchange among one another on an equal footing. In addition, a nuclear family may maintain dyadic exchange relations with families outside the network, or belonging to other networks. These dyadic ties are important because they provide the mechanism through which an outside family can be attracted to join the network, or by which a network which has outgrown its optimal size may split.

A network constituted as an extended family practises a generalized exchange of goods and services, which includes the informal pooling of resources for rent and entertainment, joint use of cooking facilities, communal child care, and many others. Each nuclear family contributes according to its ability and receives according to the availability of resources within the network. There is no accounting of any kind among the members of such a network. In a compound, on the other hand, each nuclear family has a roof and an economy of its own; yet there is an intense exchange of goods and services in the form of daily borrowings of food, tools, and money. Reciprocity here is not openly acknowledged, but it is definitely expected; each member family is supposed to provide assistance in proportion to its economic ability. Thus, if a nuclear family within a compound becomes economically more secure than the rest, it may find its resources taxed beyond the actual returns which it can expect from the network. As a result, the more prosperous families may stop asking for and offering services.

Case History

The network is of the compound type and includes two sisters A and B, who married two brothers. A non-kin neighbour is also included in the network. A third sister, C, brought in from the village to live with A, soon found work as a maid living in the house of their employer. On her days off, she visited A. A niece from the country has now joined the network with her husband. At first they lived with A, who obtained work for the husband and a room adjoining hers. All men in the network are currently working as tombstone polishers. When C became pregnant, she quit her job and went again to live with A. After she had the baby, she went back to work and left the baby in A's care during the day.

Meanwhile, B's economic status had been rising steadily. Her husband did not drink and invested in home furnishings. He also found advancement at work and became a skilled worker (in the placement of tombstones). B began to refuse to lend (or ask for) favours within the network, claiming that she had no money. Her sister, niece, and neighbour gradually stopped requesting assistance, and so did their respective husbands. After a while, B found a room just beyond the limits of the shanty town, two blocks away from her former room but with urban services. Her economic level is rated 'B'.

Sister C found a husband and moved in with him. The husband lived several blocks away from the sisters' compound, outside Cerrada del Cóndor. Yet C practically continued to live at A's place whenever her husband was away at work. When she works she leaves her child with A; when she needs money she borrows from A or from her niece. Her interaction with the non-kin member of the network is less intense; yet these neighbours have become double *compadres* in the meantime, and their exchange with both A and the niece and their respective husbands is very active. When a room became available, C began to convince her husband to move in with the network. In that case, an active exchange between C and the non-kin neighbour is anticipated.

What is Exchanged?

The following items represent the most important objects of exchange in the networks, according to my observation:

1. *Information*, including directions for migration, employment, and residence; gossip; and orientation about urban life.

2. *Training and job assistance*, including the training and establishing of a relative as a competitor. Thus, a carpet-layer or a building contractor would take his newly arrived brother-in-law along as an assistant, teach him the trade, share earnings with him, and eventually yield some of his own clientele to set him up as an independent worker.

3. *Loans*, of money, food, blankets, tools, clothing, and other goods.

4. *Services*, including the lodging and care of visiting relatives, widows,

orphans, old people; care and errand-running for such neighbours; and minding children for working mothers. Assistance among men includes help in home construction and in transporting materials. Children must lend a hand in carrying water and running errands.

5. *The sharing of facilities* such as a television set or a latrine (which the men may have built jointly).

6. *Moral and emotional support* in ritual situations (weddings, baptisms, funerals) as well as in day-to-day interactions (gossip among the women, drinking among the men). It is essential to recognize that much of the socializing in the shanty town is based on network affiliation. This constant interaction generates an overriding pre-occupation with each other's lives among the members of a network. There is little opportunity for privacy.

The ubiquity of these forms of exchange provides important evidence in support of our interpretation of shanty-town networks as economic structures which represent a specific response of marginal populations to economic insecurity in the city.

Reinforcing Mechanisms

The exchange of goods and services serves as the underpinning of a social structure: the network organization. When this exchange ceases to exist, the network disintegrates. The social structure which is erected on the basis of exchange depends on physical and social proximity of network members. Ideally, the networks are composed of neighbours related through kinship.

Actually, many networks contain non-kin members whose allegiance must be reinforced by means of fictive kinship (*compadrazgo*) and other means which will be analysed presently. Even among kin, relationships are far from secure: economic and personal differences arise frequently under conditions of extreme poverty and overcrowding. The reinforcing mechanisms to be discussed are therefore present in all networks.

Compadrazgo is widely used to reinforce existing or prospective network ties. In Cerrada del Cóndor, the *compadres* have few formal obligations toward one another as such. An informant says: 'When choosing a godfather for one's child one should look for a decent person and a good friend, if it's a couple, they should be properly married. They should be poor so no one can say that you picked them out of self-interest.' Among 426 *compadres* of baptism (the most important type of *compadrazgo* in the shanty town), 150 were relatives who lived close by, and 200 were non-kin neighbours. Another 92 were relatives who lived elsewhere in the Federal District or in the countryside, i.e., prospective network affiliates. In most cases of *compadrazgo*, the dominant factors were physical proximity and kinship. This equalitarian pattern is at variance with the frequently observed rural pattern of selecting a *compadre* above one's station in life (Forbes 1971).

The great importance of *compadrazgo* as a reinforcing mechanism of network structure is also reflected in the variety of types of *compadrazgo* that continue to be practised in the shanty town. These types are, by order of decreasing importance: baptism (426 cases), confirmation (291), communion (79), wedding (31), burial (16), Saint's Day (13), fifteenth birthday (10), Divine Child (8), Gospels (8), grade school graduation (4), habit (3), sacrament (2), scapulary (1), cross (1), and St Martin's (1). All these types of *compadrazgo* mark ritual or life-cycle occasions. The formal obligations among *compadres* can be described as follows: 'They must treat each other with respect at all times, and must exchange greetings whenever they meet.' Ideally some *compadres* should fulfill economic obligations, such as taking care of a godchild if the father dies; but these obligations are no longer taken very seriously in the shanty town.

If *compadrazgo* formalizes and legitimizes a relationship between men and women, *cuatismo*, the Mexican form of male friendship, provides the emotional content of the relationship. *Cuates* (a Nahuatl term for 'twins') are close friends who pass time together, talking, drinking, playing cards or football, watching TV, treating each other in restaurants, and having fun together; above all, they are drinking companions. Women are totally excluded from the relationship. A wife 'would never dare' to approach a *cuate* of her husband's to request a favour.

Assistance among *cuates* is ruled by social distance. Among relatives, there will be more unconditional help than among neighbours. In general, the *cuates* borrow freely from one another, help one another in looking for work, give one another a hand in fixing their homes, and stand by one another in a fight. Like *compadrazgo*, *cuatismo* is practically universal: the man who has no *cuate* and no *compadre* is lost indeed. Among a total of 10 per cent of households headed by men, the circle of *cuates* of the family head was recruited as follows: 86 were groups of *cuates* who lived in the immediate vicinity, nine were mixed groups (some *cuates* living close by and some far away), and 11 were groups whose members did not live in the immediate neighbourhood. Two heads of households had not yet made any *cuates* in the city.

It is clear that these groups of *cuates* are based primarily on the male sector of the networks described above, even though neighbours, work companions, or friends not affiliated with one's special network may be included. The existence of *cuatismo* to reinforce network affiliation is evidence that the networks are not simply built around the wives and mothers, as might be supposed from a superficial analysis. On the contrary, many networks appear to be male-dominated. If networks were based exclusively on the more visible forms of daily exchange of goods and services practised by women, the substantial overlap between networks and groups of *cuates* would be rather puzzling. Networks are constituted by nuclear families as entities; all members of each nuclear family participate actively in the relationship.

Drinking relationships among *cuates* are exceedingly important and usually take precedence over marital relationships. From a psychological point of view, drinking together is a token of absolute mutual trust which involves a baring of souls to one another (Lomnitz 1969; Butterworth 1972). From the economic point of view, cuatismo implies a mechanism of redistribution through drink which ensures that all *cuates* remain economically equal. And from a social point of view, it reinforces existing social networks and extends the influence of networks in many directions, since a drinking-circle may contain members of several networks.

The *ideology of assistance* is another important factor in network reinforcement. When questioned, most informants are reluctant to describe their own requests for assistance; yet they are unanimous in claiming to be always ready to help out their own relatives and neighbours in every possible way.

The duty of assistance is endowed with every positive moral quality; it is the ethical justification for network relations. Any direct or indirect refusal of help within a network is judged in the harshest possible terms and gives rise to disparaging gossip. People are constantly watching for signs of change in the economic status of all members of the network. Envy and gossip are the twin mechanisms used for keeping the others in line. Any show of selfishness or excessive desire for privacy will set the grapevine buzzing. There will be righteous comments, and eventually someone will find a way to set the errant person straight.

Reciprocity and *Confianza*

The types of reciprocity between shanty-town members are determined by a factor which I have called '*confianza*' (Lomnitz 1971). *Confianza* depends on cultural factors (social distance) and physical factors (closeness and intensity of exchange, as when a close friend or *cuate* enjoys greater *confianza* than a relative who lives elsewhere and is met only occasionally).

The formal categories of social distance are culturally determined. They imply a 'series of categories and plans of action' (Bock 1969, 24) which dictate expected behaviour between individuals. These categories and plans of action can only be described ethnographically: they represent an essential part of the culture or subculture of a group or subgroup. In Mexico, within the national culture, there are subcultures of each social class, each state or region, and so on down to the level of family subculture which may imply strongly particularized sets of behaviour. Two individuals are close in the scale of *confianza* to the extent that they share the same set of behaviour expectations. These expectations include a specific type of reciprocity, extending from unconditional sharing to total lack of co-operation and distrust. The scale of *confianza* measures, among other things, the extent to which these expectations are actually fulfilled. Hence, the degree of *confianza* is not rigidly determined but may vary during the evolution of the relationship.

Social Networks in the Context of Marginality

Residents of shanty towns such as Cerrada del Cóndor are often counted among the 'marginal' sector of the urban population in Latin America. The emergence of urban marginal populations is not, of course, exclusive to underdeveloped societies. In advanced industrial nations, such populations result from the displacement of certain social strata from the labour market through mechanization and automation of the means of production. These growing population sectors have no expectation of absorption into productive occupations, and become increasingly dependent on welfare. They represent *surplus* population (rather than a labour reserve) and are, therefore, an unwanted by-product of the system.

According to Quijano (1970), this situation is considerably aggravated in underdeveloped countries, because the rate and pattern of industrial development are imposed from abroad. Economic dependence introduces a factor of instability, because of the hypertrophic growth of large industrial cities at the expense of the countryside. Accessibility of sources of raw materials and cheap labour attract an overflow of hegemonic capital into formerly pre-industrial societies. As a result, (a) there is an increasing gap between 'modern' cities and 'traditional' rural areas on the verge of starvation; (b) new skills required by industrial growth are monopolized by a relatively small labour élite, while the great mass of unskilled peasants and artisans is displaced from their traditional sources of livelihood; (c) superficial modernization has caused a sudden population explosion, which increases the rate of rural-to-urban migration, thus offsetting any efforts at promoting the gradual absorption of surplus populations into the industrial labour force. Thus, the process of marginalization is not transitional, but, rather, intrinsic to the system.

Quijano specifically identifies the capitalist system in general, and the dependent industrial development observed in Latin America since 1945 in particular, as responsible for the phenomenon of marginality. Adams (1970, 89–94; 1972) generalizes this analysis to apply to any large society which is subject to a process of economic development and technological change. According to Adams, any increment in social organization is achieved at the expense of disorganization among sectors of the same society, or of dependent societies. Dialectically speaking, order is the source of disorder: work creates entropy. Starting from an undifferentiated labour force, we may build up an industrial proletariat with highly differentiated skills and a centralized form of organization; but this will generate marginalization of those populations which can no longer be assimilated or successfully utilized by the more advanced system.

In Latin America, the urban marginal strata share the following economic characteristics, which are also found among the settlers of Cerrada del Cóndor: (a) unemployment or underemployment; (b) lack of stable income; (c) generally the lowest level of income within the urban population. Most settlers of Cerrada del Cóndor are rural migrants, or the offspring of migrant parents. Most of them are unskilled workers, such as construction workers, who are hired and fired on a daily basis; journeymen and artisans who are hired for specific jobs and have no fixed income; petty traders; and people who work in menial services. They may be described as urban hunters and gatherers, who live in the interstices of the urban economy, where they maintain an undervalued but nevertheless well-defined role. They are both a product of underdevelopment and its wards.

If this vast social group lacks any economic security and has no significant support from organized welfare, how does it survive? This question was posed by Quijano (1970, 87–96), who surmises that there must be some mechanism of reciprocity operating among marginal groups, which has not been described. It is the purpose of the present section to analyse this mechanism in some detail, as a function of the socio-economic structure.

According to Polanyi (1968, 127–32) and Dalton (1968, 153), there are three forms of exchange of goods and services: (a) market exchange, in which goods and services circulate on the basis of offer and demand, without any long-term social implications attached to the exchange; (b) redistribution of goods and services, which are first concentrated in a single individual or institution from whence they flow out towards a community or society; and (c) reciprocity among social equals. Reciprocity defined in this manner is an integral part of a permanent social relationship.

The urban marginal population of Mexico is estimated at four million, a considerable part of the total urban population. While the dominant mode of exchange in the cities is market exchange and is afflicted by the internal contradictions described by Marx and Polanyi, no adequate systems of public or private redistribution have been created in response to the needs of a growing mass of urban marginals left to their own devices. Economic dependence aggravates the problem, because capital gains tend to be transferred abroad instead of becoming available for redistribution within the country.

Thus, the marginal individual cannot rely on the social system for the elementary needs of survival. He has nothing to offer the market exchange system: no property, no skills except for his devalued labour. His prospects for absorption into the industrial proletariat are slim, since marginality grows faster than the number of industrial job openings. He has nothing to fall back on: no savings, no social security of any kind. His chance of survival depends on the creation of a system of exchange entirely distinct from the rules of the market-place, a system based on his resources in kinship and friendship. This system follows the rules of reciprocity, a mode of exchange between equals, imbedded in a fabric of social relations which is persistent in time, rather than casual and momentary as in market exchange. The three basic elements of reciprocity are: (a) *confianza*, an ethnographically defined measure of social distance; (b) equivalence of resources (or lack of resources); (c) physical closeness of residence.

Characteristically, reciprocity generates a moral code which is distinct, and in some ways opposed, to the moral code of market exchange. In a reciprocity relation, the emphasis is less on receiving than on giving; the recipient is preoccupied with reciprocating, rather than with extracting a maximum personal benefit from a transaction. Both systems of exchange may be used simultaneously in different context: a member of a reciprocity network may sell his labour as worker or servant on the urban market. Yet it is the reciprocal exchange among relatives and neighbours in the shanty town which ensures his survival during the frequent and lengthy spells of joblessness. Market exchange represents the ultimate source of livelihood; but it is a livelihood at the subsistence level, without any element of security. Through sharing these intermittent resources with another six or ten people, the group may successfully survive where as individuals each of them would amost certainly fail. The networks of reciprocal exchange which we have identified in Cerrada del Cóndor are functioning economic structures which maximize security, and their success spells survival for large sectors of the population.

References

Adams, N. R. (1970) *Crucifixion by Power* (Austin: University of Texas Press).
—— (1975) 'Harnessing Technological Development', in J. H. Poggie, jun. and R. N. Lynch (eds), *Rethinking Modernization: Anthropological Perspectives* (Westport, Conn.: Greenwood Press).
Barnes, J. A. (1954) 'Class Committees on a Norwegian Island Parish', *Human Relations*, 7, 39–58.
Bock, P. (1969) *Modern Cultural Anthroplogy* (New York: Alfred A. Knopf).

Butterworth, D. (1972) 'Two Small Groups: A Comparison of Migrants and Non-migrants in Mexico City', *Urban Anthropology*, 1, 1.

Dalton, G. (1968) 'The economy as instituted process', in E. E. Le Clair, jun. and L. Schneider (eds.), *Economic Anthropology* (New York: Holt, Rinehart & Winston).

Forbes, J. (1971) 'El sistema de compadrazgo en Santa María Belén Atzitzinititlán, Tlaxcala', M.A. thesis, Universidad Iberoamericana, Mexico, DF.

Lomnitz, L. (1969) 'Patrones de ingestión de alcohol entre migrantes mapuches en Santiago', *América Indígena*, 29 (1), 43–71.

—— (1971) 'Reciprocity of Favours among the Urban Middle Class of Chile', in G. Dalton (ed.), *Studies in Economic Anthropology* (Washington, DC: American Anthropological Association).

Mangin, W. (1967) 'Latin American Squatter Settlements: A Problem and a Solution', *Latin American Research Review*, 2, 3.

Mayer, P. (1962) 'Migrancy and the Study of Africans in Town', *American Anthropologist* 64, 576–92.

Polanyi, K. (1968) *The Great Transformation* (Boston: Beacon Press).

Quijano, A. (1970) 'Redefinición de la dependencia y proceso de marginalización en América Latina', (Santiago, Chile: CEPAL, mimeo).

Turner, J., and Mangin, W. (1968) 'The Barriada Movement', *Progressive Architecture* (May).

15

Social Control and Rehabilitation in Urban China*

Martin King Whyte

RIGHT at the beginning I should state that China has not totally solved its crime problems. People are being robbed, raped, and murdered in China every day (although not usually simultaneously). I should also state that there are no useful statistics published within China that would allow one to examine trends in the crime rate, how many people are incarcerated, the rate of recidivism, or most other things we would really like to know. In such circumstances writing about how the Chinese system of social control and rehabilitation actually works may seem foolhardy indeed, but there is of course the defence that in the land of the blind those of us who have been able to see something with one eye should try to share that information with others. In the pages that follow I will try to present the general outlines of how the system of urban social controls in China has worked in recent years, drawing much of my information from interviews conducted in Hong Kong with former residents of China who had left in the years 1972–8.[1]

The impression of 'no crime in China' conveyed in the accounts of many foreigners who have travelled in China is an understandable misconception. Not only are such visitors screened from most informal contacts with the ordinary populace, but the populace knows that the penalties for robbing or assaulting a foreigner are much more severe than are those for doing the same to a fellow Chinese.[2] However, an impression that in urban China there is less crime than in America or in most cities in other societies is probably justified. There is almost nothing in the way of organized crime operating in China, major crimes against property appear rare indeed, firearms are tightly controlled, and alcoholism and drugs as sources of criminal behaviour do not constitute major problems. Most Chinese urbanites seem to feel that their cities are quite safe places in which to live. More at the level of a hunch, we would hypothesize that recidivism does

* Reprinted, in a revised form and slightly abridged, from Susan E. Martin, Lee B. Sechrest, and Robin Redner (eds), *New Directions in the Rehabilitation of Criminal Offenders* (Washington, DC: National Academy Press, 1981), by permission of the publishers.

not constitute the problem in Chinese society that it does in our own, for reasons we will specify later. We cannot be sure of any of these statements without the chance directly to research them within China, but we think it is none the less justified to assume that China has a fairly effective system for controlling crime and rehabilitating offenders; we can then proceed to examine how this system works and what its weak points are. The reader will see as we go along that the Chinese system highlights the role of other institutions besides the police, courts, and lawyers in the effort to promote orderly behaviour in the populace.

Restrictions on Urban Residence

To understand the way in which crime is controlled in China (to the extent that it is), you have to start by considering the very different structural principles upon which urban life has been transformed since 1949. First, there is a system of urban household registration that is perhaps more restrictive than any found elsewhere in the world. It is not simply a matter of having to register with a police station when you come into a city. Urban residence is a privilege, and individuals who live in rural areas, towns, or even important cities cannot move and establish residence in a larger urban place unless they have been specifically assigned there for schooling or for employment (Tien 1973, App. L). Even a person who marries someone who lives in a larger city cannot take up residence there, a policy that contributes to the not uncommon Chinese phenomenon of married couples living apart. In addition, individuals and even entire families can lose their urban registrations, temporarily or permanently, and be stuck in a smaller urban place or in the countryside. For example in the years 1968–78 a total of 17 million urban-educated young people (out of a total urban population of 160–200 million) were 'sent down' to settle in rural communes and state farms, where they became holders of agricultural (rather than urban) household registrations (Bernstein 1977). Also, in the years immediately following the Cultural Revolution (1966–9) many families were forcibly dispersed from cities as undesirable elements and sent back to their native places, which in the context of Chinese culture were often places they had never been to. The system of restrictions on urban migration and household registration seems to be fairly rigorously enforced, so that visits to a city are always possible but long-term illegal (i.e. unregistered) residence there is very difficult. Through this strict urban registration system, some older established cities have actually succeeded in reducing their populations, with Shanghai declining from 7.2 million in 1957 to 5.7 million in 1972–3 (Howe 1977).

The restrictive registration system works as effectively as it does not only due to the vigilance of the police but because so many of the necessities of life in urban areas are bureaucratically controlled and require urban household registrations (at a minimum) in order for one to have access to them. Private ownership, voluntary associations, and market forces have a very limited range of operation in supplying popular needs in urban China. In the larger cities, particularly, there is very little privately owned housing (estimates by Thompson (1975) range from under 10 per cent for national cities like Peking and Shanghai to at most 25 per cent in major provincial cities). Most residents get access to housing only by applying to their work units or directly to city housing management bureaux, and they must satisfy a complex set of criteria, of which having proper urban registration is only one, to be allocated new or better housing. There is also little in the way of a free labour market; instead, schools, labour bureaux, work units, and neighbourhood governments work in concert to allocate people to jobs (or to rural exile). In some periods individuals from outside the city have been able to get temporary jobs in fields like construction, but getting long-term employment or an urban household registration is another matter. Additionally, many of the necessities of life are rationed. One needs ration cards, books, or coupons in most cities to buy grain, cooking-oil, cotton cloth, some synthetics, sugar, pork, fish, eggs, chicken, soy-bean curd, powdered milk, soap, coal, kindling, bicycles, watches, sewing-machines, and a number of other items. To get the needed coupons to make such purchases, one needs to present one's household registration book to the local neighbourhood and police authorities and/or to one's work unit, and from the characteristics of the family and its members it will be determined what the appropriate ration amounts are (White 1977). There are free peasant markets within cities where some of these commodities (e.g. eggs, pork, fish) can be purchased without ration coupons, at a higher price, and there is a limited black market in ration coupons for more tightly controlled items, such as grain and cotton cloth. None the less, it is very difficult for a person or a family without an urban registration to live in the city for any extended period of time. Similar comments could be made about access to schooling, child care, health care, and support of the aged—for these and many other services one needs to be registered in the city in order to be eligible to benefit.

This system of urban household registration has a number of important consequences. Some of these we have already stressed. It is not possible to decide at will to move into a city, it is very difficult to arrange to be transferred there from a smaller place, and it is close to impossible to live in a city for a long time without proper registration. From the standpoint of the authorities, this means that there are not many people in cities who don't have an approved reason for being there. There are of course still visitors coming to see kinsmen and friends, people sent there on business trips, foreign tourists, and a few other categories of people, but none the less there are fewer 'strangers' lurking about than would be the case in cities almost anywhere else. Another consequence is that cities are pretty lean and productive places these days in terms of their use of human capital. For instance Shanghai had 34 per cent of its population employed in 1957, but 53 per cent employed in 1972 (Howe 1977). China has not solved its 'unemployment problem', since large numbers of young people for whom there are no jobs are simply shipped off to the countryside, and some people for whom there is no work are still permitted to stay in cities. But Chinese cities lack a large underclass of unemployed people for whom crime might constitute an alternative means of livelihood.

A further feature of urban life stems from the bureaucratic control over housing and jobs and from the fact that little in the way of new housing construction has been carried out since the 1950s in China's cities. For these and perhaps other reasons, most families tend to stay put for long periods of time, rather than moving from one home to another within cities. Most neighbourhoods, except for those that have developed in connection with new industrial developments, house people who have lived there for a long time. In addition, three-generation families are still fairly common even in China's largest cities, so that there are usually old people around in a neighbourhood even when the able-bodied are all off at work.[3] And in major cities, work units may stagger their days off, so that on any day of the week there will be residents of a range of ages present in the neighbourhood. In the evenings there is little night life or organized recreation, so the primary form of leisure is simply relaxing and chatting with friends and family where one lives before turning in. Furthermore, high-rise buildings and elevators are as yet rare in Chinese cities, and private cars non-existent, so that urban neighbours are probably in closer regular contact and less often 'escape from it all' than is the case in our own society. Taken together, it seems to us that these characteristics predispose Chinese cities to be more orderly places than are the cities we are familiar with, even without considering the institutions that are specifically designed to control the activities of people living there. The organizational systems into which urbanites are grouped turn out to be fairly distinctive as well.

The Urban Organizational System

There are two fundamental principles by which urban life is organized: the work unit and the neighbourhood. For the minority of urbanites whose work units have housing compounds attached to the work-place that house most employees, neighbourhood organizations may be irrelevant, as the work unit will supervise most facets of the lives of its employees and their families. An inclusive 'company town' setting may be fostered within a complex cityscape. The work unit of course supervises the activities of its employees on the job, and it may also run nursery schools and kindergartens, health clinics, recreational facilities, sports teams, and other activities for those within its orbit. Employees who wish to marry must receive approval from their work unit authorities to do so, as for that matter must those who wish to divorce. Work units regularly convene meetings of their employees and their dependants to discuss the government's family planning campaign, they may dictate which families are allowed to have more children and which are not, and they will pressure women who get pregnant 'over the quota' (of two children) to submit to an abortion. If the son or daughter of an employee has been assigned to settle in the countryside after graduation from secondary school, the work unit will organize meetings of the parents to pressure them to persuade the child to comply with the decision, and may even play a role in deciding what rural locale the children of employees will be assigned to and in sending personnel to monitor how the young people are doing out in the countryside. In general, problems of inadequate housing, poverty, or even the inability to find a spouse may be referred to work unit authorities in hopes of assistance. Employees who have unemployed spouses or parents living with them may see these people organized into 'dependants' committees', in which such people run small workshops, sewing-centres, and other facilities where they can make some productive contribution.

If an employee commits a crime or engages in disorderly behaviour elsewhere and is apprehended, the police will work with the security officer[4] and party officials of the unit to decide how to handle the case. If they think it useful, they may decide to bring the offender before a large meeting of employees to be criticized or 'struggled' for his waywardness. If the offence is a minor one and the employee is sufficiently repentant, they may decide to let him off with a warning, a reprimand, or a penalty of 'supervised' or 'controlled' labour. In these latter states, the employee continues to work at his regular job but is subject to special scrutiny from unit security personnel and is required to periodically report on his activities and attitudes (Cohen 1968, ch. 5). The police may decide that the offence is more serious, warranting penal confinement, with or without a criticism and sentencing meeting in the work unit. In some cases, after the offender's term has been served the work unit may agree to take him back, and in such a case the work unit would again play the major role in keeping the released person under special scrutiny, in a state not much different from supervised labour.

Work units of course vary in their size, the extent to which they provide employees with housing, the proportion of housed families that have members working in other work units, the diligence of authorities within the unit in organizing the after-hours activities of employees, and other dimensions. We would hypothesize that in the most inclusive work units, with employees and their families heavily dependent upon and involved in unit activities, social control will be at its most effective, and problems of crime and rehabilitation will be least troublesome. However, most Chinese urban residents do not live in such an all-inclusive, company town-type setting. Most work units do not provide housing for all or even most of their employees, some housing assigned by work units is dispersed within ordinary neighbourhoods, rather than being attached to the unit physically, and in most neighbourhoods individuals are employed in a variety of units spread out over the face of the city. Some may be employed in the small co-operative factories and service facilities that are administered by the neighbourhood itself, and within any one family there may be ties to several different work units (and schools). For the majority of the city residents who exist in this more complex setting, there will be two distinct social control systems operating: those of the various work units (and schools) and those of the neighbourhood. Employed individuals still have to go to their work unit with requests to marry, demands for new housing, reports on contraceptive use, and most of the other matters discussed above. But they will also spend much of their free time away from their work unit, and most of this will be spent in their residential areas. There they are to a certain extent under the authority of a set of neighbourhood organizations, and members of their families who do not have an outside work unit or school are more totally subject to this neighbourhood authority.[5] The role of these neighbourhood organizations in urban social control is again a distinctive feature of Chinese society.

Cities are generally divided into a small number of districts, and these districts are in turn divided territorially into units called neighbourhoods or streets (in fact the terminology varies somewhat from one city to another). Each neighbourhood (of, say, 2,000 to 10,000 families) is administered by a neighbourhood revolutionary committee composed of salaried state cadres, and also by a police station.[6] The neighbourhood is then subdivided into residents' committees (supervising 100 to 800 families), and these in turn are subdivided into residents' small groups (each supervising about 15 to 40 families). The small group has a single leader, the group head, while the residents' committee has several officers: a chief, assistant chief, security-defence officer, and several others. These people live in the locality and are usually appointed by the neighbourhood revolutionary committee or are nominated by them and approved by the residents in a meeting (Cohen 1968, ch. 2). Generally, only the residents' committee chief receives a small stipend, and the rest perform their duties as an unpaid social obligation. Most of the leaders at these two bottom levels are middle-aged women from good-class backgrounds,[7] but they may include a few retired or disabled male workers or educated youths allowed to stay in the city. On paper the residents' committees and their small groups are primarily subordinate to the neighbourhood revolutionary committee, but in the eyes of residents it is actually the neighbourhood police station that plays the greatest role in supervising their activities, particularly those that involve social control. It also seems to be a common practice to have a particular police officer from the neighbourhood station assigned to work with one or a few residents' committees on a day-to-day basis, and for the officers of these committees and their subordinate small-group heads to take direction primarily from this individual (Vogel 1971).

This set of organizations represents a much more penetrating organizational system than exists in cities elsewhere, and it is important to note that, while the lowest levels of it are composed of unpaid individuals and are thought of as constituting 'mass associations', they in no sense represent voluntary associations in the way this term is understood in the West. The entire structure is mandated by law, its leaders and activities are directed by higher levels, participation is required of residents, and residents are prohibited from forming their own autonomous locality associations to pursue grievances or organize activities. Residents must work through this organizational system, or try to work around it.

The activities organized by neighbourhood revolutionary committees, residents' committees, and residents' small groups vary somewhat from place to place but are always fairly broad. It is usually the neighbourhood revolutionary committee that establishes and supervises neighbourhood factories and service facilities. Many neighbourhoods will have anywhere up to a dozen or so small factories and workshops producing buttons, chopsticks, paper bags, matches, small hardware objects, and a variety of other products, some of them produced through subcontracting arrangements with larger, state-run factories. Neighbourhoods also often run small barber-shops, bicycle repair shops, snack stands, hot-water vending stations, nursery schools and kindergartens, paramedical stations, and reading-rooms (but rarely laundries), and they often license individuals in the neighbourhood to go out and sharpen knives, mend pots, sell popsicles, or work as day-labourers in transportation or construction. The mandate of neighbourhood-run economic activities is a dual one; they are designed in part to provide employment for idle labour power in the neighbourhood and in part to provide needed services for residents. Most of the small neighbourhood factories and service facilities are too small to have the sorts of resources and control over employees described earlier for state-run work units, and so their personnel (many of them former housewives, disabled people, and educated youths) are the responsibility of the neighbourhood revolutionary committee and its subordinate organizations.

Residents' committees have functions that are mainly non-economic. They work in conjunction with the police station and state commercial agencies to distribute the ration books and coupons residents require. The police will check the family members in the household registration book against their dossier on the same family and then certify the family's eligibility to receive the various ration books and coupons. When visitors come to stay with a family they are also required to register temporarily, either with a residents' committee officer or directly with the neighbourhood police station. If a small-group head or residents' committee officer suspects a local family of harbouring an unregistered individual, they notify the police station and escort the police to the home in question for a night-time search, as a result of which anyone staying there without proper papers may be hauled away to a detention station. The residents' committee and small-group leaders also work on cases of urban-educated youths who have been designated to go to the countryside but have not yet agreed to do so. They organize indoctrination groups for such youths, and they repeatedly come to their homes and pressure them and their parents until they consent to go.

Residents' committees and small groups also have important responsibilities in the areas of public health and sanitation. They notify residents of inoculation drives, they distribute fumigation chemicals to guard against pests, and if the neighbourhood is situated off a main street they collect contributions from all residents to hire a local person to sweep the streets on a regular basis (large streets are cleaned by the city sanitation department). Usually each season the city authorities will announce a clean-up campaign for all residents, and then residents' committee officers go from door to door making cleanliness inspections of people's apartments and pasting stickers saying something like 'cleanly household' on the doors of those who pass muster. These low-level leaders also play a role in the national birth-control campaign, organizing residents (primarily women who are not employed by outside units) to discuss the government's family planning goals, to fill out forms on contraceptive use, and to accept an abortion if they become pregnant 'over the quota'.

Residents' committees and small groups also regularly convene meetings of those residents not employed by outside units for the purpose of engaging in political study. These meetings will occur once a week or so in ordinary times, but more often during a political campaign or after some major national event. In these meetings the leaders transmit directives and ideas from higher authorities explaining changing official policies and interpretations of events, or they simply have residents read newspaper editorials or political study pamphlets and discuss their meaning and relation to the local situation (Whyte 1974). Often if there is a group of educated youths in the neighbourhood who have been allowed to stay in the city but have not yet had jobs assigned to them, the residents' committee will organize them to write banners and slogans for local posting and in general to serve as performers of whatever good deeds need performing. (Such youths are of course given to understand that their enthusiasm for these make-work tasks may affect the kind of job they eventually get assigned.)

On balance it seems clear that the central role of these lowest-level units concerns social control. Residents' committees organize small-group heads and local 'activists' (such as the job-aspiring youths just mentioned) to assist the police in keeping an eye on what is happening in the neighbourhood. In theory, at least, they are to report strangers whose presence is unaccounted for, neighbours engaging in suspicious activities, illegal pedlars, black marketeers, and other threats to social order. During times of disorder or for national holidays or the visits of important foreigners, the residents' committees may organize each family in the neighbourhood to take part in round-the-clock patrolling of the area. The police also periodically run campaigns in which they call meetings and put up posters to inform residents of the need for vigilance against violators of public order, class enemies, and spies, and to encourage the people to assist the police in carrying out their duties. Small-group heads are expected to remain familiar with what is going on in each family through regular, informal visits and contacts, and to use this familiarity as a basis for fulfilling their responsibility to report any untoward developments to the police. Residents are periodically convened for meetings to discuss cases of major crimes and apprehended criminals from other parts of the city, as a means of heightening their own vigilance. And if a local resident commits a serious offence, he may be brought before a mass meeting of residents for 'struggle' prior to being placed under 'mass supervision' or being sent off to penal confinement by the police. Each residents' committee has a special officer, called a security-defence officer, whose special responsibility is to assist the police in securing public order in the neighbourhood. Certain families in the neighbourhood that get designated as 'problem cases' for one reason or another will be subject to the special scrutiny of this officer as well as of the police. Individuals who have been assigned to 'mass supervision' or 'mass control' within the neighbourhood have a special set of restrictions placed upon them. They will not be allowed to participate in certain neighbourhood meetings in which national directives are communicated to the population. They have to submit periodic reports on their thoughts and activities to the security-defence officer or directly to the police station. They may be required to do periodic 'voluntary' labour stints around the neighbourhood. They may have to request special permission to make trips or to stay away from home overnight, and on occasions such as national holidays or the visits of important foreign dignitaries they will be rounded up and confined for 'study' for the duration. Released convicts are subject to a variable set of such restrictions, apparently partly reflecting the degree to which they are judged to be potentially harmful or harmless. The handling of civil disputes also fits within this framework. In earlier years there were generally mediation officers or committees as part of each residents' committee, but recently most localities lack such a formal title or office (Cohen 1971; Salaff 1971). However, when severe marital or interfamily disputes break out in the neighbourhood, the residents' small-group heads and residents' committee officers are expected to intervene and try to mediate and calm things down, although they are not empowered to enforce a particular settlement on residents. The multiple mechanisms for social control built into the urban neighbourhood structure do not mean that other, more positive neighbourhood activities are ignored, but they do illustrate the very geat concern the Chinese Communists have for social order. It can be argued that we see here a conviction that, even after 'progressive' social reforms and changes in the distribution of property and income, cities are still likely to be very unruly places unless vigorous steps are taken to make them otherwise.

Crime Control and Rehabilitation

The combination of restricted access to urban residence and highly penetrating and instrusive neighbourhood organizations creates a fairly effective environment for social control in Chinese cities, and in our judgement goes far towards accounting for the impression that these cities are fairly safe places with low crime rates (again, an impression we judge to be correct, in the absence of accurate data). Urban social control and rehabilitation in China thus depend upon a social structure in which there is almost 'no place to hide', or, in the Chinese phrase, no 'dead corners'. Urban centres are stripped of most individuals not employed or needed there, those remaining have fairly fixed locations and obligations, and a pervasive system of controls and of grass-roots organizations under the command of higher authorities makes it difficult for activities prohibited by such authorities to escape notice, as well as making it impossible for large-scale organized crime, including prostitution, narcotics, etc., to exist.

At this point it is important to specify several possible reasons for low crime that we don't see as the primary ones explaining the general orderliness of Chinese cities. We are not arguing that the change from capitalism to socialism or changes in income distribution are the solutions to urban disorders. Clearly, urban residents often live in cramped and poor conditions while high-ranking persons within sight of them have many advantages, these residents have many grievances against the restrictions within which they live, and sources of interpersonal tension and anxiety are by no means absent. Even unemployment would exist on a major scale were not much of it exported beyond the city walls. We would also be sceptical about an explanation that places primary emphasis on the use of Maoist ideology to transform urbanites into new socialist men who are unlikely to disrupt social order. Were the process for transformation of mentalities so effective, there would be no need for the involved systems of social control that we have sketched. We should also clarify that we do not see the tight-knit neighbourhood organizations creating a situation in which all residents are in fact constantly vigilant and always actively helping the police and reporting on their neighbours. It does not take such mass co-operation, but only a limited number of residents' committee officers and activists in a neighbourhood, to make it risky for anyone to engage in unorthodox or criminal activities there. Furthermore, since there are almost no written laws, and official policies are subject to change (and retroactive guilt is allowed, if not encouraged), the control system promotes a syndrome of trying to be extra cautious in order to avoid not only present but future problems.[8] Finally, the degree of local orderliness is not dependent on a neighbourhood having a high degree of solidarity and on individuals having a fear of being publicly shamed by being accused of misconduct. Even if in a particular neighbourhood everyone hates their neighbours, they will feel that disapproved behaviour is unlikely to go undetected, and in fact one could argue that precisely in such neighbourhoods people will have to be extremely cautious in their behaviour and expressions. We would argue, then, that it is primarily a tight-knit organizational and control system in Chinese cities that explains their relative orderliness, and that in this control system police play an important but not a predominant role while courts and other legal mechanisms play practically no role at all.[9]

Before turning to the weaknesses in this control network, we must consider the system of penal rehabilitation in somewhat more detail. We have argued that on even more impressionistic grounds we feel that recidivism is not a major problem in China (see also Vogel 1971, 90). At issue is whether, if this is the case, it is due to an effective method for transforming the attitudes and behaviour of inmates of penal institutions, or due to a system of controls that restricts them effectively while they are in penal institutions and after they leave. The available evidence again leads us to stress the latter explanation.

Most of those sentenced to penal confinement for either criminal or political offences are sent to labour reform or labour re-education camps in the countryside, rather than to conventional prisons. The two separate networks of camps differ somewhat in the types of inmates and their length of sentences, but their regimens are enough alike for us to discuss them together here. There are two primary activities engaged in by camp inmates: manual labour and political study. The camps vary somewhat in the kinds of work assigned, but typically inmates engage in farming, lumbering, mining, or other heavy labour tasks, often for ten hours a day or more. So far as one can determine, there is no emphasis on giving inmates new work skills or increasing their educational credentials. It is not assumed that they wound up as inmates due to lack of qualifications for an honest job, but due to poor political attitudes and disrespect for labour. In any case, inmates will not leave the camps with a set of skills or increased technical knowledge that will be of advantage to them in civilian life. Most evenings in the camps are spent in political study meetings of camp small groups, much like the residents' small groups in conception. Inmates are expected to read newspapers, study pamphlets and other materials, and discuss their own poor attitudes and failure to conform to Maoist ideals, including (ideally) any thoughts they may have of wanting to escape from the camp. It is here that the transformation of inmate attitudes is supposed to take place. However, my own research leads me to conclude that the social structure of the camps, with inmates confined and assigned to groups arbitrarily, as well as overworked and underfed, is singularly unconducive to the desired personal transformations. The successes of the camps are more of a negative sort: the same tight-knit organizational system and controls that exist in civilian society inhibit the development of any sort of 'inmate subculture' dominated by hardened criminals and thus make penal institutions less likely than prisons in our own society to function as 'schools for crime'. We would argue that the typical response to such confinement is likely to be neither personal regeneration nor hardened criminality, but controlled resentfulness and fear of running afoul of the authorities again (Bao and Chelminski 1976; Lai 1969; Whyte 1974, ch. 9).

When their terms are completed, only an unknown portion of the inmates get to return to their former homes and jobs. Both labour reform and labour reeducation camp regulations have provisions allowing the public security (police) authorities to keep released inmates on as 'free labourers' housed in communities next to the camps. It is uncertain how large a proportion of released inmates are retained in this manner, or what the specific criteria used to make the decision are. However, it is fairly clear that both the public security authorities and the inmate's former work unit and/or neighbourhood authorities must feel that he is not likely to cause future trouble in order for them to approve a return to the inmate's city of origin. If these judgements are accurate, we again see a Chinese tactic of excluding threats to social order (in this case potential recidivists) beyond the city walls.

Former inmates who do return to their original city and work unit are still subject to some special controls and scrutiny, although these are applied not by professional parole workers but by security-defence officers and local activists in the work unit or neighbourhood, as previously noted. There clearly is some stigma attached to having been incarcerated and then returned to the community. Furthermore, the immobility of Chinese society means that it is virtually impossible to leave and 'start a new life'. Even if one could change one's place of residence, the police station in one's new locale would receive notification that you were a person to be kept under special scrutiny. There may be exceptions to this pattern—a post-release exoneration, for example—but for the most part our informants describe released inmates as a very downtrodden, close-mouthed group, just trying to live from day to day and not do anything to antagonize local authorities. This general picture of how the Chinese system of penal rehabilitation works is what leads us to conclude that there is minimal personal transformation and repentance involved, and that serious problems of recidivism are avoided primarily through a very effective system of controls that leaves even disgruntled and alienated members of society with no place to hide.

Weak Spots in the Control System

We cannot fail to note, however, certain problems with the urban social control system that make it less effective than our foregoing general discussion might lead one to believe. Several of these come to mind. First there is the fact that the solidarity of a neighbourhood is to a certain extent contrived and manipulated, rather than based upon natural roots, voluntary associations, and joint ritual life. This is particularly so for those outside of housing compounds attached to large work units. In such a mixed neighbourhood, people will have ties with work units of very different types and located in different places and some families will have all their members tied into such outside units, leaving no one available to participate in residents' committee-sponsored activities. Residents are also likely to be heterogeneous in social background, family composition, and stage in the life cycle. In part this heterogeneity results from the bureaucratic system of housing allocation, which minimizes the role of market forces and personal preferences in determining who lives where. This heterogeneity and involvement in outside units tend to produce recurring tendencies for minimal involvement in neighbourhood activities and make it difficult for residents' committee personnel to get residents concerned about problems of social order that occur beyond their own doors. In many neighbourhoods residents live in very cramped conditions (generally with under four metres of living-space per person) and have to share toilets, kitchens, and other facilities with neighbours, and while this may promote mutual familiarity, as noted earlier, it may also promote conflict and resentment and weaken any sense of mutual responsibility among neighbours. We discussed earlier our guess that such neighbourhoods, which are not only heterogeneous but also filled with conflict, will spawn problems and at the same time have a high level of reporting such problems to the police. In any case, we argue that most neighbourhoods do not have characteristics that would foster a strong sense of mutual responsibility and general involvement in residents' committee activities.

A second kind of weakness in the control system is the low prestige and lack of positive resources of the neighbourhood leaders. Many of those involved are middle-aged women, often poorly educated (particularly the leaders of small groups). They do not usually have personal characteristics that would earn them public respect, and if they did they would be unlikely to be serving in such menial posts, most of them without compensation. Although they have some positive services to perform, they cannot really help residents much in solving the major problems of urban life—getting better housing, better medical care, a rewarding work post, and so forth. But they can get you into trouble—they are the ones who can call the police down on you in the middle of the night if they suspect you of something, and they are the ones who pressure reluctant youths and their parents to get the youths to go settle in the countryside. Informants feel that if you get on the wrong side of them they can make your life miserable, although they may be helpful and considerate to families they like. And if you are in conflict with them there is really no easy way out, since you cannot easily move to a new residence, and they cannot be removed by action of the residents but only if they displease the neighbourhood revolutionary committee and police station personnel. At the same time, many of these residents' committee and small-group leaders are old and not very spry, so that they have a difficult time coping with any serious (or numerous) social control problems in their neighbourhoods. The result of all of these features is that many residents may try to stay on the right side of the local leaders and say the right things in their presence, but curse them behind their backs and feel no guilt about maintaining only a minimal involvement in neighbourhood activities.

A third problem in the social control system involves the system of supplies and rationing. It has been suggested by one authority (White 1977, 1978) that there is a curvilinear relationship between the tightness of the urban rationing system and popular compliance with official controls. When ration levels are set too low, making it difficult for residents to subsist on them, urbanites may be forced to rely increasingly on black markets, speculation, and other shady activity to supply their needs. On the other hand, if ration levels are quite sufficient and if many needed food items are sold in state markets without rations, then individuals may be able to supply their needs even if they don't have proper urban registrations and ration coupons. The ideal situation, from a social control standpoint, would seem to be one in which ration levels are just sufficient to supply popular needs while markets have only a limited amount of unrationed commodities. Over the last several decades there have been fluctuations back and forth between severe shortages and a relative abundance of supplies, rather than a stable, 'optimal' level of needed food and other items.

The urban social control system can also weaken if political and other changes run counter to the control principles upon which it is based. This occurred in a major way in the decade after 1966, and we can gauge the weaknesses of the system most clearly by considering what went wrong in those years. A set of changes initiated by the Cultural Revolution, which began in 1966, contributed to a rise in urban social disorder and crime and was particularly manifested in growing juvenile delinquency. One major source of these problems was the fact that, beginning in the early 1960s, urban secondary schools began to turn out more graduates each year than the urban economy could absorb. Then, in the Cultural Revolution itself, millions of urban students were involved in Red Guard activities, rebelling against the leaders in their schools, cities, and the nation, and in the process gaining experience in political debate, physical combat, and even the handling of weapons. This period of extraordinary freedom and power for students, approved initially by Mao and other national leaders, was brought to a halt in 1968 by an effort to end the disorders of the Cultural Revolution. Red Guards were simply declared 'graduated' and then mobilized to go settle in the countryside and remould their thinking and behaviour. This campaign to resettle urban youths in the countryside continued in subsequent years, with 1 to 2 million urban secondary school graduates resettled each year. But the campaign was not very popular with most urban youths, nor with their parents. It meant that they would have to leave their families and settle in distant and poor villages, ones sometimes inhabited by minority groups speaking a different language, and they were given very little hope of ever being able to return to an urban area, much less to their native city. Urban Chinese with strong family and mobility orientations resented the 'bitter fate' this represented for their educated young people and the threat this posed of ending one's days living alone rather than with one's offspring (Bernstein 1977; Davis-Friedmann 1979). Many peasants and rural authorities also resented the intrusion of these urban youths, who did not have proper respect for rural customs and often consumed more than they produced.

Within a short period of time, it became common to have large numbers of these sent-down youths returning illegally to their cities of origin.

Although they no longer had urban household registrations, they could still live with their families and rely on family belt-tightening for food, perhaps supplemented by the rural free markets or the urban black market. Their parents were often sympathetic to their desire to come back to the city or unwilling or unable to force them to return to the village. Some residents' committee and residents' small-group leaders were sympathetic to the plight of these youths and their families (some had a child in the same boat, of course), and others were unwilling to alienate large numbers of local parents by alerting the police to these long-term 'visitors'. There might be so many of such illegally returned youths that neighbourhood authorities would have a difficult time monitoring them all, and they might also be afraid of antagonizing the youths. These were, after all, some of the same youths who had been 'steeled' in battle in the Cultural Revolution, and neighbourhood authorities might recall that returned sent-down youths of an earlier period had criticized and attacked neighbourhood leaders in the early stages of the Cultural Revolution. Returning young people often prolonged their stays by guile rather than intimidation, however. A number of dodges became popular and were used on neighbourhood leaders and police who wanted young people to return to the villages. Youths would present brigade certificates (real or forged) stating that they had permission to leave the village, they would claim they needed urban medical care or had to attend to some family crisis, or that they were planning to leave the next day. If things got too insecure in their own neighbourhood, they would go and stay temporarily with a friend living elsewhere in the city. Even if none of these devices worked, the shipping back to the village of an illegal returnee might be only temporary. A common pattern was to stay in the village for a month or two and then to head back to the city again, perhaps with some grain earned for agricultural labour and some choice food delicacies to use to curry favour with neighbourhood authorities. Youths might repeat this pattern over and over again, and generally they received no sanction except forced return to the village as long as their only offence was a desire to avoid rural hardships.

Some, perhaps most, of these illegally returning youths spent their time while in the city in a peaceful fashion—reading or studying at home, visiting friends, engaging in sports, and so on. But at least a portion of them were not so well behaved. Some began to engage in proscribed private trade activities, black marketeering, picking pockets, and petty thievery. It should be noted that there is a prohibition against employing such youths in urban work units, so any income-earning activities they developed had to be shady, if not illegal. No doubt both leisure time and a desire for excitement, as well as economic need in the absence of jobs and rations, contributed to the growing wave of youth crime that developed.

The fact that these youths had partially safe bases in their own homes made them more difficult to control than would be the case for strangers from outside the city. Even if some of these youths were apprehended and publicly made a negative example of, it is unclear that this would have much shaming or intimidating effect on other illegal returnees, who saw themselves living on the margins by their wits rather than becoming well-integrated members of the community with much to lose.

Important changes in two other sectors of society also played into the growing problem of youth crime. One of these involved the educational system. The Cultural Revolution produced important changes in urban schooling, particularly a de-emphasis on academics and on competition for selection for entrance into the university. Several months a year were spent not in the classroom but out working in factories and communes in order to learn to respect manual labour. Within the classroom the curriculum was simplified and politicized, tests, grades, and homework were de-emphasized or omitted completely, and promotion to the next grade was made automatic. However, it became impossible to go directly from secondary school to the university, and even access to upper-middle school (roughly our senior high school) was made contingent more on factors like class background, age, and political activities than on school performance. The great bulk of secondary school students knew full well that they would be sent to live in the countryside after graduation, regardless of the academic skills or industrial experience they had acquired. (More-favoured class-mates might, however, be selected to join the army or start work in an urban work unit instead.) The mass media also encouraged a spirit of students 'going against the tide' and criticizing their teachers as 'bourgeois academic authorities' (Pepper 1978).

The result of these trends was a rise in problems of discipline in schools, problems that spilled over into society. Among some students the mentality that schooling has no value became prevalent, and low motivation in school resulted. Teachers also became afraid to place strong academic demands on their students for fear of being criticized, and students felt that how they did in school could have little effect on their future lives. Parents also increasingly felt there was no point in placing high demands for school performance on their children, and were more and more at a loss as to how to help their children prepare for the future. The absence of sanctions such as failure and expulsion for poor performance in school meant that problems of truancy, vandalism, and even physical assault increased within schools, and students also got involved in petty thievery and other misbehaviour outside. In many large cities juvenile gangs emerged at this time, with bands of youths fighting with sticks and knives over territory and prestige in a pattern familiar elsewhere on the globe. City residents became increasingly aware of 'black societies', involving such youths and others, that engaged in crime and a variety of unsavoury activities (drinking, gambling, and even some reported prostitution). The students differed from the illegal rural returnees in having both urban household registrations and urban units (their schools), which gave them a more secure foothold in the city. (Young workers also played a role in the growing crime problem, as urban factories experienced some of the same discipline problems that schools did.)

The other major change affected the urban police system. During the Cultural Revolution, the public security (police) system had been attacked nationally and locally as an oppressive agency and had been thrown into some disarray. In the immediate post-Cultural Revolution years, cities were patrolled by military units supplemented by various kinds of civilian patrol units. In many cities, 'workers' picket corps' units patrolled urban neighbourhoods with clubs trying to keep order until the public security system was fully operational again. As the police began to take over the major role again, they remained under some pressure to deal with minor offences by members of the masses (those without 'bad class background' labels) by persuasion rather than coercion, which meant taking offenders to a detention station, lecturing them for a few days a week, extracting promises to behave, and then releasing them. The picket corps might turn over a suspect to a detention station for police handling, or they might mete out 'mass justice' on the spot, for example, beating a purse-snatcher severely and sending him fleeing back home. Neither of these methods was judged very effective by our informants. They saw a revolving-door phenomenon occurring, in which brazen youths would get in trouble repeatedly but would never receive serious sanctions and would not change their ways. Many citizens apparently began to feel frustrated at the lack of effectiveness of the authorities in dealing with what they saw as a growing crime wave, and some informants even report instances of ordinary residents viciously beating up a thief they caught in the act, rather than turning him over to the police for handling.

In the decade after the Cultural Revolution, then, Chinese cities were plagued by rising crime and disorder, with young people seen as the major source of the problem. Our informants uniformly perceived a marked deterioration in public safety in these years, compared with earlier periods, with increased danger of being robbed, assaulted, or even worse. Major property crimes and violent crimes were probably still uncommon compared with most other societies, but residents took increasing precautions—to stay home at night, to avoid 'gang turf', to lock their doors, and to bring property such as bicycles and drying laundry inside their apartments at night (Foreign Broadcast Information Service 1979). A combination of very large numbers of bold and adventurous young people with footholds in urban life through their families, but without opportunities for mobility within the city or major organizational ties and commitments, plus some weakening of the ability of schools and neighbourhoods to deal with disorders formed the major ingredients in this growing problem of youth crime.

In the years since 1974 a variety of steps have been taken that seem to have reduced these problems somewhat, although they are still a major concern for authorities. The sending down of youth was altered somewhat so that most urban youths were not required to settle so far from their native city, were given more economic subsidies while in the village, and were given increased hope of eventually being transferred back to an urban job. The programme thus evolved increasingly into a rotation system, rather than an exile for life (Bernstein 1977). These shorter and less distant rural sojourns were somewhat more palatable, and more urban youths were motivated to perform well in the countryside in hopes of being transferred back to the city sooner, or to a more desirable job. Changes have also taken place in the educational system, with tests, homework, failures, and other accoutrements of highly competitive academics back in style in a big way. Strict classroom discipline and respect for teachers are also mandated now, and stints working in factories or communes have been reduced to nominal periods. Beginning in 1977 it became possible for urban secondary school graduates to do well on a placement exam and be selected to go directly to the university, thus avoiding a rural labour stint after graduation entirely. After 1980 the rural sojourns were discontinued entirely, and major steps were taken to provide a greater number and variety of urban jobs in order to cope with the large number of unemployed urban youths. And the picket corps and other informal policing agencies have been phased out increasingly while the police have begun major crackdowns on crime and have expanded their youth reformatories to handle more youths on a long-term basis (Butterfield 1979). Informants report that in the last few years urban social order has improved as a result of these changes but is still not back to what it was prior to 1966.

Cultural Continuities and Discontinuities

To what degree is this system of social controls a distinctive product of Chinese traditions and customs? The social control system we have sketched has a number of roots and precedents in traditional Chinese culture, but it none the less constitutes something new in very important ways. Population registration systems existed over long periods of Chinese history and at some times required houses to have signs posted by the door listing all family members and their characteristics. Intermittently in various dynasties, families were formed into groups of families that were supposed to have mutual responsibility for the orderly behaviour of all their members (Leong and Tao 1915, 61–3). There is also a long tradition in China of expecting the rulers to set moral standards for all members of the community, and of regarding the ability of the government to maintain public order as a primary indication of that government's legitimacy. By the same token, unorthodox thoughts and actions were more likely to be interpreted as a threat to public order, and less likely to be seen as healthy diversity or the exercise of natural rights, than has been the case in our Western tradition (Munro 1977).

Yet the population registration and mutual responsibility systems of past dynasties were not the same as those of today. Large-scale movements of people back and forth between rural and urban locales were common throughout Chinese history. There is no lengthy Chinese heritage of not being able to change one's job or place of work at will. There was a healthy tradition of voluntary associations and providing of social needs from kin and other natural groupings, rather than from the government (Leong and Tao 1915; Skinner 1977). The bureaucratic system of control over jobs, food, housing, and most other necessities is really quite a departure from traditional norms, and for many other reasons those people raised in the 'old ways' often find it difficult to fit into post-1949 China (Whyte 1974, ch. 3). Chinese cities were also not such orderly places in the past. This is after all a society known for its urban secret societies, opium dens, and unemployment. Perhaps some elements of the current system of controls have traditional roots, but still on balance this system represents a major change in the way society is organized.

Notes

1. The research project, entitled 'Urban Life in the People's Republic of China' and supported by a grant from the National Science Foundation, is being carried out in collaboration with William L. Parish of the University of Chicago. During our year-long stay in Hong Kong, we conducted semi-structured intensive interviews with 133 individuals who had lived in cities large and small in various parts of China (but with a heavy overrepresentation of cities in the southern province of Kwangtung, which is adjacent to Hong Kong). The interviews consumed 1,382 hours in total and resulted in 2,369 pages of single-spaced typed notes covering many different aspects of urban life, including those dealt with in this paper. The urban system as described here is the one that existed in the 1970s. Since that time some, but not all, features of this system have been modified. For information on the impact of these modifications, see the monograph produced by our research (Whyte and Parish 1984).
2. In the spring of 1978, for instance, a man later described as deranged attacked a

party of foreigners in Peking that included the actor William Holden, and stabbed one individual in the shoulder. Within the week, the attacker had been taken out and shot. Our informants also stated that they were repeatedly made aware of the fact that foreigners were to be treated with extreme caution and respect, and that the penalty for an offence against a foreigner would be especially severe.

3. There are also few 'broken homes' in the sense that these are written about in literature on juvenile delinquency in the West, since official policies strongly discourage divorce and require repeated mediation of marital disputes. However, two other family situations may not be as favourable in terms of the family's role in fostering orderly behaviour of its members. A certain number of homes are 'broken' because of job assignments to places away from the parents or because of marriage of people with household registrations in different communities. Also, a very high percentage of younger mothers are fully employed, as are fathers. Thus there are a fair number of 'latchkey children' who don't have parents around to supervise them after school, and if no grandparent is at home and the school makes no special effort to supervise them, they may not receive much supervision except informally, from neighbours.

4. The security officer or committee is in charge of handling general social order problems and serious deviance cases in the work unit and co-operating with police agencies outside. This is a much broader role than that played by security guards in our own society, but this person is part of the work unit administration, rather than an employee of the police. There is a comparable role within neighbourhood organizations, the security-defence officer, that will be discussed below.

5. Generally speaking, families that have at least one adult member who is not employed or enrolled in school outside must send a representative to residents' committee and residents' small-group meetings and must participate in other activities organized by these bodies. Families all of whose members have ties to outside units do not have to send representatives to these meetings generally, but they would be involved in neighbourhood clean-up campaigns, household inspections, and other local activities described below, and would generally get their rations through the neighbourhood organizations, rather than through their work unit. In some periods, efforts have been made to foster more active participation of these people tied to work units in neighbourhood activities, but never with much success, given the heavy time and energy involvements people maintain in their work units (Salaff 1971).

6. The term 'revolutionary committee' emerged during the Cultural Revolution, and after 1979 it was phased out in favour of the term that had been used previously, simply 'neighbourhood office.'

7. Class background refers to labels affixed to families based on their economic position just before 1949, such as landlord, rich peasant, poor peasant, capitalist, worker, etc. These labels were maintained in later years and passed on patrilineally, even though they no longer had any clear connection to the economic position of the family (Kraus 1977). Only in 1979 did the government indicate its desire to phase out this system of labels, through reclassifying those

with 'bad labels' (e.g. landlord, rich peasant) under more neutral terms (e.g. commune member).

8. Much of our discussion throughout has been based on the notion that there are certain objectively specifiable behaviours that can be identified as crimes and that do not fluctuate sharply over time or depend upon unpredictable whims of enforcement agencies for their identification. This view is of course an over-simplification in our own society, as theorists of labelling, Marxist, and other perspectives point out, and it is also an over-simplification for China. Activities and even thought and speech that are not proscribed at one time become crimes at another; class background labels, political position, and other factors influence whether one is likely to be arrested for a particular act; and in many cases there are no documents or laws one can cite to defend oneself against unjust treatment by the authorities. For a discussion of some of these problems, see Greenblatt (1977).

9. In 1979 a policy change occurred in China when a set of new legal statutes, including a criminal law code and a code of criminal procedure, were enacted for the first time (to take effect as of 1 Jan. 1980). However, there is an acute scarcity of personnel to man the new legal institutions, and many of the existing personnel spent long years being oriented against reliance on written legal standards. It remains to be seen how substantial an effect these changes will have on social control and citizen behaviour in Chinese cities.

References

Bao Ruo-wang, and Chelminski, R. (1976) *Prisoner of Mao* (New York: Penguin).

Bernstein, T. (1977) *Up to the Mountains and down to the Villages* (New Haven, Conn.: Yale University Press).

Butterfield, F. (1979) 'Peking is Troubled about Youth Crimes', *New York Times*, 11 March.

Cohen, J. A. (1968) *The Criminal Process in the People's Republic of China 1949–1963* (Cambridge, Mass.: Harvard University Press).

—— (1971) Drafting People's Mediation Rules', in J. W. Lewis, (ed.), *The City in Communist China* (Stanford, Calif.: Stanford University Press), 29–50.

Davis-Friedmann, D. (1979) 'Old People and their Families in the People's Republic of China', Ph.D. dissertation, Boston University.

Foreign Broadcast Information Service (1979) *Daily Report: People's Republic of China* 15 Jan. K-3; 12 Feb. G-6; 14 Feb. M-1; 21 Mar. O-5; 22 Mar. R-2; 23 Mar. O-1; 26 Mar. O-1.

Greenblatt, S. (1977) 'Campaigns and the Manufacture of Deviance in Chinese Society', in A. Wilson, S. Greenblatt, and R. Wilson (eds), *Deviance and Social Control in Chinese Society* (New York: Praeger), 82–120.

Howe, C. (1977) 'The Industrialization of Shanghai, 1949–1970s,' Paper presented at a conference on Shangai since 1949, Cambridge University.

Kraus, R. (1977) Class Conflict and the Vocabulary of Social Analysis in China,' *China Quarterly*, 69 (Mar.), 54–74.

Lai Ying (1969) *The Thirty-Sixth Way* (Garden City, NY: Doubleday).

Leong, Y. K., and Tao, L. K. (1915) *Village and Town Life in China* (London: George Allen & Unwin).

Munro, D. J. (1977) Belief Control: The Psychological and Ethical Foundations', in A. Wilson, S. Greenblatt, and R. Wilson (eds), *Deviance and Social Control in Chinese Society* (New York: Praeger), 14–36.

Pepper, S. (1978) 'Education and Revolution: The "Chinese Model" Revised', *Asian Survey*, 18(9), 847–911.

Salaff, J. (1971) 'Urban Residential Committees in the Wake of the Cultural Revolution, in J. W. Lewis (ed.), *The City in Communist China* (Stanford, Calif.: Stanford University Press), 289–324.

Skinner, G. W. (ed.) (1977) *The City in Late Imperial China* (Stanford, Calif.: Stanford University Press).

Thompson, R. (1975) 'City Planning in China', *World Development*, 3, 595–606.

Tien, H. Y. (1973) *China's Population Struggle* (Columbus, Ohio: Ohio State University Press).

Vogel, E. F. (1971) 'Preserving Order in the Cities', in J. W. Lewis (ed.), *The City in Communist China* (Stanford, Calif.: Stanford University Press).

White, L. T. (1977) 'Deviance, Modernization, Rations, and Household Registers in Urban China', in A. Wilson, S. Greenblatt, and R. Wilson (eds), *Deviance and Social Control in Chinese Society* (New York: Praeger), 151–72.

—— (1978) *Careers in Shanghai*. (Berkeley, Calif.: University of California Press).

Whyte, M. K. (1974) *Small Groups and Political Rituals in China* (Berkeley, Calif.: University of California Press).

—— and Parish, W. L. (1984) *Urban Life in Contemporary China* (Chicago, Ill.: University of Chicago Press).

VII

Patterns of Political Integration and Conflict

INTRODUCTION

AT the heart of Third World politics lies a sharp contrast. Political decisions and administrative procedures affect large sectors of the population profoundly—to a much greater extent than in affluent market economies. From arbitrary arrest by a police officer to the systematic killings of political opponents, from the trading licence of a street vendor to the import licences of large corporations, political decisions and their implementation affect the lives of all but subsistence farmers. At the same time the mass of the population typically has little impact on decision-making and frequently is deprived of effective legal protection in its dealings with government. In other words, Third World countries are highly stratified; the mass of the population is dominated by a small élite whether the regime is committed to a free market economy or to socialism.

Elections supposed to give voices to the entire population rarely serve that purpose. Many countries have not witnessed elections for a long time. In many other countries the masses find only a limited voice in what amount to plebiscites for one-party regimes. Elsewhere patron–client networks deliver votes to patrons who pursue élite interests whatever faction they may belong to. Nearly everywhere electoral fraud is common, in quite a number of countries intimidation of opponents part of electoral strategy.

Rural populations invariably enjoy little effective participation in the political process. Where they do take part in elections their vote is frequently controlled by local élites through patronage or outright coercion. Powell (1980, 202), in a comprehensive review of rural voting patterns, concludes that there are hints almost everywhere, when one penetrates beneath the surface of political behaviour in rural areas, of a style of politics that can be labelled a 'rural mafia' system. The rural masses are thus usually without a voice, and their fate is neglect. They may rebel because land has been taken away from them, taxes raised, or the prices paid for their crops lowered, but such upheavals tend to be isolated and rarely constitute an effective threat to the élites in the cities. The country at large is affected when peasants withdraw from the market and revert to subsistence farming. But the most serious consequence of the neglect of the rural masses is the stream of rural–urban migrants who cannot be absorbed into the urban economy productively and who put severe pressure on urban resources. They swell the ranks of the urban masses strategically poised at the centres of local, regional, and national decision-making. The politics of Third World countries are by and large played out in these urban arenas.

The Cuban revolution challenged the presumption, established for over a century, of US hegemony over the Western Hemisphere. Various branches of the US government, including the Defence Department, as well as private foundations, sponsored a large body of research on the urban poor in the 1960s. There was little concern to define and describe urban poverty, but rather a preoccupation with the political implications of poverty. The research subsided in the 1970s, not because urban poverty was on the decline, but because the proposition that the urban poor are not vanguards of revolutionary movements was gaining wide acceptance (Eckstein 1976). For a comprehensive review of such research in Latin America as well as elsewhere in the Third World see Nelson (1979). The following chapters focus on patterns of patronage that muffle the voice of the masses, ethnic, religious, and caste identities that divide the masses, and three forms of mass action that at times impact on the urban political arena.

Patron–client relationships are found throughout the world in the most diverse settings, from national politics to the politics of academia, but their significance varies. In many Third World countries they play a central role in fashioning political integration and delineating lines of conflict. In a detailed study of Mushin, a suburb of Lagos, Nigeria, Barnes (1986) shows that most political participation is based on patron–client relationships, that a clientele that can be mobilized for various forms of action—voting, influencing, disturbing—provides the basis for power, and that this power in turn is used to strengthen and expand client support. For a recent comprehensive discussion of patron–client relationships that also provides a review of the literature on such relationships in a large number of Third World countries, see Eisenstadt and Roniger (1984).

Susan Eckstein and Michael Johnson, in Chapters 16 and 17, describe and analyse two sharply contrasting systems of patronage. In Mexico, after the devastations of a long civil war, political power was effectively concentrated at the apex of the state, in the hands of the president and his close advisers. Co-optation throughout the political and administrative system has effectively channelled and scaled down demands and paralysed nearly all potential opposition. The monopoly of power of the Institutionalized Revolutionary Party, ruling under a variety of 'revolutionary' names since 1928, has not been threatened in any of the regularly scheduled elections until very recently. A party that legitimizes itself on the basis of a revolution, yet consistently flouts the officially proclaimed revolutionary goals, has managed to stay in power for more than two generations. Eckstein, in a study of three low-income neighbourhoods in

Mexico City, shows how local leadership is co-opted and integrated into the power structure.

In Lebanon the origin and some of the persisting peculiarities of patronage can be traced to a feudal past. Johnson presents a case-study of Sunni Beirut where until the 1970s political bosses operated machines that usurped much of the powers of the independent state established in 1943. They distributed economic and other services to their clients in exchange for consistent political loyalty. Alliances with strong-arm neighbourhood leaders served to recruit local communities into these machines and to control them, by force if necessary. The political bosses maintained control because they were in a position to play arbiter: to shield their followers, and in particular their strong-arm retainers, from the authorities, or to call in the police to punish the recalcitrant. The structure of the machine, based on local neighbourhoods and clans, prevented the emergence of self-conscious, horizontally linked social categories such as interest groups or classes. The electorate was furthermore fragmented because most favours remained in the gift of the political bosses and went to individuals. If the political bosses largely controlled both the allocation of state resources and the state's coercive potential, their position was predicated on a state providing resources and holding the means of coercion in the police and the army. The collapse of the state in the civil war in 1975 deprived the political bosses of their base and they lost control over their strong-arm retainers, some of whom emerged as prominent militia leaders over the course of successive rounds of fighting. Initially a progressive coalition was established, but it broke down, not least because of foreign interventions, and anarchy resulted. Solidarities of neighbourhood and clan remained as the only defence in an age of violence.

Primary alignments based on region of origin, religion, or caste play a central role in the politics of many Third World countries. The salient characteristics of such identities are articulated in juxtaposition to the identities of other groups. In much of Africa, and in major parts of Asia, most migrants in the cities and their children categorize themselves and others in terms of region of origin. To refer to such a pattern as 'tribalism' is unfortunate because of the pejorative connotations the term 'tribe' has acquired. It is also misleading. Such identities in the urban setting, and their cultural content, usually bear little relationship to traditional societies and their culture. The past provides the raw materials, but ethnic identities are fashioned in the confrontations of the urban arena. In parts of Africa and Asia adherents of two or even three world religions live within the same city boundaries. At times they have clashed with each other, most dramatically Hindu and Muslim in South Asia. But conflict can also crystallize around internal divisions within a religion, such as the Muslim sects in Syria. Caste is of central importance in the Hindu religion, and for the majority of Indians it remains the primary referent. Determined by birth and supposedly immutable, it also changes its guise according to context. For a comprehensive review of such cleavages and their relation to national politics in major Third World regions, see Young (1976).

Chapter 18 is drawn from Abner Cohen's classic study of the Hausa in Ibadan, Nigeria. Enjoying political control over their internal affairs under British rule, the advent of independence brought the prospect that they, a small minority, would lose all such control and with it their monopoly of the long-distance trade in cattle and cola nuts. In the face of that threat they reaffirmed their cultural distinctiveness from Yoruba Muslims by joining a different Muslim sect, a move that led to a veritable ethnic renaissance and effectively excluded the Yoruba from breaking into trades built on trust among co-ethnics.

Alignments of shared origin, religion, or caste commonly reinforce political and economic grievances. An ethnic group may move ahead of other groups because its land offers exceptional opportunities, a religious group may have the edge in education, or a caste may enjoy advantageous occupational opportunities. Once some members of such a group have established themselves in a privileged position, they wield considerable influence over the opportunities open to others. To the extent that their patronage goes to kinsmen, fellow villagers, or any 'brother', an entire group can be seen to enjoy privilege, others to be excluded. Furthermore, the continuous modifications of identities of origin, religion, and caste typically increase the congruence between such identities and political and economic interests. Still, such divisions usually obscure stratification at the national level and defuse class conflict.

Mass movements, rather than elections, have been the means by which the masses in a number of Third World countries have been able to extract concessions from the political system or indeed to transform it. Among such actions three types stand out: squatter movements, trade unions, and insurrectionary movements. In many Third World cities the most conspicuous political action of the urban poor has been the illegal occupation of land. Frequently it has been organized on a scale so large as to persuade the authorities initially to condone it and eventually to grant the squatters legal title. The presence and form of squatter settlements varies from country to country, even among cities within one country, and it changes over time, depending on the political context. A large body of research has accumulated, especially for Latin America. Leeds and Leeds (1976) pioneered comparative analysis when relating squatter movements to political systems in Brazil, Peru, and Chile. Manuel Castells, in Chapter 19, compares squatter movements in Lima, Peru, Mexico City and Monterrey, Mexico, and Santiago de Chile, drawing on an even larger body of research.

Successful squatter movements account for improved housing conditions for large numbers of city-dwellers in the Third World. All the evidence indicates, however, that the political role of such movements remains narrowly circumscribed by political élites, and that established squatters are particularly prone to be drawn into patron–client relationships. Unlike squatters, urban workers have real leverage. Any country's economy is immediately dependent on the sustained labour of its work-force. The leverage different sectors of the labour force can exert varies greatly though. Unskilled workers can be easily replaced by eager recruits from among the unemployed and underemployed, but many skills are in short supply. At the same time, underdeveloped economies tend to be heavily dependent on the effective operation of a few key sectors. A strike by railroad or dock workers brings the entire economy to a halt in many a country; a work stoppage by miners can threaten a foreign exchange crisis. Those who operate such crucial economic resources, and whose skills are not easily replaced, potentially enjoy considerable political power.

By now nearly all Third World countries have a history of worker protest. It is the subject of a rapidly growing body of literature. Thus Bergquist (1986) provides accounts of close to a century of labour struggles in Chile, Argentina, Venezuela, and Colombia. In contrast, there has been surprisingly little serious research on the role of organized labour in contemporary Third World politics. Two notable recent studies, however, signal a shift of scholarly interest to organized labour. Humphrey (1982) and Roxborough (1984) both have chosen a Latin American country. Brazil and Mexico respectively. And both have focused on the same industry, automobile production. In both countries this is a new industry. The major role it has come to play in Brazil and Mexico, the huge capital investments it entails, and the skills it requires, give considerable leverage to its recently formed, young labour force.

Throughout the Third World workers have increasingly become aware of revolutionary changes wrought in some countries, of reforms achieved by their counterparts in many others. Consciousness of both historical precedent and actual condition is facilitated by rising levels of educational experience. The wide gap in living conditions separating the labouring masses from local élites, foreign advisers, and tourists is less and less taken for granted. Still, the emergence of a broad, class-conscious movement of workers is very much the exception in Third World countries.

The organization of the labour force, and its role in the political arena, is most advanced in parts of South America. Paul Drake, in Chapter 20, demonstrates its resilience even under repressive military regimes. His comparison of four countries—Chile, Argentina, Uruguay, and Brazil— deepens our understanding of the processes at work and is made all the more convincing because it addresses the economic, the legal, and the political context in which organized labour operates.

In the final chapter I argue that contemporary revolutions require urban insurrections to succeed in this day and age. The Chinese revolution was the last revolution to be based on a peasant army. The intelligence, communications, transport, and destructive capabilities of modern armies are such as to preclude the establishment of revolutionary peasant armies. And while rural guerrilla movements may induce colonial rulers to depart, they are unlikely to thus persuade indigenous élites.

References

Barnes, Sandra T. (1986) *Patrons and Power: Creating a Political Community in Metropolitan Lagos*. International African Library (Manchester: Manchester University Press; Bloomington and Indianapolis: Indiana University Press).

Bergquist, Charles (1986) *Labor in Latin America: Comparative Essays on Chile, Argentina, Venezuela, and Colombia* (Stanford: Stanford University Press).

Eckstein, Susan (1976) 'The Rise and Demise of Research on Latin American Urban Poverty', *Studies in Comparative International Development*, 11, 107–26.

Eisenstadt, S. N., and Roniger, L. (1984) *Patrons, Clients and Friends: Interpersonal Relations and the Structure of Trust in Society*. Themes in the Social Sciences (Cambridge, London, New York, New Rochelle, Melbourne, and Sydney: Cambridge University Press).

Humphrey, John (1982) *Capitalist Control and Workers' Struggle in the Brazilian Auto Industry* (Princeton, NJ: Princeton University Press).

Leeds, Anthony, and Leeds, Elizabeth (1976) 'Accounting for Behavioral Differences: Three Political Systems and the Responses of Squatters in Brazil, Peru, and Chile', in John Walton and Louis H. Masotti (eds), *The City in Comparative Perspective: Cross-national Research and New Directions in Theory* (New York, London, Sydney, and Toronto: Halsted Press), 193–248.

Nelson, Joan M. (1979) *Access to Power: Politics and the Urban Poor in Developing Countries* (Princeton, NJ: Princeton University Press).

Powell, John Duncan (1980) 'Electoral Behavior among Peasants', in Ivan Volgyes, Richard E. Lonsdale, and William P. Avery (eds), *The Process of Rural Transformation: Eastern Europe, Latin America and Australia*. Comparative Rural Transformation Series (New York, Oxford, Toronto, Sydney, Frankfurt, and Paris: Pergamon Press), 193–241.

Roxborough, Ian (1984) *Unions and Politics in Mexico: The Case of the Automobile Industry* (Cambridge, New York, New Rochelle, Melbourne, and Sydney: Cambridge University Press).

Young, Crawford (1976) *The Politics of Cultural Pluralism* (Madison and London: University of Wisconsin Press).

16

The Politics of Conformity in Mexico City*

Susan Eckstein

W H Y does poverty persist even when poor people organize in legitimately recognized groups publicly concerned with their welfare? How can income distribution become more inegalitarian as national wealth increases and a mass-based party officially commands the reins of government? Poor people's organizational effectiveness, as shown below, may be limited if the groups with which they associate, or leaders of those groups, are formally co-opted *and* if informal pressures impede those co-opted from using the status they thereby acquire to primarily serve their own ends. This article highlights the informal processes that impede co-opted groups from utilizing formally gained power for their own purposes. Individual and group co-optation are near-universal, but the particular form organizations take and the particular processes through which poor people are regulated are country-specific.

Types of co-optation and informal regulatory processes, and structural conditions which account for them, are discussed in turn below and illustrated with data on Mexican urban poor. The study suggests that Mexican poor have access to the symbols of power, not power itself, through affiliation both with the official party—the Institutionalized Revolutionary Party (PRI)—and with government-linked groups. Above all the analysis reveals that:

1. Organizations are not likely to serve as instruments of political power when members are organized into groups lacking institutionalized access to power; organizations of the poor may help legitimate the status quo, extend the government's realm of administration, and reinforce existing social and economic inequities if the poor are co-opted—overtly or covertly, collectively or through group leaders—into institutions affiliated with the official power structure; however, members are not necessarily aware of the effect their affiliations have and do not necessarily organize with these ends in view.

* Reprinted, in a revised form, from the *British Journal of Sociology* 27 (2), 1976, by permission of Routledge & Kegan Paul. I am grateful to Michael Useem for comments on an earlier version.

2. Formal power does not become an effective base of power when hierarchical, class, and other informal societal forces inhibit members from using the groups to serve their own interests.

3. Organized poor are co-opted largely because structural forces induce group leaders to establish ties with the official political-administrative apparatus.

Methodology

This account focuses on organizational relations in (a) a centre city slum, with a history antedating the Conquest; (b) a now-legalized squatter settlement formed by an organized land invasion in 1954; and (c) a low-cost government-subsidized housing development which opened in 1964 in Mexico City. The three areas were purposively selected to represent different kinds of lower-class residential settlements, in order to ascertain whether organizations and their effects differ in contrasting types of dwelling environs housing urban poor.

The data derive from a multi-method study I did in 1967–8 and in 1971–2 in the three areas. Statements about leaders and group activities are based on 'semi-structured' interviews with heads of all locally operating formally chartered groups and agencies,[1] participant-observations, and written documents. The leaders were located by visiting all visible local political, administrative, social, economic, and religious agencies and organizational headquarters, asking those in charge of the operations the names and addresses of other agencies, institutions and organizations in the areas, speaking with the persons in charge of the places so named and asking them, in turn, the same question.[2] I continued this procedure until I learned of no new groups. In the case of divisions of city-wide and national groups I interviewed not only the heads of the local affiliations but also the non-local persons to whom they formally were responsible in their organizational hierarchy, since they too shaped relations within the communities.

The positional leaders interviewed included all parish priests (7); all leaders of parish-wide lay groups (4); all local ministers (9); directors of all local parochial schools and church-linked hospitals (3); non-resident religious personnel who worked locally (4); administrators of government-run facilities, i.e. directors of at least one shift of all public primary and secondary schools (20), all medical and other social service agencies (6), all sports facilities (6), and all markets (10); all *subdelegados* (the highest-ranking local functionaries of the territorial-based organization of the municipal government) (4), and their non-local hierarchical superior; the administrator of the housing development; personnel in government agencies concerned with the areas, e.g. the agencies which designed the housing development and handled problems in the squatter settlement relating to legalization of land claims and provisions of basic social services; six of the ten highest-level local representatives of political parties; all persons in charge of territorial-based politically affiliated social groups (2); local union representatives of market vendors (10 of 15); and all formally organized nominally autonomous social groups (3).

I supplemented and cross-checked information I obtained through these interviews both with observations I made in the communities, including at local group meetings held during the time I did my field research, and with information from newspapers, government files, and group documents. The documents included formal statutes of groups; letters of correspondence between local individuals and groups on the one hand, and district, city-wide and national political government agencies on the other; and records kept by local groups.[3]

Formal Co-optation and Incorporation

In the three areas there were over two dozen locally operating formal groups. Most of them were either formally incorporated into national PRI or government-affiliated organizations or exposed to PRI and government influence through heads of the groups who overtly or covertly had ties with such national organizations.[4] The local groups included territorial-based divisions of political parties, party-affiliated groups, and the government's administrative apparatus; economic associations; athletic and other social associations.

Local groups initially addressed themselves to local problems. Since the problems residents of each area had differed somewhat, divisions of the same parent organization at first did not necessarily concern themselves with the same sets of issues. In this respect the organizations adapted to members' interests.

Independently of the specific conditions which initially led to the formation of most of the groups, the groups came to claim a common and general concern about the welfare of members. For example, while the first groups in the squatter settlement began as loosely structured 'followings' around particular leaders who attended to the problems residents had securing legal title to their illegally held land and basic social and urban services, the heads of these groups presently proclaimed a general commitment to the moral, social, and economic defence of residents.

When originally affiliating with national PRI and government groups members received a variety of benefits they coveted, including schools, roads, pavements, markets, and public transportation. In the squatter settlement they thereby also secured property rights. Occasionally the local divisions also successfully convinced government authorities to replace corrupt, exploitative hierarchically appointed leaders.[5] However, their ability to secure goods and services in most instances proved to be limited and short-lived, even when they were still in need. In particular, school and medical facilities in the areas remained inadequate: for example over 2,000 children in the squatter settlement could not attend school for lack of space. Despite remaining needs and previous success in extracting benefits from the state, they pressured agencies for little. Above all, they never collectively pressured for income subsidies or other benefits which would have improved their socio-economic standing at the expense of wealthy Mexicans. Also, even though they at times had unpopular functionaries removed from office, they had no way of preventing their successors from abusing them too. Their restricted organizational effectiveness stemmed mainly from the fact that they had no institutionalized authoritative power or budgetary discretion. They were dependent on their capacity to convince decision-makers, with no guarantee that their interests would be protected.

Congressmen were among the only popularly elected *políticos* with official institutionalized power. Yet even they rarely used their authority to advance residence interests. They did not view their main task as one of representing constituent interests in the legislature, as evidenced by the way they divided their time, defined their responsibilities, and voted in Congress. None the less, they did not entirely ignore their electoral constituents, for they extra-officially solicited government agencies for community facilities and helped some residents secure jobs. Furthermore, their district headquarters offered such services as medical care and classes in sewing, homemaking, and typing which residents, if they so chose, could make use of.

The groups, whether they had institutionalized access to power or not, placed constraints on members. After affiliating with national groups, associations which once had highly active memberships became largely inactive, and members increasingly came to deal with their group leaders on an individual basis about individual problems rather than collectively about common concerns. Moreover, those who continued to partake in group-sponsored activities generally did so in return for or in anticipation of favours from the group leaders. Such opportunistic participation reflected administrative involvement,[6] not civic concern. In the very process of affiliating with PRI and the government and periodically participating in civic and political activities though, residents helped legitimate the regime and extend the government's administrative apparatus to new areas.

Incorporation into national groups with popular objectives served as an instrument of social and political control because the local divisions consequently were demobilized,[7] made to impose reduced demands on the regime, and indebted to PRI and the government for the benefits they secured. However, independently of such participatory constraints, when members of any group secured the benefits which initially induced them to join, or when they became completely disillusioned with the unfulfilled promises of their leaders, they also lost interest in the groups. Although occasionally participation remobilized when non-local functionaries allowed local functionaries to mediate specific concerns of residents, e.g. housing rights, the leaders used the opportunity to extend political-administrative order by involving persons interested in the issue at hand in civic festivities.

Even social groups which did not formally affiliate with national institutions were subject to political-administrative constraints. Although these groups, unlike PRI and government-affiliated groups, technically had authoritative power, because their constituents were poor they had limited financial resources with which to satisfy members' concerns. Therefore, their leaders sought help from *políticos* and in so doing were subject to socio-political controls. In addition, leaders of these groups, either because they individually formally affiliated with PRI or government-linked groups or because they anticipated political patronage, organized their groups comparably to those which were formally incorporated. They introduced hierarchical, territorial-based organizations, with representatives on each block, urged members to attend civic and political manifestations, and spoke of political-administrative concerns at their group meetings. Thus, when the head of a Church-initiated social group privately affiliated with a PRI-linked association, he began to give political and civic matters high priority in his own association. He subsequently reported regularly on the activities of the PRI group and allowed the leader of the PRI group to monopolize a number of rallies the social group organized. As a result, the group was diverted from its original social and economic and deliberately non-political objectives. Similarly, the head of a large, formally autonomous soccer league in the city area allowed PRI to offer social services at the group's headquarters, required members to partake in civic celebrations, and organized his group by blocks, as PRI did, so that, as he phrased it, he would be ready should the Party call on him.

Clearly, residents' inability to secure many benefits from the government or to improve their socio-economic standing relative to other classes stem not from any 'organizational incapacity' or 'political incompetence', as some commentators on Mexico claim.[8] Nor does their inability stem from affiliation with groups lacking a professed concern with their social welfare, or from their failure to work through legitimate channels. Their organized efforts met with limited success because their groups rarely entitled them to institutionalized access to power and because affiliation with PRI and government-linked groups weakened their organizational effectiveness. Groups that had been co-opted were left without real power, and subjected to political and administrative controls. Moreover, when local groups affiliated with national political and administrative institutions they helped, knowingly or not, to legitimate the regime. In general, their organizational effectiveness was hampered by informal as well as formal processes, to the extent that they came to manifest some of the characteristics of the so-called culture of poverty. Residents because apathetic; yet they were not always that way. The change came about when local groups, or leaders of local groups, established 'extra'-local ties and when relations within the groups, in turn, were routinized, for informal pressures operating in the society at large consequently affected relations locally.

Informal Processes Constricting the Political Power of Organized Residents

Since formal power potentially serves as a basis of effective power, co-optation and incorporation alone do not account for the limited political effectiveness of organized residents. My field-work suggests the following additional hypothesis: when inter- and intra-group relations are hierarchically structured, when class biases discriminate against the poor, when the government monopolizes the use of force, poor people cannot readily use formal power for their own ends.

The hierarchical way in which *políticos* were appointed and removed from office and the hierarchical channels through which urban and community services were obtained limited residents' ability to use local groups to advance their own interests. These conditions constrained leaders to conform with 'rules of the game' that they themselves did not establish. They felt that they otherwise could not advance politically or secure benefits for their constituents. The prospects of removal from office also compelled local subordinates to conform with the expectations, or perceived expectations, of higher-ranking functionaries. Thus, 'appropriate' local concerns were delimited and heads of local groups were encouraged to be subservient, although they were not necessarily rewarded upon being so. As an illustration, a politically ambitious head of a local division of a government agency complained, 'If you do more than your boss you're in real trouble.'

Furthermore, those in charge of local groups could not readily consolidate power within the ranks of their branch affiliations because they were periodically reassigned or removed from office. This kept most local officeholders from becoming entrenched in local interests. In this connection, one market administrator explained to me that the head of the Division of Markets regularly reassigned administrators to different markets so that no administrator gets entrenched in local politics to the extent that he becomes subservient to local interests and fails to fulfil his obligations. As this case suggests, higher-ranking functionaries may periodically replace low-level functionaries because they are concerned with the effectiveness of their particular group, not because they wish to manipulate and exploit local residents. But irrespective of intent, such actions do serve to restrict residents' ability to use existing groups for their own ends.

In addition, the hierarchical promotion system drained the local areas of their best-trained leaders. Local leaders were pressured to partake in activities in other parts of the city, in order to secure or maintain the support of higher-ranking functionaries. Thus, during the last election the leading *políticos* engaged, almost exclusively in political activities in other electoral districts. Again, in so doing, higher-ranking functionaries did not deliberately attempt to undermine the organizational effectiveness of the local communities but their own opportunistic concerns had such an effect.

The hierarchical system also prevents groups from organizing 'horizontally'. At times local groups were isolated from, and in competition with, other divisions of their same parent organizations. In the case of unions, local divisions were not all incorporated into the same sector of PRI. Since benefits were allocated to party sectors and groups within each sector separately and unequally, the unionized workers did not share a common fate, even though they were affiliated with the same party. Furthermore, the hierarchical structure of groups served to divide workers' loyalty *within* occupational groups, and informal constraints exposed rank-and-file members to non-local, non-economic Party and government influences. In response to commands from regional union officers, local leaders encouraged workers to participate in civic and Party rallies and vote for PRI, but to be otherwise politically quiescent.

Independently of the hierarchical structure of political and administrative groups, class biases which informally operated in the society at large permeated relations within the three areas, constricting the political effectiveness of local groups. Access to local leadership posts, opportunities for political advancement, and the ability of local groups to secure benefits for members—particularly once relations within the groups were routinized—depended more on the socio-economic status of group members and leaders than on the grass-roots structure and formal objectives of the national groups.

While local functionaries were not the local economic élite, they tended to be disproportionately recruited from the most economically successful resident populations, and increasingly so over the years. Furthermore, local leaders' prospects of political mobility became progressively limited by their class standing. Leaders of humble class origin stood the greatest chance of promotion when members of their groups were highly mobilized, which rarely occurred once national political and administrative institutions established local divisions and relations within the areas were routinized.

Although the functionaries might have served the interests of resident poor even when they themselves were not of low socio-economic status, because those working locally were primarily interested in their own advancement they tended to be hierarchically rather than community-oriented. Local leaders rarely collaborated to promote programmes for the benefit of residents. In fact, the leaders had much more contact with persons of higher rank in their own organizations, outside the local communities, than they had with one another.

Not merely the leaders but common residents as well were rewarded by the government inversely to their economic need, once they ceased being highly 'mobilized'. Residents of the housing development, the most middle class of the three areas, were most successful in getting social services from the administration, even though they always had the best facilities. Moreover, *within* the development the sections with the largest proportion of middle-class residents and middle-class leaders benefited most from government assistance. Their effectiveness stemmed not from specific formal group affiliations, for local divisions of the same PRI and government-affiliated groups in the other two areas were not equally successful in securing assistance from non-local authorities. Rather, their good fortune derived both from higher-ranking non-local functionaries favouring their middle-class subordinate staff and the middle-class interests they 'represented' and from local middle-class functionaries generally having the best contacts with non-local personnel who could help them.[9]

The personal manner in which *políticos* were appointed and removed from office, goods and services distributed, and local demands and conflicts articulated and resolved, further limited residents' ability to use local groups to serve their own ends. The groups operated in a highly personalized manner because the structure of power induced politicians to act accordingly. This structurally-induced *personalismo* made residents—particularly in the two newer areas—feel dependent on and indebted to the government for material benefits, including land and pavement for which they actually paid. The benefits generally were personally 'given' to them by high-ranking functionaries at public rallies.

Moreover, in the very process of securing personal 'favours' from the government the collective effectiveness of residents, seemingly paradoxically, weakened. Particularly when legalizing illegal land claims local functionaries collaborated with non-local functionaries to establish social order and routinize local organizational life in a section of the city where community relations initially were not institutionalized. For such reasons, the allocation of goods and services, in effect, serves as a regulating mechanism.

In addition, local leaders reported that they were instructed by higher-ranking functionaries to personally request, not demand, goods and services and to petition for benefits either individually or in small groups, not *en masse*. Thus, whereas leaders in the squatter settlement initially mobilized large followings, they now go to government offices alone or in small delegations. In so doing residents, in effect, put less pressure on the government to provide them with benefits.

Furthermore, because politics is personalized, conflict is also. When residents' expectations were not fulfilled they blamed individuals, and competition for political and economic spoils pitted local leaders against one another. The hierarchical system of appointments induced leaders to compete for local followings and influence, even when the groups they headed were ultimately affiliated with the same parent institution and when the groups concerned themselves with the same sets of issues. Such petty jealousies, in turn, weakened the potential collective strength of local inhabitants because divisions thereby were generated between different local leaders and their respective constituents. These divisions directed residents' energies inward towards local personalities, not outward against the institutions ultimately responsible for the conditions giving rise to local conflicts. Also, by personalizing conflict residents never criticized political and administrative institutions in ways that could have helped to undermine the legitimacy of the inegalitarian social order.

Any political power which residents might have enjoyed through collective organization was further weakened by the sheer multiplicity of groups operating locally. This multiplicity was a by-product of the fact that numerous vertically structured groups operated nationally in affiliation with PRI, and the government,[10] and that higher-ranking functionaries tacitly supported the array of groups by periodically distributing gifts to different associations, by attending local group meetings, and by seeking collaboration from the groups in civic events.

However, at the same time that the government encouraged groups and allocated certain goods and services to constituents it regulated residents. All groups were required to be officially registered and a so-called law of 'social dissolution' gave the administration the right to intervene in local groups. The very existence of the laws, and the government's monopoly of the legitimate use of force, probably inhibited residents from forming groups and engaging in activities which might be officially condemned. Moreover, the laws served as a useful device by which controversial local leaders were repressed, even when they espoused the same objectives and concerns as leaders of officially recognized groups. Yet incidences of violent government repression were rare. Co-optation and informal constraints generally effectively induced conformity.

Structural Conditions Inducing Leadership Co-optation and Formal Political and Governmental Affiliation

The preceding discussion highlighted how and why organizations espousing populist ideologies regulated their lower-class memberships. But why did resident poor affiliate with organizations that precipitate such effects? The situation within the three areas suggests that leaders are prompted to establish ties with national political and administrative groups, even when they do not in the process gain institutionalized access to power, if they have the opportunity and feel such ties are personally expedient.

My personal discussions with local leaders and my observations of local activities suggest that the leaders of groups symbolically co-opted other group leaders or incorporated entire groups and, in turn, were co-opted and incorporated by others, mainly for opportunistic reasons—not because of shared ideological commitments, or organizational 'needs'. Viewed from the local level, leadership co-optation and group incorporation occurred because those in local command posts perceived challenges to their claims to leadership or opportunities to enrich themselves and extend their sphere of influence. Rank-and-file members generally gained some benefits in the process, particularly in the short run, but the leaders above all were guided by personal desires to secure political and economic spoils.

Local leaders generally were induced to establish formal ties with national political-administrative groups because they felt, in view of the hierarchical, centralized nature of the regime, that their own political and economic mobility prospects would thereby be enhanced and that they would be able to obtain goods and services for their constituents. They did not by choice establish ties which left them without institutional access to decision-making. Rather, they perceived no viable alternative.

On occasion, however, local leaders deliberately established informal rather than formal ties: when they felt their constituents or potential constituents were antagonistic to the idea of being directly associated with the government or PRI. For example the politically ambitious head of the large soccer league in the centre city area never formally affiliated with a national government or PRI group because members of his group felt they would be compromised if they had such affiliation. Yet, as he admitted to me, because of his own political ambitions he collaborated with political and government functionaries, and ran his group in ways which he felt would please those functionaries. Aware that prior leaders of the league and prominent local athletes in the past had been awarded posts in PRI and the government, he hoped to secure political patronage on the basis of his organization. Similarly, the head of the Church-sponsored social group in the housing development who privately affiliated with a PRI-affiliated group informed me that he never allowed his group to be formally associated with a PRI or government-affiliated group because he thought he would be forced to resign if he did. Yet he deliberately affiliated with the PRI-linked group individually, as he felt he could thereby secure the necessary material assistance to maintain a following.

The higher-ranking non-local functionaries with whom I spoke told me that they formally co-opted local leaders and formally incorporated local groups as branch affiliations of their organizations because they felt that still higher-ranking functionaries would be impressed by the increased size of their following and by their ability to establish social and political order in the areas. In turn, they felt that their chances of promotion would be increased. Since they themselves did not have institutionalized decision-making power they neither wanted nor were able to extend institutionalized power to local leaders or groups.

Conclusions and Implications

The experience of the low-income communities demonstrates that organization may contribute to the persistence of poverty and inequality when organized poor are co-opted into groups which both reinforce and extend the legitimacy of a regime and provide no institutionalized access to power.[11] Under such circumstances organizations serve regulatory functions, even when not formed with this end in view and when other-than-poor do not deliberately manipulate group members with this end in view. Neither the mass base of the groups nor the 'popular' ideology which the groups propagate offset the effects of formal and informal constraints. The constraints derive both from forces fundamental to organizations and from forces in the society in which the organizations are embedded.

Co-optation resulted from the opportunity structure which induced leaders, largely for opportunistic reasons, to affiliate with national political and administrative groups, either individually or together with other members of their groups. The PRI's populist doctrine and grass-roots organization facilitated acceptance of the official party and the government and enabled the government to stand as the champion of local residents, while the co-optive processes enhanced the stability of the regime.

In so doing, local organizations contributed to the extension of formal democracy. Locally operating groups were more democratic in theory than in practice, as they gave residents access to the public symbols of power but only limited, informal, and indirect access to government resources and decision-making.

None the less, co-optation did result in a certain commitment by the government to divert occasional resources to local residents. Because of collective efforts and persevering leaders, residents now enjoy a number of social and urban facilities and property rights which they otherwise might not. However, they secured most when they initially were least acquiescent but not antagonistic towards the government—when they were organized informally and most 'mobilized'. Moreover, they actually paid for many of the urban services they ostensibly secured through organization.

The findings imply that Mexican urban poor are not likely to gain much through territorial or class-based organizations linked formally or informally to the government-party apparatus. Aware of this structural constraint, some low-income neighbourhood groups, under the guidance of militant university-educated students, have, in other areas of Mexico, resisted establishing ties to the state and they have pressed for rights, not favours. This has occurred, for example, in the northern industrial city of Monterrey. There, some neighbourhood groups have rejected state help. They preferred, instead, to steal materials and illegally obtain water, electricity, and other urban services.[12]

If the state's legitimacy is seriously challenged on the national level, if the state's material resources become increasingly inadequate to provide low-income groups with sufficient goods and services to convince them that loyalty to the state pays off, or if more powerful sectors pressure the state to allocate increasingly less scarce public revenues to the poor, the poor in the neighbourhoods I studied and elsewhere may indeed follow the example of the Monterrey groups. Should they organize autonomously of the state and resist co-optation they may, in turn, contribute to an erosion of state hegemony. However, as long as important segments of the capitalist and middle classes remain the politically most influential groups, the redistributive impact of such autonomous organization is not likely to be great. Indeed, the state may respond with repression rather than material rewards and reform.

Notes

1. Throughout the text I use the terms group, organization, and association interchangeably. I specify whether the terms apply to nominally autonomous units or ones affiliated with national or 'supra'-local organizations or institutions.
2. The 'decision-making' approach proved inapplicable because authoritative decisions were not made at the 'grass-roots' level. The 'reputational' approach also proved of little use, as few interviewees were well informed about the communities. On these approaches see Nelson Polsby, 'How to Study Community Power: The Pluralist Alternative', *J. Politics*, 22 (Aug. 1960),

474–84; Floyd Hunter, *Community Power Structure: A Study of Decision Makers* (Chapel Hill: University of North Carolina Press, 1953).

3. For a more detailed discussion of the methodology see Susan Eckstein, *The Poverty of Revolution: The State and the Urban Poor in Mexico*, Princeton, Princeton University Press, 1977, 2nd edition, 1988.

4. None of the other political parties had an organizational apparatus approximating that of the 'official' PRI and government-affiliated groups.

5. Similarly, Cornelius found that residents in the low-income areas be studied did not influence government policy formation, although they did effectively informally demand benefits from the government. Since the demand-making he describes is not a formally prescribed mode of political participation outlined in the statutes of political and administrative groups, it corroborates the point made here that the formal political apparatus provides resident poor with no guaranteed, institutionalized means by which to influence the decision-making process. Wayne Cornelius, jun., 'Urbanization and Political Demand-Making: A Study of Political Participation among the Migrant Poor', *Amer. Pol. Sci. Rev.* 68 (Dec. 1974). On relations between the state and urban poor, see also Jorge Montaño, *Los Pobres de la Ciudad en los Asentamientos Espontáneos* (México, DF: Siglo Veintiuno Editores, 1976).

6. In contrast, Almond and Verba assume that involvement in local politics reflects civic responsibility and competence. Gabriel Almond and Sidney Verba, *The Civic Culture* (Princeton: Princeton University Press, 1965).

7. In contrast, Deutsch and Lerner assume that participation is an ever-increasing phenomenon. Karl Deutsch, 'Social Mobilization and Political Development' in Jason Finkle and Richard Gable, *Political Development and Social Change* (New York: John Wiley & Sons, 1966).

8. Oscar Lewis, *Five Families: Mexican Case Studies in the Culture of Poverty* (New York: New American Library, 1959); Susan Purcell and John Purcell, 'Community Power and Benefits from the Nation: the Case of Mexico', *Latin American Urban Rev.* 3 (1973); Almond and Verba, *The Civic Culture*; Roger Hansen, *The Politics of Mexican Development* (Baltimore: Johns Hopkins Press, 1971).

9. Johnson argues that on the national level particular *camarilla* or political clique memberships are generally of greater political consequence than formal political or administrative career patterns. The limited access residents have to such cliques informally limits their political power. Kenneth Johnson, *Mexican Democracy: A Critical View* (Boston: Allyn and Bacon, 1971). Judith Adler Hellman, in *Mexico in Crisis* (New York: Holmes & Meier, 1978) and Nora Hamilton, in *The Limits of State Autonomy: Post-Revolutionary Mexico* (Princeton: Princeton University Press, 1982), also highlight the informal class basis of the Mexican state.

10. In countries where two or more parties effectively compete for lower-class support, parties rather than groups affiliated with a single party often create division among urban poor. On the divisive effect of political parties among urban poor in other Latin American countries, see Talton Ray, *Politics of the Barrios of Venezuela* (Berkeley and Los Angeles; University of California Press 1969), 98–127, and Raymond Pratt, 'Parties, Neighbourhood Associations,

and the Politicization of the Urban Poor in Latin America', *Midwest J. Pol. Sci.* 15 (Aug. 1971).

11. Yet protest also provides no guarantee that group interests will be protected: Stevens argues on the basis of an analysis of three major strikes that the Mexican government to date has succeeded in limiting, discouraging and manipulating participation in decision-making and demand-making among protest groups. Evelyn Stevens, *Protest and Response in Mexico* (Cambridge, Mass.: Massachusetts Institute of Technology Press, 1974).

12. Manuel Castells, *The City and the Grassroots: A Cross-Cultural Theory of Urban Social Movements* (London: Arnold; Berkeley, Calif.: University of California Press, 1983), 194–9. This section is reprinted in Chapter 19 in the present volume.

17

Political Bosses and Strong-Arm Retainers in the Sunni Muslim Quarters of Beirut*

Michael Johnson

IN the mid-twentieth century Lebanon's service-based economy developed such that Beirut became the banking and trade centre for the Arab Middle East. The relative decline of Lebanese agriculture, coupled with the emergence of capitalist farming in what were previously neo-feudal estates, led to an increasing migration to the city. A small and dependent industrial sector remained subordinate to the interests of commerce and finance, and did not provide work for significant numbers of immigrants who simply swelled the ranks of the semi-employed sub-proletariat.[1] The overwhelming majority of the labour force was employed in the service sector. Just as there was peasant individualism in Lebanese villages, so Beirut's society was characterized by sub-proletarian and petty-bourgeois individualism. There was little or no class consciousness, and political parties (with one exception, the Christian Phalangists) were weak to the point of insignificance, as were trade unions and other interest groups. Religion divided Beirut, as it did the rest of Lebanon, into two roughly equal confessional groups, the Christians and the Muslims, which were further subdivided into a plethora of sects.

Partly as a response to the fragmented and individualistic electorate, Lebanon's democratic political system became dominated by locally powerful leaders called 'za'ims' (Arabic: *za'im*, pl. *zu'ama*) who in the cities developed sophisticated machines to recruit a clientele. This paper starts by describing how these machines operated in the Sunni Muslim quarters of Beirut before the outbreak of civil war in 1975, and emphasizes the role of neighbourhood strong-arm men who recruited and policed the clienteles of the za'im.[2] The strong-arm retainers were called 'qabadays' (Arabic: *qabaday*, pl. *qabadayat*), a word which implies the use of physical force to defend and promote honour and leadership. In the latter part of the article it is shown how the Sunni qabadays of Beirut began to assert

* This chapter is based on field-work conducted by the author in West Beirut during 1972 and 1973 as part of a Manchester University project, funded by the British Social Science Research Council, on Lebanese politics and society.

themselves as leaders in their own right during the early 1970s, and how they became completely independent of the za'ims in the Lebanese civil war.[3]

Although a few published sources on Lebanon's political economy are cited in the footnotes, most of the information in the article was collected by the author in periods of field-work in the Sunni quarters of Beirut. Thus the 'clientelist system' and its breakdown, which are analysed here, are specific to Sunni Beirut. There is, however, some evidence to suggest that similar changes took place in similar political systems in other parts of urban Lebanon.

Political Bosses

At the end of World War I, to the delight of Lebanese Christians, the League of Nations awarded the Mandate for Syria to France. At first the Sunni notables of Beirut, accustomed to being the agents of Ottoman rule, refused to have anything to do with what they regarded as a foreign and Christian imposition on a Muslim majority. But Sunni za'ims gradually began to participate in the new democratic institutions, and they eventually co-operated with the Christians in founding the independent state of Lebanon in 1943. Parliamentary democracy provided access to governmental patronage and, in order to maintain a large clientele, it became essential for za'ims to be regularly elected as deputies and appointed as ministers. As part of the elaborate system of confessional checks and balances in Lebanon, the office of president was reserved for a Maronite Christian. But the Sunnis were given the premiership, and conflict between Sunni za'ims in Beirut was mainly concerned with competition for this powerful office.

Most of these Sunni za'ims were descended from the notable families of the early twentieth-century Ottoman period. They inherited the wealth and clientele of their fathers, forming the basis of their electoral support in independent Lebanon. The clientele was bound to the za'im by a network of transactional ties, where economic and other services were distributed to the clients in exchange for consistent political loyalty. This political support usually took the form of voting for the za'im and his allies in parliamentary elections, but the clientele could be required to support the za'im in other political conflicts, and could even be expected to take up arms in disputes with other za'ims. A considerable amount of ritual support was involved in the patron–client relationship, and clients had to demonstrate their loyalty in a variety of ways: on feast days, clients visited their za'im to wish him the compliments of the season; and when a za'im returned from a journey, his supporters usually turned out to welcome him home. Such occasions often involved the supporters slaughtering sheep and holding a great reception, during which the za'im's armed retainers would fire their machine-guns into the air—a popular method of expressing loyalty, political strength, and jubilation.

The za'im maintained his support in two important ways: first by being regularly returned to office, so that he could influence the administration and continuously provide his clients with governmental services; and secondly, by being a successful business man, so that he could use his commercial and financial contacts to give his clients employment, contracts, and capital. Depending on the wealth and influence of his clients, the za'im provided public works contracts, governmental concessions, employment in the government and private sectors, promotion within the professions and civil service, free or cheap education and medical treatment in government or charitable institutions, and even protection from the law. In order to survive electoral defeats and periods in opposition, the za'im had to be rich, or have access to other people's wealth, so as to buy the support of the electorate as well as the acquiescence of ministers and officials responsible for particular governmental services. Although election to the Assembly was of considerable advantage, it was not always essential, and a za'im could survive temporary periods of opposition and political weakness by using the credit he had built up in the past. Thus, for example, a judge, who owed his original appointment to a particular za'im, would probably continue to be lenient to the za'im's criminal clients even if the za'im were no longer in office; and a business man, who had been enriched by the za'im granting him a contract or concession, could be expected to donate some money to the campaign fund when the za'im was in opposition. In both cases, the grateful clients might eventually be bought off by rival za'ims. Such changing allegiances were not uncommon, and lower down the social hierarchy large-scale defections from the clientele took place when a za'im failed to deliver the goods.

Za'ims were not elected on the basis of a programme, but on their ability to provide their clientele with services. In this sense, the clients' support was a transactional obligation, rather than a form of moral, ideological loyalty. But although national and programmatic appeals were not usually made by za'ims, they did make some moral appeals, and sought to woo their electors by posing as local champions and confessional representatives. Thus a Beiruti Sunni za'im appealed to Beirutis *qua* Beirutis, and to Muslims *qua* Muslims. Za'ims did win the support of constituents who were not of their religion. Indeed, the electoral system compelled them to do this.[4] But the major part of their support usually came from their own sect. During the 1950s, for example, Sunni za'ims in Beirut put forward demands for increased Muslim representation in the administration, and increased influence in what they described as a Christian-dominated state. Similarly, by providing governmental services for his constituency, the za'im was able to demonstrate that he was an active local representative. During the 1960s, when the reformist 'Shihabist' regimes of Presidents Shihab (1958–64) and Hilu (1964–70) diverted governmental resources to the poorer, outlying regions of the country, Beirutis complained that contracts and jobs were going to non-Beirutis at their expense. Sa'ib Salam,[5] consistently excluded from the premiership during this period, was thus able to capitalize on this feeling, portray his competitors as traitors to their city, and maintain the support of a large part of his clientele even though he was politically weak for such a long period of time.

Strong-arm Retainers

A defining characteristic of the za'im was his close coercive control of the clientele. All Lebanese politicians entered into transactional relationships with their constituents, but what distinguished the za'im was his sophisticated machine and his willingness to use force, not only in attaining political objectives, but also in maintaining the loyalty of his clientele. A za'im did not necessarily set out to bully his clients into accepting his leadership. But in building his apparatus to recruit and control the machine, he made alliances with strong-arm neighbourhood leaders or qabadays, whose support was based, in part at least, on coercion.

The word *qabaday* is generally supposed to be derived from the Arabic verb 'to grasp or hold' (*qabada*). But the same word is found in Turkish (*kabadayi*) where it means 'swashbuckler, bully, tough'. In Lebanon *qabaday* is an ambiguous word which can have positive or negative connotations, although it is usually a sign of approval and is used to describe someone who is quick-witted, physically strong, heroic, or possessed of some other supposedly masculine attribute. Negative connotations include bullying, throwing one's weight around, and other forms of unruly and obstreperous behaviour. In its most specific sense, *qabaday* is a title given to a street or quarter boss who combines both positive and negative attributes into a leadership role. Here the qabaday recruits a following on the basis of his reputation as a man of the people, as a helper of the weak and the poor, as a protector of the quarter and its inhabitants, and, most important, as a man who is prepared to defend his claims to leadership by the use of force. All these characteristics might apply to the za'im as well as the qabaday, but prior to the outbreak of civil war in 1975 the two leaders could be distinguished by their social origins and levels of leadership. Whereas the za'im was born of a rich notable family and was recognized as the leader of a large following, the qabaday's parentage was of low socio-economic status and his political influence was limited to his immediate neighbourhood. Typically, the qabaday was a criminal involved in protection rackets, gun-running, hashish smuggling, or other similar activities. The za'im provided him with protection from the police and the courts in return for his political loyalty and services. These services included recruiting and controlling the za'im's clientele, organizing mass demonstrations of support, and, if necessary, fighting for the za'im in battles with his rivals.

The qabaday was particularly well suited for ensuring the loyalty of the za'im's clients because although he could use his relative wealth and his underworld connections to help the inhabitants of his quarter, he could also use physical force to control the clientele. One informant described in the following way how a qabaday established himself:

A qabaday starts by throwing his weight around, picking quarrels and beating up those who don't treat him respectfully. For example, if he is sitting on the street corner playing *tawila* [a type of backgammon], and someone from the quarter walks by without wishing him a good day and enquiring after his health, he might beat him. Later he'll probably shoot someone. Perhaps he'll kill another criminal, perhaps someone in a quarrel. The important thing is not so much who he kills, but the way he does it. Anyone can shoot someone. A qabaday has to do it openly and be willing to accept responsibility, and possibly go to prison for about four years.

In other words, the importance of the killing was the symbolic nature of the act. A qabaday was someone prepared to promote his leadership claims by an open murder in complete disregard for the law. He accepted the risk of imprisonment and a protracted vendetta, but in committing the act openly he won the respect of his fellows. Because physical strength and notions of honour were (and are) highly prised values, the qabaday attracted a local following and could therefore expect the protection of a za'im and a reduced prison sentence for his crimes. He acted outside the law, but was governed by another code of norms accepted by the society in which he lived. Some Beirutis, particularly the highly educated, did not accept this code, and for them the qabaday was *az'ar*, a criminal, thug, or murderer. But in the poorer quarters, where the qabadays were particularly powerful, *az'ar* was a term reserved for a man who killed dishonourably, for a robber of the poor and, most significantly, for a qabaday who worked for an opposing za'im. One man's qabaday was thus another's *az'ar*, but most people had ambiguous feelings about their local strong-arm men. They supported them partly out of respect for their honourable qabaday qualities, and partly out of fear of their willingness to use force.

One way Muslim Beirutis expressed and attempted to resolve this ambiguity was to indulge in myth-making, particularly about those qabadays who operated in the early part of this century. In these stories the qabaday aided the weak and the poor, fought against Christian attacks on Muslims, and protected the Sunni quarters against the excesses of the Turkish and French governments. Although the myths admitted the criminal activities of many of these leaders, they always emphasized the qabaday's basically honourable character and the services he performed for his neighbours. Of course, the stories changed according to who was the narrator and who the listener, and it is interesting to note that, as a 'Christian foreigner', I was often told the qabaday was a repository of all Arab virtues, that he was hospitable to strangers, and that Muslim and Christian qabadays were friends who regularly visited one another's houses. So according to the myth, the paradigmatic qabaday was the strong, honourable quarter leader of the Ottoman period. A coffee shop that I used to visit in the early 1970s was owned by just such a leader. As an old man of some 90 years, he was able to lay claim to the 'true qabaday status'. He never smoked cigarettes, nor drank alcohol; he wore traditional dress, and much regretted the decline of traditional moral standards. His predominantly youthful customers accorded him great respect, greeting him with elaborate courtesy and ritual, including kissing his hand. Although they broke most of the old man's rules of morality, they always favourably compared him with what they called the 'new qabadays':

The real qabaday has disappeared. Nowadays the new qabadays drink in night-clubs, smoke hashish and deal in cocaine and heroin. In comparison with the 'Hajj' [the coffee-shop owner had made the pilgrimage to Mecca] they are all weaklings. Look at him; even at his age he is still physically strong. These new men are only strong because they carry a pistol.

Nevertheless, despite their opinion, which they shared with many other people, most of the young men respected the new qabadays, valued their friendship, and even had qabaday aspirations themselves. This was largely due to the qabaday's proven ability to defend and champion his local community. Periodic confrontations between the Muslim and Christian halves of the city encouraged the local population to see the qabaday as their communal protector. Most of the more powerful qabadays in the early 1970s had established themselves during the civil war of 1958 when, as young men, they had demonstrated their ability to mobilize armed bands of *shahab* (young bloods) to fight for the insurrectionist za'im, Sa'ib Salam, against the government of President Chamoun. Their bravery in battle, and their ultimate victory over the Chamoun regime, turned them into heroes. Often uneducated men from relatively humble backgrounds, they emerged as popular leaders of their quarters and, as such, were extremely useful to the za'ims. The za'im, born of a rich notable family, had great social and economic status, and he won support partly because of that. But by working through the qabaday, he was able to show his willingness to come down closer to his low-status clientele. While he gained increased moral support through his alliance with a popular leader, the za'im also derived considerable pragmatic advantage, because the qabaday, as local boy and local hero, was in an ideal position to recruit and maintain a loyal clientele.

The Qabaday as Part of the Za'im's Apparatus

The most important part of the za'im's apparatus was the core of qabadays. During elections, they acted as 'election keys' (*mafatih al-intikhabat*) and ensured that the za'im's clientele voted for him and his allies. Between elections they recruited supporters, channelled requests for services, and organized mass demonstrations of support. All politicians used intermediaries to recruit and maintain support, but the za'im's strength lay in his ability to give protection and assistance to the popular leaders of the street. Other politicians were either unable to offer this protection, because they were new to the game, or unwilling because of their commitment to some form of social change.

Successful za'ims recognized the necessity of keeping their organization as simple as possible. Even if the za'im could find the time to see each of his clients when they asked for services, he would have found it impossible to ensure their loyalty. In the predominantly Sunni constituency of Beirut III, a Sunni za'im could expect between 12 and 16 or even 18 thousand votes. A proportion of these voters would be clients of other politicians on the za'im's electoral list (a temporary alliance of candidates in multi-member constituencies), but the majority would in some way be beholden to the za'im. In constructing his machine, the za'im sought to organize these voters into manageable groups, where he could leave the responsibility of controlling the clientele to his lieutenants.

The za'im kept the number of lieutenants to a minimum, and usually had something in the region of 15 to 20 qabadays in his core group. Typically, a member of the core was an established political boss who controlled a network or gang of lesser qabadays. Two major types of core qabadays can be distinguished: quarter bosses, who looked after the za'im's interests in particular localities, and family qabadays,[6] who were often also organized by quarters.

Perhaps the most important type of qabaday was the boss of a quarter. The boundaries of Beirut's administrative districts are virtually the same as those of the nineteenth-century quarters which were established as Beirut developed beyond the walls of the old city. The confessional war of 1860 prompted thousands of Christians to settle in the relative safety of Beirut, and the expansion of trade attracted members of all confessions to seek their fortune in the service sector of the economy. The different confessions tended to settle in particular districts. The Christians settled in the east, the Muslims in the west, and these two regions were subdivided into quarters dominated by particular sects. Native Beirutis, generally Sunnis, also moved to a pleasanter environment outside the walls, and while immigrant Muslims (usually Shi'ites) tended to settle in the inner and outer rings of the western city, the native Sunnis were predominant in the quarters of the middle ring. The Sunni mercantile families were the first to move, but the large residences of rich merchants were soon surrounded by the homes of poor and middle-class Sunnis. Although this meant that the Sunni quarters were not inhabited by homogeneous occupational groups or classes, they did acquire some distinguishing characteristics. Basta, with its port workers and stevedores, had a more popular character than Musaytiba where established mercantile and sheikhly families tended to settle; and the quarters of the inner and outer rings (respectively, Bashura and Tariq al-Jadida) were considerably poorer than the generally middle-class quarters of Mazra'a. The process of urbanization gradually changed the character of these quarters. The middle ring, in particular, became much more heterogeneous than before, and the boundaries between quarters became less precise. Nevertheless, there is still a sense of inhabitants 'belonging' to a quarter. There are differences between quarter dialects, and particular families or clans continue to be associated with particular districts.

In pre-civil war Beirut, the za'im could conveniently organize his electoral clients by quarters, and he established local bosses to control them. Sometimes these bosses were associated with the traditional occupation of the quarter. In Basta, for example, one qabaday inherited a family business in the port where there was a qabaday tradition of gang fights between competing families and factions over such prizes as government concessions to manage the barges and to transport cargo from the ships. Some quarter bosses were protection racketeers; others were alleged to be political assassins or agents in the pay of foreign embassies; but most were outwardly respectable business men. The important thing was that they were powerful local leaders who knew their quarter and its inhabitants, were always ready to see and help the za'im's clients, and could mobilize lesser qabadays to fight for the za'im.

The other important type of core lieutenant was the family or clan qabaday. In a sense, every qabaday represented both his family and his neighbours in the quarter, but some qabadays came from such large clans that they were particularly important for ensuring the loyalty of their kin. In Sunni Beirut the 'Itanis formed perhaps the largest family, with around 4,000 voters. Traditionally, they supported the Salams, both Sa'ib and his father Salim. But because the family was so large, and spread all over the

Sunni quarters of Beirut, it needed a certain amount of organization if it was to form an efficient unit in Sa'ib Salam's clientele. Other za'ims recognized the value of the family's votes, and various attempts were made to woo the 'Itanis away from their traditional za'im.

So as to maintain his hold on the family, Salam established a number of 'Itani qabadays to control their kin. This was not a novel nor a unique tactic. What usually happened was that the za'im chose particular individuals who had a certain amount of support in the family, and then built them up to a position of monopoly leadership. Thus according to some informants, Salam let it be known that any 'Itani living in Musaytiba, who wanted a service from him, should first see Hashim 'Itani, his chosen representative and the 'official' 'Itani qabaday of the quarter. In 1958 Hashim had been a minor qabaday fighting under Salam's leadership. Over the years, he increased his local influence, and it was said that when Salam recognized Hashim's strength, he gave the qabaday considerable economic assistance. It is certainly the case that from being the owner of a small coffee shop, Hashim 'Itani extended his business interests such that he became part-owner of two of Beirut's biggest cinemas, and one of the directors of a company that owned restaurants, bars, and cafés in the fashionable quarters of Ras Bayrut. Although the amount of governmental patronage that Salam could dispense was limited in the 1960s, he was able to capitalize on his close relations with right-wing Christian politicians, including his former enemy, ex-President Chamoun, and give help to his clients in the fields of business and commerce. Thus a number of informants told me that Salam helped to find Hashim 'Itani Christian capital and expertise, and that in doing so he created an indebted ally. These same informants claimed that by refusing services to those 'Itanis not vetted by Hashim, Salam could control his 'Itani clientele in Musaytiba through one loyal and grateful lieutenant.

The example of Hashim 'Itani emphasizes the moral, as opposed to transactional, relationship between the za'im and his qabadays.[7] All za'ims had a moral core of lieutenants, who over time received so many transactional benefits from their patron that the relationship acquired a degree of permanency. Members of the core remained loyal to the za'im not simply because of the expectation of future services. They also had a debt of gratitude for past services. This debt changed the character of the za'im–qabaday dyad from a patron–client exchange to a leader–follower relationship, which was often further transformed into a condition of friendship. Perhaps Sa'ib Salam's greatest strength was his ability to make such a transformation with a few key people. Reference has already been made to his exclusion from high office during the so-called Shihabist regimes of Presidents Shihab and Hilu. During that period, the agents of the Deuxième Bureau (intelligence apparatus) skilfully operated the za'im–qabaday system against Salam and other za'ims opposed to Shihab. They suborned criminal qabadays who needed government protection, and they established their own powerful qabaday network to organize the clientele of Shihabist za'ims. But although Salam lost a large part of his clientele to the Shihabist qabadays and their masters, he was able to maintain his apparatus by giving commercial and financial assistance to his lieutenants, and by forming close moral relationships with them. Thus a man like Hashim 'Itani, who did not rely on criminal activities for his economic well-being, was not subverted by the Deuxième Bureau.

The core qabaday's role should be contrasted with that of the notable (*wajih*). Notables were accorded high status because of their wealth, philanthropy, and reputation for religious piety. But they were not usually significant as election keys, as they were not prepared to spend time dealing with the individual problems of the poor, and were not able to coerce the clientele during elections. If a notable did use his contacts in government to build up a large client following of his own, he was likely to become a politician in his own right and ultimately pose a threat to the za'im's dominance. For this reason, the notable could not be trusted. The qabaday, on the other hand 'came from the masses', was protected or enriched by the za'im, and was much more easily controlled. In addition, because he usually lacked education and the statesmanlike qualities required of a politician, he was not likely to stand as a candidate in parliamentary elections.

The Qabaday's Functions in the Machine

In using the qabaday as an intermediary, the za'im cut down the risk of performing services for his rivals' supporters, left the policing of the machine to a few individuals over whom he had close control, fragmented the electorate into politically artificial entities of quarter and family groups, and further bound his individual clients into the patron–client debt relationship.

The za'im's patronage resources were not infinite, and it was necessary to supervise carefully the recruitment as well as the loyalty of his clientele. Thus he and his secretaries operated a selective procedure when a potential client requested a service. Most attention, and a personal audience with the za'im, were granted to rich clients, clients from large families, and other voters who could provide the za'im with significant monetary and electoral resources in return for patronage. A second category of clients were those poorer and politically less important voters who made up the majority of the za'im's support base, but individually could only pledge their own votes, the votes of their nuclear family, and possibly the votes of a few friends. Usually such people had relatively simple requests and were dealt with by the za'im's secretarial assistants. On behalf of the za'im, secretaries wrote a note or made a telephone call to the relevant government department or whatever institution or individual was concerned with the client's case. A third category of clients included those who voted in another constituency and those who were not enfranchised. An example of the latter group was the Palestinian community. Most Palestinians were not granted full civil rights in Lebanon, did not have the vote, and, as a result, had nothing to offer the za'im. Such people were refused services, while Lebanese citizens from other constituencies were told to approach their own za'ims.

The qabaday was particularly important with regard to the second category. As scores of people came daily to the za'im's house or office to ask for services, the za'im and his secretaries could not easily distinguish which potential clients were voters in the constituency, and which of those were consistent supporters. Elaborate records were kept, but the best way of establishing a client's credentials was to insist on his first seeing a quarter or family qabaday. The qabaday had extensive knowledge of his local domain, and was well placed to vet the client's loyalty and political reliability. He could see which clients regularly presented themselves on feast days, and other similar occasions, to demonstrate their ritual support and continuing allegiance to the za'im, and he could usually ensure that clients voted in the way they were instructed.

The use of the qabaday also served to fragment the electorate and prevent the emergence of self-conscious, horizontally linked social categories such as interest groups or classes. By forcing clients to approach him through their quarter or family qabaday, the za'im encouraged the individual client to see himself as a member of a particular quarter or family grouping, and discouraged the formation of other social categories that might have posed a real threat to the status quo. Although urbanization tended to lead to the break-up of the quarter as a homogeneous unit, and the extended family gave way to the nuclear family, the za'im capitalized on his clients' mythological conceptions of social organization, and in some cases actually created artificial quarter and family identification.

But although the za'im used family and quarter as units of organization, he did not deal with them as corporate groups with their own elected spokesmen. He did not, for example, make a contract with the president of the 'Itani Family Association to provide the family with a sum of money, or a certain fixed number of services, in return for the whole family's support. The electorate was further fragmented because the contracts were made between the za'im and *individual* members of the family or quarter. The qabaday played a crucial role in establishing and policing these contracts.

Although the qabaday independently performed some services, such as mediation of disputes, protection of the quarter, and charitable distributions of small sums of money, his access to governmental patronage was limited. He therefore tended to act as a broker who did not speculate in his own right, but effected and facilitated the formation of contracts between the individual clients and the za'im. Claims that election keys in Beirut delivered vote blocs should be treated with some scepticism. When informants told me that a qabaday was worth at least 2,000 votes to a particular za'im, this did not mean that if the qabaday defected the za'im would lose all 2,000 votes. He would inevitably have lost some, but he would only lose a large number if he failed to find another qabaday to take over the defector's role. Even if a client went first to his street qabaday, then to the quarter boss and only after this went to the za'im's house, where he perhaps saw a secretary rather than the za'im himself, the contract was made between the client and the za'im, not between the client and the qabaday. The important function of the qabaday was to reinforce this za'im–client relationship. When a client received a service from the za'im or his secretary, he was making an agreement with a notable who was socially distant and only occasionally seen in a face-to-face situation. But the transaction was channelled through the qabaday whom the client saw in the quarter possibly every day, and this regular contact between client and qabaday served as a continual reminder of the client's debt to the za'im.

The Retainers become Bosses

Until the outbreak of civil war in 1975, the clientelist system of the za'ims worked remarkably well. It is true that the lack of political parties and national political structures meant there were few arbitrating mechanisms which could control local conflicts. This resulted in costly local feuds between notables, and contributed to the outbreak of a full-scale civil war in 1958. But the system was flexible enough for enemies to become reconciled. From 1958 to 1964 the reformist President Shihab dealt with some of the socio-economic and political problems which had given rise to civil war; and his successor, President Hilu, carried through a number of Shihabist reforms. In addition, the factional squabbles between za'ims usually served the interests of the dominant social class, for a relatively weak state could not interfere in trade and finance. The most serious danger to the Lebanese commercial-financial bourgeoisie was the prospect of social unrest among the urban sub-proletariat. So long as the za'ims could control these people, the persistence and reproduction of the social formation was not endangered.

The success of the clientelist system was, however, predicated on the continued availability of patronage, particularly with regard to the provision of employment, promotion, and career opportunity. Furthermore, given the strong social attachments to religious primordial loyalties, the distribution of this patronage had to be fairly shared between the confessions. In the early years of independence, these conditions appeared to be being met, but as time wore on fundamental problems emerged which led ultimately to the revolutionary climate of the 1970s. The rate of unemployment rose, and it became increasingly obvious that the Muslim community—and especially the Shi'ite sect—was underprivileged as compared with the Christians. With the development of social and communal conflict, the clientelist structures of Sunni Beirut came under increasing pressure and, during the civil war, eventually collapsed into a murderous anarchy.

During the latter 1960s, two groups of qabadays could be distinguished in Sunni Beirut. One was composed of strong-arm business men loyal to Sa'ib Salam, and these included Hashim 'Itani in the quarter of Musaytiba. The second group was Shihabist and was protected or paid by the Deuxième Bureau. These leaders of the street were more committed than the Salamists to the popular ideology of Arab nationalism. They were at least nominally Nasserist and had remained loyal to Cairo after Sa'ib Salam had changed his own allegiance to Saudi Arabia. The most important and influential qabaday in this group was Ibrahim Qulaylat from the popular quarter of Tariq al-Jadida. A somewhat mysterious 'Robin Hood' character, Qulaylat was generally considered to be an agent in the pay of Egyptian intelligence or the Deuxième Bureau, and probably both. In the late 1960s, he was unsuccessfully prosecuted for the assassination of a right-wing and pro-Saudi newspaper editor, and many observers thought that pressure had been applied on the court from higher echelons of the Shihabist regime.[8]

The election of President Frangieh in 1970 brought about a change of regime. Sa'ib Salam became prime minister and, with his president's blessing, set about purging the Deuxième Bureau. Principal officers were put on trial and others were removed from positions of influence. The new internal security agency was staffed by officers loyal to the Frangieh–Salam regime, and although these men were competent they lacked access to the intelligence networks which had been assiduously built up since 1958. As a result, the regime lost control of the political underworld.

During the Shihabist period (1958–70), the Deuxième Bureau had been able to tolerate some minor excesses on the part of the racketeers, while at the same time keeping them under a tight political control. After the purge of the Bureau, these Shihabist qabadays were able to operate more independently. While some pro-Frangieh za'ims were able to recruit certain criminal qabadays and bring them under state control by offering a continued protection from the law, other street leaders were able to build links with powerful patrons outside the Lebanese clientelist system. These patrons were principally the Palestinian organizations and the Libyan and Iraqi governments.

By the beginning of the 1970s, a number of Palestinian gangs, largely recruited from the refugee camps, had developed smuggling and protection rackets of their own and were emerging as competitors of the indigenous Lebanese gangs. Through their connections with commando organizations, the Palestinian criminal networks were well armed; and as the Cairo Agreement of 1969 allowed the Palestinians to enforce their own laws in the refugee camps of Lebanon, the gangsters had a secure base from which they could operate. The competition between these new gangs and the Lebanese qabadays led to armed clashes. The za'im protectors of the qabadays were powerless to prevent these, and were often unable to bring the Palestinians to justice. Revenge was exacted through vendettas which exacerbated the existing state of tension.

Some of the Muslim racketeers dealt with the new situation by working in partnership with Palestinian gangs, and it appears that the commando movement itself took control of some qabaday networks. Palestinian and Muslim Lebanese gangs were also brought closer together as a result of their common opposition to some of the Christian gangs. There was a traditional rivalry between the communal champions of Muslim and Christian Beirut—a rivalry that was usually contained by the mediation of za'ims and notables from both communities. It seems that in the 1970s, one of the objections of Christian parties like the Phalangists to the armed Palestinian presence in Lebanon was that it was upsetting this delicate balance of the political underworld.

Of perhaps greater concern was the fact that some Muslim qabadays were becoming more overtly political than the normal type of racketeer. A prominent example of this trend was Ibrahim Qulaylat, who after the death of President Nasser had switched his allegiance to Arab paymasters more radical than President Sadat and had founded his own organization called the Independent Nasserist Movement. The populist ideology of Nasserism was attractive to such 'primitive rebels',[9] appealing particularly to their sense of Arabism and to their political identification with Islam. It was during the 1950s that it had first provided a populist expression of Sunni discontent, and many of the more powerful qabadays had established themselves in the 1958 civil war as the communal champions of their quarters. Qulaylat was 16 years old at the time and he, like other Sunni qabadays, fought for the 'Nasserist' insurrectionary za'im, Sa'ib Salam. The adoption of Nasserism by Salam and other za'ims reduced the already limited revolutionary potential of the ideology, incorporating it and the qabadays into the clientelist system. After Salam broke with Egypt in the

1960s, the Nasserist qabadays were brought under the supervision of the Deuxième Bureau. Again the radical element of the ideology was controlled, possibly more effectively this time by the apparatus of a quasi-police state. Manipulated first by the za'im representatives of the state, and then by the state itself, Nasserism provided a framework for socializing the urban poor and their qabaday leaders into an acceptance of the status quo. The end of Shihabism, however, provided an opportunity for freeing the ideology from the control of the state and liberating its populist appeal.

The radicalization of popular Nasserism was a gradual process, and it became a significant political force only when state authority finally collapsed in Beirut during the fourth 'round' of fighting in September 1975. Throughout the first three rounds of the civil war, the quarters of Sunni Beirut remained relatively free from the fighting. Barricades were erected around some quarters and there were cases of kidnappings, mutilations, and murders, but there was no heavy fighting as there was in the suburbs. The Sunni establishment only lost control of the situation in September when the Phalangist militia began to bombard the western half of Beirut's commercial district in an apparent attempt to force the government to send in the Lebanese Army to restore order.[10] This brought the fighting into the centre of the city and, by the end of October, Ibrahim Qulaylat's militia, the 'Murabitun', had become involved in a strategic battle which earned Qulaylat the right to be considered as one of the foremost leaders of the National Movement (the coalition of 'progressive' forces fighting the predominantly Maronite rightist militias).

In October fighting had again broken out along the 'confrontation line' which divided West and East Beirut, and later spread westwards along the coast from the commercial district towards Ras Bayrut. The Christian rightist militias were attempting to control access to the port, and as part of their strategy they occupied the high-rise building of the Holiday Inn. Opposing them were a number of leftist and Muslim militias under the overall leadership of the Murabitun. The Muslim militias inflicted a significant defeat on the Christians, and the Murabitun demonstrated that it was a relatively well-disciplined and powerful fighting force. In the fifth round of fighting, in December, Ibrahim Qulaylat again achieved notoriety when he led another attack on the Christian strongholds in the hotel district, and forced the Maronite militias out of the Saint George and Phoenecia hotels.[11]

Before the hotel battles of October and December, Qulaylat's Independent Nasserist Movement had been just one of many Nasserist and populist groups in Sunni Beirut. As a result of the bravery and success of its fighters at the end of 1975, however, the Murabitun militia became an influential force in Lebanon's political mosaic. This period marked the final collapse of the za'ims' clientelist system in Sunni Beirut. Men like Ibrahim Qulaylat would never again be beholden to their traditional patrons. Although Sa'ib Salam, for example, made persistent communal appeals to the Muslim community, forming a new political party based on the 'teachings of Islam' and condemning what he called the 'destructive left',[12] the initiative had already passed to the new Nasserist leaders. In contrast to 1958, the war of 1975 brought about the establishment of a popular movement in Sunni Beirut completely independent of the za'ims and their clientelist system.

After March 1976, when the Murabitun and its allies finally took the Holiday Inn, the National Movement and the Palestinians extended their control to take over almost 80 per cent of Lebanese territory, including part of the Maronite stronghold in Mount Lebanon. The rightists faced imminent defeat and were rescued only by shipments of weapons from Israel and by the intervention of Syrian troops in June. The latter invasion, and the rightist counter-attack which it allowed, inflicted severe losses on the Palestinian and progressive coalition and led to a military stalemate which amounted to a *de facto* partition of the country. Whether the left could have brought about a new political order in Lebanon is a matter for speculation. All that can be said is that once the Syrian troops had prevented a leftist victory, the progressive parties were not very efficient at imposing such an order on their much-reduced sphere of influence.[13]

In East Beirut the Phalangists controlled their 'street' and made it illegal to carry a gun without a permit. Criminals were arrested and sometimes summarily executed, and taxes were raised and services provided. Such an order was built on the fact that the Phalangists had either eliminated their rivals or brought them into a unified military force under the command of Bashir Gemayel. Gemayel's rise to power was destructive and bloody,[14] but the rough discipline he imposed was considered by many inhabitants of Christian Lebanon as preferable to the chaos which prevailed in West Beirut.

One major division amongst the progressive forces in Muslim Beirut had occurred when a number of organizations, such as the pro-Syrian Ba'th faction, formed a separate front sympathetic to Damascus. Relations between this new coalition and the Lebanese National Movement (LNM) were especially tense during the Syrian invasion, but improved a little in the latter part of 1977 when the pro-Syrian Ba'th and the Progressive Socialist Party (predominantly Druze Muslim) issued a joint proclamation as a step towards an improvement of relations between Damascus and the LNM. Events outside Lebanon also had an influence on the making and breaking of alliances. The 1978 Camp David agreements between Egypt and Israel promoted a *rapprochement* between Syria and Iraq, which was reflected in Lebanon by close relations between the two wings of the Ba'th Party. But hostilities broke out once more when Syria supported Iran in its war with Iraq. Similarly, the Shi'ite 'Amal' militia supported Iran while

Arab nationalist forces tended to support Iraq. Such divided loyalties led to non-Lebanese conflicts being fought by proxy in West Beirut and the suburbs, and these conflicts became inextricably wound up with the parochial rivalries between street and quarter gangs.

The number of fighting forces in Muslim Beirut seems to have increased after the 1975-6 civil war. Syria, Iraq, Libya, and even Iran and Saudi Arabia, either directly or indirectly funded and supplied militias, and there were probably other countries involved as well. Some Arab regimes such as Syria and Libya supported more than one militia, and by the late 1970s there were around 30 separate fighting units in West Beirut alone. Many were the clients of one particular regime, while others received material aid from a number of different sources; some were financed by wealthy Lebanese to protect their businesses and property, and many were connected to protection rackets and smuggling. There were disciplined political organizations amongst them, but a large proportion were little more than adolescent street gangs—inheritors of a qabaday tradition which had degenerated into its most anarchic form. In a complicated process of fission and fusion, such gangs and militias fought and allied according to the nature of the dispute. Often they fought street to street over the extent of their territories and protection rackets, but faced with a common foe they would band together, as when pro-Syrian groups united to fight supporters of Iraq. During the Israeli siege of Beirut in the summer of 1982, the various groups tended to co-operate in readiness for an invasion of their quarters, but even then there were some battles over who would defend which street. The most petty disputes could give rise to gun-battles: motorists shot at each other after quarrels over a right of way or parking-space; dissatisfied customers gunned down shopkeepers; jilted lovers killed their rivals; and because many people could count on the support of their local gunmen, such incidents often widened into artillery duels between one street and another. The Syrian Army, already unpopular because of its intervention on the side of the Maronite rightists, was unwilling to move too hard against the Muslim street. Instead, Syria preferred to work with a new clientelist system and seek out favoured clients through whom it could manipulate the violent and unstable politics of West Beirut.

Conclusion

In a relatively short article it is impossible to trace the course of the Lebanese civil wars that erupted again in the 1980s. After the Israeli invasion of 1982, it looked as though some form of political order could be re-established. Sa'ib Salam became more prominent and played an important role as a mediator between the Palestinians and the US envoy who negotiated the withdrawal of the commandos from Beirut. For a brief period a reconstituted Lebanese Army patrolled the streets of West Beirut and there were hopes for a lasting peace. But disputes between the Druze Muslims and Maronites of Mount Lebanon led to more fighting which eventually resulted in Shiʻite militias taking control of large parts of Sunni Beirut in 1984 and destroying the Murabitun in 1985.[15] During ten years of civil war, ideological politics had gradually retreated before the powerful primordial forces of confessional allegiance. It had become impossible to talk of a political left and right in Lebanon, and what remained of the country was a patchwork of territories, in and around Beirut, under the control of the different confessional militias and their foreign backers.

The clientelist system which had maintained political order before 1975 had given way to anarchy in some areas and to enclaves of confessionally based authority elsewhere. In Sunni Beirut the relatively ordered structure of the zaʻims' political machines had long been replaced by unbridled gangsterism and, in the mid-1980s, communal conflict between Sunni and Shiʻite Muslims became more prominent than ever before. In response, Sunnis increasingly turned inward on the quarter and family solidarities which had been manipulated so effectively by the pre-war zaʻims. Without a central authority which could enable the zaʻims to control the qabadays, such primordial solidarities provided some sense of community and security in a highly competitive and violent society.

Whereas quarter and family solidarities in the 1960s and early 1970s can be described as essentially 'mythological' and often were artificially created by zaʻims to maintain control of the clientele, they had become powerful forces by the 1980s. The divisions between quarters and clans, fuelled by rivalries between local militias, had contributed to a fragmentation of the Sunni community. The pre-war zaʻims had been remarkably effective at maintaining political order, but their system had relied on their access to the superior force of the Lebanese Army and security apparatus which could be used to control rebellious qabadays. Criminal qabadays had been protected by their zaʻim, but if they stepped out of line they could be handed over to the police. The break up of the army and police during the civil war permanently weakened the Sunni zaʻims and left their community open to parochial conflicts and divisions. In particular, the attachment of many Beiruti Sunnis to the values of Arabism meant that the divisions of the Arab world became replicated in damaging intra-confessional conflicts between Sunni militias, thus leaving them in the 1980s with a weak sense of communal solidarity at a time when a confessional balance of forces between Maronites, Druze, and Shiʻites seemed to be the most likely framework for an end to the civil wars.

Notes

1. One of the best analyses of the Lebanese economy in the modern period is S. Nasr, 'Backdrop to Civil War: The Crisis of Lebanese Capitalism', *MERIP Reports* 73 (1978.) For more details, see C. Dubar and S. Nasr, *Les classes sociales au Liban* (Paris, 1976).
2. The material on pre-civil war political machines is drawn from M. Johnson, 'Political Bosses and their Gangs: *zu'ama* and *qabadayat* in the Sunni Muslim Quarters of Beirut', in E. Gellner and J. Waterbury (eds), *Patrons and Clients in Mediterranean Societies* (London, 1977).
3. Material on the early 1970s and the civil war of 1975–6 is drawn from M. Johnson, 'Popular Movements and Primordial Loyalties in Beirut', in T. Asad and R. Owen (eds), *Sociology of Developing Societies: The Middle East* (London, 1983). For the late 1970s and the civil wars of the 1980s, see M. Johnson, *Class and Client in Beirut: The Sunni Muslim Community and the Lebanese State, 1840–1985* (London, 1986), ch. 8. The latter book elaborates on the 'clientelist system' which operated in Sunni Beirut and elsewhere in Lebanon, and explains how the system broke down and was changed in Lebanon's civil wars.
4. The seats in the Assembly were allocated to the various confessional communities according to their supposed size in the population. For the elections of 1960, 1964, 1968, and 1972, in the constituency of Beirut III, four of the five seats were reserved for Sunni Muslims, and one for a Greek Orthodox Christian. In order to win, a Sunni candidate usually had to recruit the support of Christians as well as Muslims.
5. After 1958 Sa'ib Salam was the most powerful Sunni za'im in Beirut, and he remained influential even after the outbreak of civil war in 1975. Other Sunni za'ims, such as 'Uthman ad-Dana, lost virtually all influence during the wars of the 1970s and 1980s.
6. The word 'family' refers here to a clan of people bearing the same surname and having the same confession, who trace their genealogy to common ancestors. Often such clans have family benevolent associations, open to all members of the clan, and these give some organizational solidarity to the group.
7. Such concepts as 'moral', 'transactional', and 'core group' are derived from F. G. Bailey, *Stratagems and Spoils: A Social Anthropology of Politics* (Oxford, 1969), esp. 34–49.
8. Johnson, *Class and Client in Beirut*, ch. 3, case-study 7.
9. E. J. Hobsbawm, *Primitive Rebels: Studies in Archaic Forms of Social Movement in the 19th and 20th Centuries* (Manchester, 1959).
10. K. S. Salibi, *Crossroads to Civil War: Lebanon 1958–1976* (Delmar, NY and London, 1976), 97 ff.
11. Ibid. and *MERIP Reports* 44 (1976), 14.
12. *MERIP Reports*, 44 (1976), p. iv.
13. For further information on the 1975–6 civil war and its aftermath, see, for example, W. Khalidi, *Conflict and Violence in Lebanon: Confrontation in the*

Middle East (Cambridge, Mass., 1979). The material in this article on the 1970s and 1980s is drawn from Johnson, *Class and Client in Beirut*, chs. 7 and 8.

14. J. Randal, *The Tragedy of Lebanon: Christian Warlords, Israeli Adventurers and American Bunglers* (London, 1983), 109 ff.

15. See Johnson, *Class and Client in Beirut*, ch. 8.

18

The Politics of Ethnicity in African Towns*

Abner Cohen

The Making of an Ethnic Polity in Town

Sabo, the Hausa quarter in the city of Ibadan, developed in conjunction with the growth and organization of Hausa monopoly in long-distance trade in kola and cattle between northern and southern Nigeria. The Hausa have been able to overcome the technical problems which are encountered in this trade by the development of an ethnic monopoly over the major stages of the trade. This has involved the development of a network of Hausa migrant communities in the Western Region of Nigeria. Sabo thus came into being as a base for control over parts of the southern end of the chain of the trade.

But in the process of achieving such control, the Hausa have come face to face with increasing rivalry, competition, and opposition from various Yoruba individuals and groups. From the very beginning, economic competition led to political encounters with members of the host society. The Hausa, confronting mounting pressure from the Yoruba majority, were forced to organize themselves for political action. With the growth of the trade, the increase in the number of settlers in the Quarter, and the expansion of the host city, Hausa political organization became more complex and more elaborate in two different, but closely related, spheres. First, the Hausa developed and maintained their tribal exclusiveness. Second, they built an internal organization of political functions: communication, decision-making, authority, administration, and sanctions, and also political myths, symbols, slogans, and ideology. The principal aims of the whole system are (a) to prevent the encroachment of men from other ethnic groups into the trade, (b) to co-ordinate the activities of the members of the community in maintaining and developing their economico-

* Reprinted, in a revised form, from the concluding chapter of Abner Cohen, *Custom and Politics in Urban Africa: A Study of Hausa Migrants in Yoruba Towns* (London: Routledge & Kegan Paul; Berkeley and Los Angeles: University of California Press, 1969), by permission of the publishers.

political organization, and (c) to maintain mechanisms for co-operation with other Hausa communities in both the South and the North, for the common cause.

During the period of indirect rule by the British many of these functions were officially recognized and constituted part of the formal organization of power which had been set up by colonial rule. The Hausa were recognized as a distinct 'tribal' group and were given a well-defined residential base and a recognized 'tribal' chief. The authority of the chief was ultimately supported by the power of the administration.

This formal recognition of Hausa political organization enabled the people of Sabo, not only to consolidate their gains in the control of trade, but also to capture more economic fields, and the actual Quarter itself, with its buildings, sites, and strategic position within the city, became a vast vested interest for the community.

With the coming of party politics in the 1950s, as the Nigerian nationalist movement arose and later with independence, the whole formal basis of Hausa distinctiveness was undermined. Sabo was no longer officially recognized as an exclusive 'tribal' grouping and the support which had been given by the power of the government to the authority of its chief was withdrawn. The weakening position of the chief affected not only the organization of the functions of communication, decision-making, and co-ordination of action, but also the very distinctiveness of the Quarter because it was no longer possible for the chief to force individuals to act in conformity with the corporate interests of the community.

In the meantime, the ethnic exclusiveness of Sabo was being threatened by increasing social interaction between Hausa and Yoruba in two major social fields: in party political activities and in joint Islamic ritual and ceremonial. Interaction of this kind was likely to result in the creation of primary, moral relations between Hausa and Yoruba, under new values, norms, and symbols.

The adoption of the Tijaniyya[1] by the Quarter brought about processes which halted the disintegration of the bases of the exclusiveness and identity of Sabo. The reorganization of the Quarter's religion was at the same time a reorganization of the Quarter's political organization. A new myth of distinctiveness for the Quarter was found. The Quarter was now a superior, puritanical, ritual community, a religious brotherhood, distinct from the masses of Yoruba Muslims in the city, complete with its separate Friday mosque, Friday congregation, and with a separate cemetery.

The localization of ritual in the Quarter inhibited the development of much social interaction with the Yoruba.[2] On the other hand, the intensification and collectivization of ritual increased the informal social interaction within the Quarter, under Hausa traditional values, norms, and customs.

The principle of intercession which the Tijaniyya introduced, and the concentration of all the mystical forces of the universe in Allah, vested a great deal of ritual power in the malams. The malams became the sole mediators between laymen and the supernatural powers of Allah. Through their services as teachers, interpreters of the dogma, ritual masters, diviners, magicians, spiritual healers, and officiants in rites of passage, the malams developed multiple relations of power over laymen and, through the hierarchy of ritual authority instituted by the Tijaniyya, this power is finally concentrated in the hands of the big malams.

Through their manifold relationships with the business landlords and the chief, the big malams have become part of the 'Establishment'. They act as advisers to the landlords and to the chief and they formally participate in the formulation of problems, in deliberation, and decision-making, and in the co-ordination of action in matters of general policy. They also play significant roles in the processes of communication and co-ordination in the course of the routine administration of the Quarter.

The Hausa of Sabo are today more socially exclusive, or less assimilated into the host society, than at any other time in the past. They thus seem to have completed a full cycle of 'retribalization'. They speak their own language even in their dealings with the Yoruba, and they dress differently and eat differently from their hosts. Hausa customs, norms, values, and beliefs are upheld by a web of multiplex social relationships resulting from the increasingly intense interaction within the Quarter. On the other hand, the absence of intermarriage with the Yoruba, and the ritual exclusiveness brought about by the Tijaniyya, have insulated the Hausa from much social interaction with the Yoruba and thus inhibited the development of moral ties and loyalties across the lines of tribal separateness. Finally, with the withdrawal of the British from Nigeria, the two ethnic groupings came into a sharp confrontation and the cleavage between them became deeper and more bitter.

Sabo has acquired more social and cultural distinctiveness as a result of marked social and cultural changes among the Yoruba in Ibadan. During the past few decades the Yoruba have developed cash crops, trade in European goods, and some light industry. They have adopted a relatively great measure of Western education and developed a fair degree of occupational differentiation and specialization in their society. The adoption of the city as capital of the Western Region, and its development as an administrative centre, together with the building of a large university[3] and a university hospital in it,[4] have brought further differentiation within its population. The formation of different kinds of voluntary associations,[5] the intensified activities of political parties, the emergence of a Western-oriented élite and of a new economically privileged class in it, have created a web of links and cleavages cross-cutting one another, and have thus changed the structure of Yoruba society.

In sharp contrast with all this change among the Yoruba of Ibadan, Sabo society and culture remain basically unaffected, like an island of continuity in a sea of change. Its economy remains stable and is today not much more sophisticated in its organization than it was 20 years ago. Its education remains purely 'Arabic', almost untouched by Western education.[6] The ambition of a Sabo man is success in trade, higher Islamic learning, and, as the crowning of success in both endeavours, pilgrimage to Mecca. While many of the Yoruba are culturally oriented towards European or American civilization, the Hausa of Sabo remain oriented towards the North and the North-east, towards the interior of Africa and the civilization of Islam.

The Economics of Ethnic Groupings

This 'conservatism' on the part of the Hausa is not the result of their situation as migrant strangers in the Yoruba city. Other migrant strangers in Ibadan have changed socially and culturally along with the Yoruba. This can be clearly seen among the Western Ibo who began to settle in Ibadan in the early 1920s and who in 1964 numbered 2,000–2,800.[7] They, too, were originally regarded as strangers and were concentrated in a special residential area.

But, as Okonjo shows, their residential segregation has completely broken down and today ethnic mixture in the compounds where they live 'is the rule rather than the exception'. They do have a tribal association, 'The Western Ibo Union, Ibadan', but it is a weak association, meets once a month, and has often suffered from the embezzlement of its funds and from frequent quarrels among its members. Like many other tribal associations in Africa, the Western Ibo Union in Ibadan has aimed not at the development of ethnic exclusiveness but on the contrary at promoting the successful adaptation of its members to modern urban conditions. Indeed some writers show that affiliation to such tribal associations is often only a temporary measure taken by new migrants to the city to get help to integrate within the new social milieu. Second-generation Western Ibos in Ibadan speak Yoruba 'without accent' and have Yoruba as their playmates. As Okonjo shows, the Western Ibos in Ibadan can be found scattered in places of work all over the town. They are occupationally differentiated, ranging from university lecturers, through mechanics, clerks, and printers, to workers of all sorts.

This integration of the Western Ibo in Ibadan on the one hand, and the apparent insensitivity of the Hausa of Sabo to the great changes that have been taking place around them in the town on the other, has been attributed by some writers to strong 'achievement motivations' among the Ibo (and the Yoruba) and to particularly 'conservative' and 'traditional' traits in the basic structure of the Hausa personality.[8] Thus in a comparative study of 'achievement motivations' among the Hausa, Yoruba, and Ibo in Nigeria, LeVine writes: 'Hausa traders are everywhere in West Africa . . . but their pattern of trade is traditional and no matter how long they stay in modern cities like Accra and Lagos, they remain conservative with regard to education, religion and politics and aloof from modern bureaucratic and industrial occupations.' According to LeVine, this traditionalism among the Hausa, as against the spirit of being 'go-ahead' among the Ibo and the Yoruba, is the result of '. . . conservatism versus modernism, authoritarian versus democratic ideology, and Islamic obedience versus Christian individualism.'[9]

I do not intend to discuss here the methodological and theoretical assumptions in these arguments, nor is this the occasion to question the value and the validity of comparing large ethnic stocks, like the Hausa, Yoruba, and Ibo, in this way when the multiplicity and complexity of cultural, historical, and social variables have not yet been sufficiently analysed. What I want to point out is that if a comparison of this sort is to be attempted at all between members of these different ethnic groups, it must be carried out within the same, or similar, situations, such as those that I have been discussing in this book where Yoruba have confronted Hausa over the capture of strategic positions in the organization of long-distance trade.

It is quite true that Hausa organization of long-distance trade is 'traditional', but what LeVine seems to overlook is that in the present circumstances this organization is the most rational, the most economic, and hence the most profitable. This is not a 'petty trade' of the type indulged in by masses of men, women, and children all over West Africa, including Ibo and Yoruba; it is serious business involving large sums of money and yielding wide margins of profit and steady incomes to the men who are engaged in its far-flung organization. If it were not so profitable, the Yoruba competitors would not have gone to such a great deal of trouble, during all these years, in order to gain a foothold in it. So far it has been economically more viable than the numerous small-scale modern enterprises which have sprung up everywhere in Southern Nigeria following independence, and have ended in bankruptcy after a short time. The Hausa landlord knows of the existence of banks, and sometimes even makes use of their services, but he will still keep large sums of cash in his house and thus run the risk of losing it through theft, not because of the blind force of custom or of ignorance but because of a number of practical, rationally calculated considerations. Similarly, the Hausa cattle-dealer knows of the advantages of having his cattle transported from the North to the South by rail, instead of having them driven on the hoof. The journey on the hoof takes about 40 days and by the time the cattle arrive at the southern markets they will have lost much weight, contracted disease, and thus depreciated in value. On the other hand, the journey by rail takes only two days, after which the cattle arrive in a healthy condition and thus fetch a higher price. Nevertheless, the dealer will often drive his cattle rather than send them by rail. Here again he does this not in a blind adherence to tradition but only because, after the calculation of costs and risks, he finds that driving the beasts will be more profitable than sending them by rail. The Hausa is here making a choice between alternative courses of action and his decisions are rational and are aimed at the 'maximization of profits'. If this were not the case, i.e. if there were more economical methods of organizing the business so as to reduce costs and raise profits, then the question should be asked, why have the more 'go-ahead' Yoruba not adopted them and thus succeeded in throwing the Hausa out of business? Indeed an attempt of this kind was made a few years ago by a number of enterprising Europeans who intended launching a modern organization for securing beef supplies for southern Nigeria and for export, by slaughtering cattle in the North and then transporting the meat to the South by air. The experiment failed and the whole enterprise was given up as uneconomic, and the Hausa continue until today to dominate the trade and to run it in the 'old ways'. The Hausa are not ignorant about the possibilities of air transport or air travel and anyone who has travelled on the internal West African air lines would have noticed the steady traffic of Hausa traders using this method of transport daily. But they have not used it for transporting meat because they can deliver the meat more cheaply by the traditional ways. They understand the operation of the widely spread network of relations in which the trade is conducted, between breeders, dealers, middlemen, brokers, financiers, speculators, drovers, and scores of other intermediaries in different communities of the network which extends from the remote expanses of the savanna, where the Fulani raise their cattle, down to the southern parts of the forest belt, where the beef is finally consumed. To rationalize the trade and to put it on modern bases will require a complete social and economic revolution covering almost every stage in the chain of the trade and this at present is not feasible.

Thus in the field of long-distance trade, the Hausa cannot be said to be inferior to the Yoruba in 'achievement motivation'. Indeed, the Hausa can be said to have outmanœuvred and outwitted the Yoruba in many situations. This is not a matter of superior or inferior psychological make-up but a question of political developments and political organization.

And this brings us back to the phenomena of ethnicity and to the process of ethnic political grouping.

Ethnic Grouping as a General Political Process

The Hausa in Ibadan are more 'retribalized' than the 'Western Ibos', not because of their conservatism, as LeVine suggests, and not because of special elements in their traditional culture, as Rouch and others contend, but because their ethnicity articulates a Hausa political organization which is used as a weapon in the struggle to keep the Hausa in control of the trade. Ethnicity is thus basically a political and not a cultural phenomenon, and it operates within contemporary political contexts and is not an archaic survival arrangement carried over into the present by conservative people.

The development and functioning of Sabo, as an autonomous polity is far from being a unique case of 'retribalization' in contemporary African society. The same processes have operated in the formation of other Hausa communities in other Yoruba towns. This is confirmed both by records and by partial observations which I made of a number of such communities. Similar processes have also been evident in the formation of other communities of this type in southern Ghana where I made some limited enquiries, and elsewhere in Ghana as indicated in the literature. Each one of these Hausa communities is unique in the sense that it has developed under a particular combination of historical, geographical, demographic, economic, cultural, and political circumstances. But the general pattern and the sociological interconnections between the major variables, and particularly between the economic and the political ones, are similar.

Nor is the phenomenon of the Hausa diaspora in West Africa unique in this respect, for there are other ethnic diasporas in other parts of the subcontinent. If one goes further afield, one will meet similar situations in the organization of Lebanese and Syrian communities in West Africa, and in the development of a network of Indian migrant communities in East and South Africa. If one leaves the African continent, one will find similar examples in the organization of Chinese migrant communities in different parts of the Far East and of South-East Asia, in the network of Arab Muslim trading communities in non-Arab lands during the Middle Ages, of Jewish trading communities around the Mediterranean, and so on.

But the processes of political 'retribalization' are not confined to the formation of ethnic diasporas in pre-industrial societies. Everywhere in the world today there are ethnic groupings which are engaged in a struggle for power and privilege with other ethnic groupings, within the frameworks of formal politcal settings. Recent studies by sociologists and anthropologists of various communities in the USA reveal the dynamic nature of ethnic groupings in contemporary society. In a study of such groupings in New York city, Glazer and Moynihan find that: 'New York organizational life today is in large measure lived within ethnic bounds.'[10] These ethnic groups are not a survival from the age of mass immigration, but new social forms. In many cases members who are third-generation immigrants, or descendants of even earlier immigrants, have lost their original language and many of their indigenous customs. But they have continuously re-created their distinctiveness in different ways, not because of conservatism but because these ethnic groups are in fact interest groupings whose members share some common economic and political interests and who, therefore, stand together in the continuous competition for power with other groups.

Political ethnicity has been particularly striking in the newly emerging states of the Third World because under colonial rule some ethnic groups succeeded in gaining a great deal of power while others became underprivileged.

With independence, the underprivileged closed their ranks to redress the balance whilst the privileged had to close their ranks to retain their privileges. In the course of the ensuing struggle, many of these ethnic groups exploited some of their traditional values, myths, and ceremonial in order to establish an elaborate political organization. And because within the formal structure of the new state there is no place for the formal organization of these ethnic groups, this organization has been informally developed by being articulated in terms of some of these traditional symbolic forms.

This process is dramatically manifested in post-independent British West Africa, but it is also present, though only at certain levels of political organization even in those parts of Africa which have been characterized by a great deal of political detribalization. This can be seen in the now classic example of the industrial centres in Central Africa. Here, as Epstein shows in his study of the industrial town Luanshya,[11] the tribally heterogeneous African labour force, confronting the monolithic, bureaucratic industrial organization which had been set up by white employers, aligned their forces together in a struggle for higher wages and better working conditions. But, within the unions themselves, competition over positions of power was conducted on ethnic lines. What this means is that on this level of organization, ethnicity must have been political and each ethnic group must have organized itself for political action. A leader of such an ethnic group will do his best to emphasize ethnic distinctiveness and to mobilize power relations within the group to support him. Studies on this level of political organization have not yet been made in Central Africa. Mitchell has recently drawn attention to this gap in our knowledge:

. . . the interaction of African townsmen in industrial and commercial environments has been little studied. Industrial sociologists have shown that in Europe and

America informal relationships among workmen modify and augment the formal pattern of relationships among them. We would expect this to be true also of African workers. From a theoretical point of view it would be interesting to know whether such factors as tribalism and kinship play a more important role in informal relationships in the work situation in Africa than they do in Europe and America. . .[12]

This means that even in the industrial towns of the Copperbelt tribalism *is* a live political and economic issue and is not just a method of categorization to help the African migrant to deal with the bewildering complexity of urban society or to regulate for him such 'domestic' matters as marriage, friendship, burial, and mutual help. It has for long now been recognized by social anthropologists that political organization in a society does not exclusively consist in formal political institutions, but includes many institutions that are not formally political, like marriage, friendship, and security for old age.

Epstein must have been aware of all this because soon after indicating that tribalism *was* a live political issue within the unions he introduced into the picture of the African camp a new variable, that of social stratification, to indicate that this was a countervailing force against political tribalism. But, here, too, our information is scanty and a great deal depends on whether the new lines of stratification cut across, or overlap with, tribal cleavages.

If status cleavages will cut across ethnic divisions, then the manifestations of ethnic identity and exclusiveness will tend to be inhibited by the emerging countervailing alignments of power. The less privileged from one ethnic group will co-operate with the less privileged from other ethnic groups against the privileged from the same ethnic groups. The privileged groups will, for their part, also close their ranks to protect their interests. If the situation continues to develop in this way, tribal differences will be weakened and will eventually disappear and the people will become politically detribalized. In time, class division will be so deep that a new subculture, with different styles of life, different norms, values, and ideologies, will emerge and a situation may develop which is similar to that of 'the two nations' of Victorian Britain.

However, the situation will be entirely different if the new class cleavages will overlap with tribal groupings, so that within the new system the privileged will tend to be identified with one ethnic group and the underprivileged with another ethnic group. In this situation cultural differences between the two groups will become entrenched, consolidated, and strengthened in order to express the struggle between the two interest groups across the new class lines. Old customs will tend to persist, but within the newly emerging social system they will assume new values and new social significance. A great deal of social change will take place, but it will tend to be effected through the rearrangement of traditional cultural items, rather than through the development of new cultural items, or, more significantly, rather than the borrowing of cultural items from the other tribal groups. Thus to the casual observer it will look as if there is here stagnation, conservatism, or a return to the past, when in fact we are confronted with a new social system in which men articulate their *new roles* in terms of traditional ethnic idioms. This is why a concentration on the study of culture as such will shed little light on the nature of this kind of situation, and it is for this reason that Gluckman and others insisted that ethnicity should be studied within the social context of the town.[13]

Notes

1. The Tijaniyya order is an Islamic sect that was joined by the overwhelming part of the Hausa of Sabo in 1951–2.
2. Some Yoruba became Tijanis, but because of the localization of ritual under local Mukaddams no interaction with Tijani Hausa could take place.
3. It was initially established as University College associated with the University of London, and was meant to cater for the whole of British West Africa. It was converted into an independent university in 1963.
4. Nearly half of the doctors of Nigeria were concentrated in Ibadan.
5. Ibadan has been prolific in the formation of numerous types of voluntary associations, particularly since World War II.
6. Except for the small group of men who learnt English privately
7. C. Okonjo, 'The Western Ibo', in P. C. Lloyd, A. L. Mabogunje, and B. Awe (eds), *The City of Ibadan* (Cambridge University Press, 1967).
8. R. A. LeVine, *Dreams and Deeds* (University of Chicago Press, 1966), 84.
9. Ibid., 93.
10. N. Glazer and D. P. Moynihan, *Beyond the Melting Pot: The Negroes, Puerto Ricans, Jews, Italians, and Irish of New York City* (MIT Press and Harvard University Press, 1963).
11. A. L. Epstein, *Politics in an Urban African Community* (Manchester University Press, 1958).
12. J. C. Mitchell, 'Theoretical Orientations in African Urbanization Studies', in M. Banton (ed.), *The Social Anthropology of Complex Societies*. ASA. Monograph no. 4 (Tavistock Publications, 1966), 51–2.
13. M. Gluckman, 'Tribalism in Modern British Central Africa', in *Cahiers d'Études Africaines* 1: 55–70.

19

Squatters and the State in Latin America*

Manuel Castells

THE conditions of urbanization in Latin American societies force an increasing proportion of the metropolitan population to live in squatter settlements or in slum areas. This situation is not external to the structural dynamics of the Third World, but is connected to the speculative functioning of some sectors of capital as well as to the peculiar patterns of popular consumption in the so-called informal economy.[1] On the basis of their situation in the urban structure, the squatters tend to organize themselves at the community level. Their organization does not imply, by itself, any kind of involvement in a process of social change. On the contrary, most of the existing evidence points to a subservient relationship with the dominant economic and political powers.[2] Nevertheless, the fact of a relatively strong local organization is itself a distinctive feature which clearly differentiates the squatters from other urban dwellers who are predominantly organized at the work-place or in political parties, when and if they are organized at all.

Furthermore, the state's attitude towards squatter settlements pre-determines most of their characteristics. Thus the connection between the squatters and the political process is a very close one. And it is precisely in this way that urbanization, and its impact on community organization, becomes a crucial aspect of political evolution in Latin America. Let us, therefore, explore the relationship between squatters and the state in three major Latin American countries: Peru, Mexico, and Chile.

Squatters and Populism: The Barriadas of Lima[3]

Lima's spectacular urban growth has mainly been due to the expansion of *barriadas*, peripheral[4] substandard settlements, often illegal in their early

* This is an edited version of ch. 19 in Manuel Castells, *The City and the Grassroots: A Cross-Cultural Theory of Urban Social Movements* (London: Edward Arnold; Berkeley and Los Angeles: University of California Press, 1983), reprinted by permission of the publishers. References to other parts of the book and a photograph have been omitted, notes renumbered.

stage, and generally deprived of basic urban facilities. The population of the *barriadas* came, on the one hand, from the slums of central Lima (*tugurios*) once they had reached bursting-point or when they were demolished, and on the other hand, from accelerated rural and regional migration,[5] the structural causes of which were the same as for all dependent societies.[6] And in Peru, this particular form of urbanization—the *barriadas*—cannot be explained without reference to the action of political forces as well as to the state's policies.[7] Given the illegal nature of land invasion by the population of the *barriadas*, only institutional permissiveness or the strength of the movement (or a combination of both) can explain such a phenomenon. More specifically, given the way power has been unevenly distributed within Peruvian society until very recent times, the land invasion must be understood to have been the result, in part, of policies that originated from various dominant sectors. Very often landowners and private developers have manipulated the squatters into forcing portions of the land onto the real estate market, by obtaining from the authorities some urban infrastructure for the squatters, thus enhancing the land value and opening the way for profitable housing construction. In a second stage, the squatters are expelled from the land they have occupied and forced to start all over again on the frontier of a city which has expanded as a result of their efforts.

Nevertheless, the main factor underlying the intensity of urban land invasion in Lima has been a political strategy consisting of protection given for the invasion in exchange for poor people's support. Table 19.1, constructed by David Collier on the basis of his study of 136 of Lima's *barriadas*, clearly shows the political context for the peak moments of urban land invasion between 1900 and 1972.[8] The political strategies and their actual effects on the squatters were really very different from one land invasion to another.

The most spectacular stage in the history of land invasions corresponds to the initiative of General Odría's government in 1948–56. At a time of political repression against the Communist Party, and particularly against the Alianza Popular Revolucionaria Americana (APRA), which was trying to seize power to implement an 'anti-imperialist programme', Odría's populism was a direct attempt to mobilize people on his side by offering to distribute land and urban services. The aim was to dispute APRA's political influence by taking advantage of the urban poor's low level of political organization and consciousness, and by mobilizing people around issues outside the work place where the pro-APRA union leaders would be more vulnerable. Nevertheless, APRA's reaction was very rapid: they demanded that Odría keep his promises to the squatters' organizations by accelerating the invasion, and the final outcome was a political crisis and the downfall of Odría's government.

TABLE 19.1   Number and Population of *barriadas* (Squatter Settlements) Formed
in Lima in each Presidential Term, 1900–72

President	Number of cases	Percentage of cases	Population	Percentage of population
Before Sánchez Cerro (1900–30)	2	1.5	2,712	0.4
Sánchez Cerro (1930–1, 1931–3)	3	2.2	12,975	1.7
Benavides (1933–9)	8	5.9	18,888	2.5
Prado (1939–45)	8	5.9	6,930	0.9
1947: Information uncertain	5	3.7	24,335	3.2
Bustamante (1945–8)	16	11.8	38,545	5.1
Odría (1948–56)	30	22.1	203,877	26.9
1956: Information uncertain	2	1.5	11,890	1.6
Prado (1956–62)	30	22.1	93,249	12.3
1962: Information uncertain	2	1.5	22,377	2.9
Pérez Godoy (1962–3)	2	1.5	1,737	0.2
Lindley (1963)	3	2.2	11,046	1.5
Belaúnde (1963–8)	15	11.0	93,407	12.3
Velasco (1968–1972 only)	10	7.4	217,050	28.6
Total	136	100.3	759,018	100.1

Source: David Collier, *Barriadas y élites: de Odría a Velasco* (Lima: Instituto de
Estudios Peruanos, 1976; updated 1978).

We can understand, then, why Prado's government, supported by
APRA, continued to be interested in the *barriadas* in order to eliminate
the remaining pro-Odría circles and to widen its popular basis. Instead of
stimulating new invasions, Prado launched a programme of housing and
service delivery for the popular neighbourhoods, trying to integrate these
sectors into the government's policy without mobilizing them. In a
complementary move, APRA started formally controlling the organizations
of squatters (the *Asociaciones de Pobladores*) in order to expand its
political machine from the trade unions to the social organizations centred
on residential issues.

Belaúnde's urban policy was very different. Although he also looked for
some support from the squatters, allowing and stimulating land invasions,
he did not limit his activity to the struggle against APRA, but tried a
certain rationalization of the whole process. His 'Law of *Barriadas*' was the
first attempt to adapt urbanization to the general interest of Peruvian
capitalist development without adopting a particular set of political
interests. The activity of his party, Accion Popular, was aimed at
modernizing the *barriadas* system and facilitating an effective connection
with the broader interests of corporate capital. The social control of the
squatters was then organized by international agencies, churches, and

humanitarian organizations, which were closely linked to the interests of the American government. Belaúnde's strategy was quite effective in weakening APRA's political influence among the *pobladores*, but it was unable to solidly establish a new form of social control. This situation led to a very important change in the government's strategy after the establishment of a military junta in the revolution of 1968. At the beginning, the military government tried to implement a law-and-order policy, repressing all illegal invasions and putting the *asociaciones de pobladores* under the control of the police. Nevertheless, its attitude towards the *barriadas* changed dramatically on the basis of two major factors: first, the difficulty of counteracting a basic mechanism that determines the housing crises in the big cities of dependent societies, and second, the military government's need to obtain very rapidly some popular support for its modernizing policies once these policies had come under attack from the conservative landlords and business circles.

The turning-point appears to have been the *Pamplonazo* in May 1971. An invasion of urban land in the neighbourhood of Pamplona was vigorously repressed and provoked an open conflict between the Minister of Interior, General Artola, and Bishop Bambaren, nicknamed the '*Barriadas* Bishop', who was jailed. The crisis between the state and the Catholic Church moved President General Velasco Alvarado to act personally on the issue. He conceded most of the *pobladores*' demands, but moved them to a very arid peripheral zone close to Lima, where he invited them to start a 'self-help' community supported by the government. This was the beginning of Villa El Salvador, a new city which in 1979 housed up to 300,000 inhabitants recruited among Lima dwellers and rural migrants looking for a home in the metropolitan area.

The military government learned a very important lesson from this crisis: not only did it discover the dangers of a purely repressive policy, but it also realized the potential advantages of mobilizing the *pobladores*. Using the Church's experience, the military government created a special agency, the *Oficina Nacional de Pueblos Jóvenes* (New Settlements' National Office), charged with legalizing the land occupations and with organizing material and institutional aid to the *barriadas*. At the same time, within the framework of Sistema Nacional de Movilizacion Social (SINAMOS), the regime's 'social office', a special section was created to organize and lead the *pobladores*. Under the new measures, each residential neighbourhood in the *barriadas* had to elect its representatives who would eventually become the partners of the government officials, controlling the distribution of material aid and urban facilities. At the same time, the new institution relied on the existing agencies and voluntary associations (most of them linked to churches and international agencies) to tailor their functions and co-ordinate their activity to the parameters set by state policy. This policy would develop along several paths: economic (popular savings institutions, production and consumption co-operatives); legal (laws recognizing the squatting of urban land); ideological (legitimizing of the *pobladores'* associations, propaganda centres for the government); and political (active involvement in the Peruvian 'revolution' through SINAMOS). The *barriadas* became a crucial focus of popular mobilization for the new regime.

As a consequence of the successive encouragements given to squatter mobilization by the state, as well as by the political parties, the *barriadas* of Lima grew by extraordinary proportions: their population increased from 100,000 in 1940 to 1,000,000 in 1970 and became an ever larger proportion of the population in most of Lima's districts.[9]

Nevertheless, it would be wrong to conclude that all forms of mobilization were identical save for different ideological stances. In his study, Etienne Henry makes clear some fundamental differences in practice. Odría's and Prado's policies expressed the same relationship to the *pobladores*, patronizing them to reinforce their political constituencies. In the case of Belaúnde, the action to integrate people was subordinated to the effort of rationalizing urban development. The military government's policy between 1971 and 1975 represented a significant change in urban policy. It was not an attempt to build up partisan support for a particular political machine but was, in fact, a very ambitious project to establish a new and permanent relationship between the state and urban popular sectors through the controlled mobilization of the *barriadas* now transformed into *pueblos jóvenes* (new settlements). This transformation was much more than a change in name: it expressed the holding of all economic and political functions of the *pobladores'* voluntary associations by the state in exchange for the delivery and management of required urban services. The goal was no longer to obtain a political constituency but to build a 'popular movement' mobilized around the values promoted by the revolutionary regime. In this sense, the *barriadas* became closely linked to Peruvian politics and were increasingly reluctant to adapt to the new government's orientation resulting from the growing influence of the conservative wing within the army.

The picture of the Lima squatters' movement appears as one of a manipulated mob, changing from one political ideology to another in exchange for the delivery (or promise) of land, housing, and services. And this was, to a large extent, the case. The *pobladores'* attitude was quite understandable if we remember that all politically progressive alternatives were always defeated and ferociously repressed. So, as Anthony and Elizabeth Leeds[10] have pointed out, the behaviour of the squatters was not cynical or apolitical, but, on the contrary, deeply realistic, and displayed an awareness of the political situation and how their hard-pressed demands could be obtained. Thus it appears that the Peruvian urban movement was, until 1976, dependent upon various populist strategies of controlled mobilization. That is, the movement was, in its various stages, a vehicle for carrying the social integration of the urban popular sectors in the same direction as the political strategies of the different political sectors of the dominant classes.

Now this process, like all controlled mobilizations, expresses a contradiction between the effectiveness of the mobilization and the fulfilment of the goals assigned to the movement. When these goals are delayed as a result of the structural limits to social reform and when people's organization and consciousness grow, some attempts at autonomous social mobilization occur. A sign of this evolution was, in the case of Lima, the organization of the *Barriada Independencia* in 1972. When the autonomous mobilization expanded, the government tried to stop it by means of violent repression as it did, for example, in March 1974. In spite of repression, the movement continued its opposition, making alliances with the trade unions and with the radical left, as was revealed in the *barriadas*' massive participation in the strikes against the regime in 1976, 1978, and 1979. After the dismantling of SINAMOS by the new military president, Morales Bermudez, the political control of the *barriadas* rapidly collapsed. Ironically, Villa El Salvador became one of the most active centres of opposition to the state's new conservative leadership.

This evolution supports a crucial hypothesis. The replacement of a classic patronizing relationship, ruling class to popular sectors, by controlled populist mobilization expands the hegemony of the ruling class over the popular sectors which are organized under the label of 'urban marginals'. But the crisis of such an hegemony, if it does happen, has far more serious consequences for the existing social order than the breaking of the traditional patronizing ties of a political machine. In fact, it is this type of crisis that enables the initiation of an autonomous popular movement, the further development of which will depend on its capacity to establish a stable and flexible link with the broader process of class struggle.

Our analysis of the Lima experience, although excessively condensed, provides some significant findings:

1. An urban movement can be an instrument of social integration and subordination to the existing political order instead of an agent of social change. (This is, in fact, the most frequent trend in squatter settlements in Latin America.)

2. The subordination of the movement can be obtained by political parties representing the interests of different factions of the ruling class and/or by the state itself. The results are different in each case. When the

movement has close ties with the state, then urban policies become a crucial aspect of change in dependent societies.

3. Since urbanization in developing countries is deeply marked by a growing proportion of squatter settlements (out of the total urban population), it appears that the forms and levels of such urbanization will largely depend upon the relationship established between the state and the popular sectors. This explains why we consider urban politics to be the major explanatory variable of the characteristics of urbanization.

4. When squatter movements break their relationship of dependency *vis-à-vis* the state, they may become potential agents of social change. Yet their fate is ultimately determined by the general process of political conflict.

Between *Caciquismo* and Utopia: The *Colonos* of Mexico City and the *Posesionarios* of Monterrey

Mexico's accelerated urban growth is a social process full of contradictions.[11] An expression of these contradictions during the 1970s were the ever-increasing mobilizations by the popular sectors and urban squatters to obtain their demands in the *vecindades* (slums) and *colonias proletarias* (squatters settlements on the periphery) of the largest Mexican cities.[12] The potential strength of this urban mobilization must be seen in the context of a political system that was perfectly capable of controlling and integrating all signs of social protest.[13]

In traditional squatter settlements on the periphery of big cities, the key element was a very strong community organization under the tight control of leaders who were the intermediaries between the squatters (*colonos*) and the administration officials. In its early stages, this form of community organization may be considered to be dominated by *caciquismo*, that is, by the personal and authoritarian control of a leader, himself recognized and backed by local authorities. Therefore illegal land invasion, by itself, did not present a challenge to the prevailing social order. Indeed, economically, it represented a way to activate the capitalist, urban land market; politically it was a major element in the social control of people in search of shelter. What must be emphasized is that *caciquismo* was not an isolated phenomenon, but had a major function to fulfil within both the political system and the state's urban policies. The local leaders were not neighbourhood bosses living in a closed world: they were representatives of political power through their relationships with the administration and with the Partido Revolucionario Institucional (PRI)—the government's party—from which they obtained their resources and their legitimacy. So Mexican squatters have always been well organized in their communities, and this organization has performed two major functions: on the one hand, it has allowed them to exert pressure for their demands to stay on the land they have occupied and to obtain the delivery of urban services; on the other hand, it has represented a major channel of subordinated political participation by ensuring that their votes and support goes to the PRI. Both aspects have, in fact, been complementary, and the *caciques* (the community bosses) were the agents of this process. They were not, however, the real bosses of the squatters, since they exercised their power on behalf of the PRI. To understand this situation we must remember the historical and popular roots of the PRI, and the need for it to continuously renew its role of organizing the people politically while providing access to work, housing, and services in exchange for loyalty to the PRI's programme and leadership.

Thus the new urban movements that developed in Mexico during the 1970s derived from a previous network of voluntary associations existing in the *vecindades* and *colonias* which were, at the same time, channels for expressing demands and vehicles of political integration with PRI. Taking into consideration the ideological hegemony of the PRI and the violent repression exercised against any alternative form of squatter organization, how can we explain the upheaval caused by autonomous urban movements since 1968? And what were their characteristics and possibilities?

Two major factors seem to have favoured the development of these movements:

1. President Echeverria's reformism (1970–6) to some extent recognized the right to protest outside the established channels, while legitimizing aspirations to improve the living conditions in cities.[14]

2. Political radicalism among students, after the 1968 movement, provided militants who tried to use the squatter communities as a ground on which to build a new form of autonomous political organization.

This explains how the evolution of the new urban movements came to be determined by the interaction between the interests of the squatters, the reformist policy of the administration, and the experience of a new radical left, learning how to lead urban struggles.

From the outset of urban mobilization, radicals tried to organize and politicize some squatter settlements, linking their urban demands to the establishment of permanent bases of revolutionary action and propaganda within these settlements. These attempts were often unable either to overcome the squatters' fears of reprisal, or to uproot the PRI's solid political organization. When the radicals did succeed in their attempt to organize a squatter settlement as a revolutionary community, the state resorted to large-scale violence, having taken care to undermine the movement by claiming it had subversive contacts with underground guerrillas. The most typical example was the *colonia* Ruben-Jaramillo in the city of Cuernavaca, where radical militants organized more than 25,000 squatters, helped them to improve their living conditions, and raised the level of their political awareness. The *colonia*'s radicalism prompted a violent response by the army which occupied it and put it under the control of a specialized public agency to deal with squatter settlements.

Nevertheless, other settlements resisted police repression and survived by maintaining a high level of organization and political mobilization. The best-known case is the Campamento 2 de Octubre in Ixtacalco in the Mexico City metropolitan area. Four thousand families illegally invaded a piece of highly valued urban land where both private and public developers had considerable interests. Students and professionals backed the movement and some of them went to live with the squatters to help their organization. The squatters kept their autonomy *vis-à-vis* the government, and used their strong bargaining position to call for a general political opposition to the PRI's policies. They became the target of the most conservative sectors of the Mexican establishment. After a long series of provocations by paid gangs, the police attacked the *campamento* in January 1976, starting a fire and injuring many of the squatters. Some days later, several hundred families returned to the settlement, reconstructed their houses, and started to negotiate with the government to obtain the legal rights to remain there. But if repression could not dismantle the *campamento*, it succeeded in isolating it by making it too dangerous an example to be followed by other squatters. When the Ixtacalco squatters tried to organize around themselves a Federacion de Colonias Proletarias to unite the efforts of other settlements, they obtained little support because of their image of extreme radicalism. In fact, their demands were relatively modest, consisting of the legalization of the settlement and a minimum level of service. But the repression was very severe because the government saw a major danger in the movement's will to autonomy, its capacity to link urban demands and political criticism, and its appeal to other political sectors to build an opposition front, bypassing the political apparatus of the PRI within the communities. At the same time, police action was made easier by the political naïveté of some of the students, who at the beginning of the movement thought of the settlement as a 'liberated zone' and spent much of their energy on verbal radicalism. In this sense, Ixtacalco was an extraordinarily advanced example of autonomous urban mobilization, but was also a very isolated experience which went forward by itself without considering the general level of urban struggle in elsewhere Mexico City.

In fact, the most important urban movements in recent years have taken place in northern Mexico, particularly in Chihuahua, Torreón, Madero, and above all in Monterrey, where the movement of the *posesionarios* (squatters) was perhaps one of the most interesting and sizeable in Latin America. Let us examine this experience in some detail.

Monterrey, the third largest Mexican city, with a population of 1,600,000, is a dynamic industrial area with an important steel industry. It is dominated by a local bourgeoisie with an old and strong tradition, cohesively organized and closely linked to American capital. The so-called Monterrey Group is a modernizing entrepreneurial class, politically conservative and socially paternalistic. It has always opposed state intervention, often criticized the PRI, and succeeded with its workers through a policy of social benefits and high salaries. In Monterrey, the powerful Confederacion de Trabajadores Mexicanos (CTM), the major labour union controlled by the PRI, is relatively unimportant since most workers have joined the *sindicatos blancos* (the white unions) which are manipulated by company management. The city, which is proud of maintaining the highest living standards in Mexico, has experienced a strong urban growth-rate since 1940: 5.6 per cent annual growth 1940–50 and 1950–60, and 3.7 per cent in 1960–70. Urban immigration has been the result of both industrial growth and accelerated rural exodus caused by the rapid capitalist modernization of agriculture in northern Mexico. This urban growth has not, however, been matched by the increase in housing and urban services. The big companies have provided housing for their workers, but for the remaining people (one-third of the population) no housing is available. The consequence has been, as in other Mexican cities, the invasion of surrounding land and massive construction of their own housing by the squatters. Three hundred thousand *posesionarios* have settled there. The underlying mechanisms were similar to those already described: speculation and illegal development on the one hand, and the role of the PRI's political machine as intermediary with local authorities on the other.

It was against this background that the student militants acted, trying to connect urban demands to political protest in the same mould as the university-based radicalism which began in 1971. The students led new land invasions contrary to the agreement with the administration. To differentiate themselves from the former settlements, the students called the new ones *colonias de lucha* (struggle settlements). In 1971 they founded the first settlement, Martires de San Cosme, in the arid zone of Topo Chico. The police immediately surrounded them, but withdrew after a month of violent clashes. Then the squatters built their houses and urban infrastructure and established a very elaborate social and political organization. The same process was renewed in the following years, and the cumulative effect made it extremely difficult to use repression as a means of halting their progress. The participants of each land invasion included not only its beneficiaries but also squatters already settled elsewhere who considered the new invasions as part of their own struggle. The timing of each invasion was extremely important as a means of averting repression: one of the most courageous invasions, in San Angel Bajo, succeeded in occupying good open space close to the municipal park without suffering reprisals because it was carried out on the eve of President Lopez Portillo's arrival in Monterrey during his electoral campaign of 1976. Once the invasion was accomplished, people raised the Mexican flag, running the red flag up a few weeks later. In similarly ingenious ways Tierra y Libertad, Revolución Proletaria, Lucio Cabañas, Genaro Vásquez, and 24 other settlements were born and eventually combined in an alliance, the Frente Popular Tierra y Libertad, representing, at the time of our field-work in August 1976, about 100,000 squatters.[15]

A key element in the success of the movement was its ability to take advantage of the internal contradictions of the ruling élite. For example, the Monterrey bourgeoisie openly opposed President Echeverria's reformism, and launched a major attack against the governor, who replied by trying to obtain support from the people. Using the themes of the governor's populist speeches as justification of their actions, the squatters made open repression against them more difficult. Nevertheless, the embittered local oligarchy, which controlled the city police, reacted by organizing continuous provocations. In one of the police actions, on 18 February 1976, six squatters were killed and many others wounded. The movement's protest was impressive, and signs of solidarity came from all over the country. There were street demonstrations in Monterrey for 15 days, some of them attracting over 40,000 people, organized jointly by squatters, students, and workers. For two months, the squatters occupied several public places. Finally, they were personally entertained by President Echeverria in Mexico City. Victims' relatives obtained economic compensation, an official inquiry was opened, the city police chief was ousted, and the government provided strong financial support for the revolutionary squatter settlements.

So, in a critical moment, the movement clearly displayed its strength and political capacity. But it also revealed its limits. To understand this crucial point, we must consider the organizational structure and the political principles of the Monterrey squatter's movement.

The basic idea, shared by all the squatters' leaders, was that struggles for urban demands were meaningful only as far as they allowed people to unite, to be organized, and to become politically aware, because (according to these leaders) such political strength was the only base from which to successfully make demands. On the other hand, they wanted to link the squatters' actions to a collective theme aimed, in the long term, to the revolutionary transformation of society. Only if these principles are remembered can some surprising aspects of the movement be understood. For instance the squatters strongly opposed the legalization by the government of their illegal land occupation. Their reasons were threefold: economic, ideological, and political. Economically, legalization implied high payments for a long time under conditions that many families could not afford. Ideologically, the movement could be transformed into a pressure group *vis-à-vis* the state instead of asserting their natural right to the land. Above all, politically: legalization, by individualizing the problem and dividing the land would create a specific relationship between each squatter and the administration. Thus, the movement itself could be fragmented, lose its internal solidarity, and be pushed towards integration with the state's machinery. Therefore, to preserve their solidarity, cohesiveness, and strength (which they considered to be their only weapons), the squatters refused the property rights offered by the state, and expelled from the settlements those squatters who accepted the legal property title. A similar attitude was taken towards the delivery of services.

The squatters believed in self-reliance and rejected the state's help in the first stages of the movement. They did not, however, avoid contact with the state, since they were continuously engaged in negotiation, but wanted to preserve popular autonomy in a Mexican context where the political system is quite capable of swallowing up any initiative by a grass-roots organization. So they stole construction materials or obtained them by putting pressure on the administration, but they collectively built the schools, health services, and civic centres, with excellent results (unlike most Mexican squatter settlements). Houses were built by each family but in lots of a collectively decided size, in proportion to the size of the family and following a master plan approved by the settlement's General Assembly. Water, drainage, and electricity were provided by illegal connections to the city systems. It is interesting to note that several settlements refused electric power in order to avoid television because it was considered a source of 'ideological pollution'. To overcome transport problems, the squatters seized buses on several occasions, finally forcing the bus company to adapt to the new urban structure. Schools were integrated into the general educational system and paid for by the state, but were controlled and managed by the parents' association in collaboration with the children's representatives. A similar organization managed health services. There was also in each settlement an honour and justice committee which passed judgement on conflicts, the most serious of which were handled by the General Assembly. Alcohol and prostitution were strictly forbidden. Settlement leaders organized vigilante groups to protect the squatters. The general organization was based on a structure of block delegates which nominated the settlement committees which, in turn, reported to the General Assembly. There were a variety of voluntary associations, the strongest of which were the women's leagues and the children's leagues. The ideology of collective solidarity was reinforced. On Red Sundays—in 1976, every Sunday—everybody had to do collective work on shared urban facilities. There was also a high level of political and cultural activity run by 'activist brigades'.

Nevertheless, in spite of this extraordinary level of organization and consciousness, the *posesionarios'* movement in Monterrey suffered from the shortcomings of its isolation, geographical, social, and political. Geographically, it was the only urban movement of such size and character in the whole country. Socially, the squatters' population consisted almost entirely of unemployed, migrant peasants, having little contact with Monterrey's industrial workers. Politically, the leaders had no national audience and were only important at a local level.

The movement's leaders were well aware of this situation and of the danger of closing themselves into a new kind of communal Utopia. To break this isolation they tried to launch a series of actions to support 'fair causes': for instance each time a worker was unjustifiably fired, the squatters occupied the manager's homeyard until the worker was reappointed. Each individual repression was faced by the whole movement, and so it became increasingly politicized. But such political radicalism based only on the squatters' support carried two major risks: first, increasing repression, chiefly from the army; and second, political infighting within the movement.

Two crucial elements emerge from the analysis of this extraordinary experience:

1. The speed and development of an urban movement cannot be separated from the general level of organization and consciousness in the broader process of political conflict.

2. The relationship to the state is not exhausted either by repression or integration. A movement may increase its autonomy by playing on the internal contradictions of the state. Monterrey was able to go further than Ixtacalco mainly because of the type of relationship which the *posesionarios* were able to establish with the state.

Such political, urban movements as Monterrey or Ixtacalco are only able to stabilize if the power relationships between social classes change in favour of the popular classes. But this does not seem to have been true of Mexico, so that the survival of these community organizations ultimately required some alliance with a sector within the state. Thus the experience of Mexico shows, again, the intimate connection between urban movements and the political system. We will now turn to the most important political squatter movement in recent Latin American history—Chile during the Unidad Popular—so that we may study this relationship in detail.

Urban Social Movements and Political Change: The *Pobladores* of Santiago de Chile, 1965–73[16]

The historical significance of urban movements in Chile between 1965 and 1973 has been surrounded by a confused mythology. Our respect for the Chilean popular movement requires a careful reconstruction of the facts, as well as a rigorous analysis of the experience.

The squatter movement in Chile was closely linked with class struggle and its political expressions,[17] and this explains both its importance and shortcomings. While invasions of urban land had always happened in Chile,[18] they changed their social implications when they became entrenched with the political strategies of conflicting social classes: urban popular movements reached a peak as a consequence of the failure of the Christian Democratic programme for urban reform.[19] The reform, initiated under Eduardo Frei's presidency in 1965, relied on three elements:

1. A programme of distribution of urban land (Operación Sitio) combined with public support for the construction of housing by the people.

2. The formation of voluntary associations of *pobladores* and of house-wives (*centros de madres*) linked to a series of public agencies, organized around the government's Department of Popular Promotion.

3. The decentralization of local governments after the creation in 1968 of advisory neighbourhood councils (*Juntas de Vecinos*), elected by the residents of each neighbourhood.[20]

In fact the programme of urban reform failed because of two constraints: the first from the structural limits of the system (the difficulty of redistributing resources without affecting the functioning of private capital),[21] and the second from the pressure of interest groups (mainly the Chilean Chamber of Private Builders, and the Savings and Loan institutions) which used the programme as a means of producing profitable housing for middle-class families.[22]

As a consequence of this failure, the Christian Democrats lost control of the *pobladores'* movement and the neighbourhood councils became a political battlefield.[23] The movement then started to put pressure on the government in two ways: on the one hand, the residents of the popular neighbourhoods started asking for the delivery of promised services, and on the other thousands of families living with relatives or in shanties gathered to form Committees of the Homeless (*Comités Sin Casa*). These committees, in the late 1960s, took the initiative of squatting on urban land to force the government to provide the housing and urban services promised in the reform programme.[24] In the first period of the movement, between 1965 and 1969, the government responded by repressing the invasions, even causing a massacre (Puerto Montt, March 1969) and partially succeeded in stopping the process.[25] But the presidential elections were scheduled for September 1970, and in the Christian Democratic Party the left had won endorsement for its leader, Tomic, against the wishes of the incumbent President Eduardo Frei. Therefore, an open repression of the *pobladores* might have been politically costly among the urban popular sectors, whose vote had been crucial for the electoral victory in 1964. So, when in 1970 the police were restricted in the use of violence, mass squatting was launched in most cities of the country, taking advantage of the new leniency to establish a new form of settlements called *campamentos* to symbolize their political ideology (see Table 19.2).

TABLE 19.2   Illegal Invasions of Urban Land, Chile, 1966–71, by Year. (Units are Acts of Land Invasion, Regardless of the Number of Squatters Involved.)

	1966	1967	1968	1969	1970	1971	Sept. 1971– 31 May 1972	1 Jan. 1972– 31 May 1972
Santiago	0	13	4	35	103	n/a	88	n/a
Chile (including Santiago)	n/a	n/a	8	23	220	560	n/a	148

Source: FLACSO Survey on Chilean Squatters, 1972 (for Santiago); Dirección General de Carabineros, cited by FLACSO (for Chile 1968–71); and Ernesto Pastrana and Monica Threlfall, *Pan, techo, y poder: el movimiento de pobladores en Chile 1970–3* (Buenos Aires: Ediciones SIAP, 1974) who also relied on the Dirección General de Carabineros (for Chile 1972).

When the newly elected socialist President Salvador Allende took office in November 1970, more than 300,000 people were living in these *campamentos* in Santiago alone. At the end of 1972, by which time the number of urban invasions had stabilized, more than 400,000 people were in the *campamentos* of Santiago, and 100,000 or more in the other cities.[26] The main characteristic of these *campamentos* was that from the beginning they were structured around the *Comités Sin Casa* that led the invasions, each of which were in turn organized by different political parties[27] so much so that we can say that the Chilean *pobladores'* movement was created by the political parties. Of course, to do so they took into consideration the people's urban needs, and they were instrumental in organizing their demands and supporting them before the government. But we can by no means speak of a 'movement' of *pobladores*, unified around a programme and an organization; it was not, for instance, like the labour movement, which in Chile was unified and organized in the Central Única de Trabajadores (CUT), in spite of political divisions within the working class.

The majority of the *pobladores* were organized by the Comando de Pobladores de la Central Única de Trabajadores (the urban branch of the trade unions), linked to the Communist Party, and by the Central Única del Poblador (CUP), dependent on the Socialist Party. A very active minority constituted itself as the Movimiento de Pobladores Revolucionarios (MPR), a branch of the radical organization Movimento de Izquierda Revolucionaria (MIR). Almost 25 per cent of the *campamentos* were still under the control of the Christian Democratic Party, and a few settlements were even organized by the National Party (radical right). This whole situation had two major consequences:

1. Each *campamento* was dependent upon the political leadership which had founded it. Political pluralism within the *campamentos* was rare, except between Socialists and Communists (for instance, the largest *campamento*, Unidad Popular, had a joint leadership of both parties).

2. The participation of the *campamentos* in the political process very closely followed the political line dominant in each settlement. We should actually speak of the *pobladores'* branch of each party, rather than of a 'squatters' movement'. While all the parties always spoke of the need for unifying the movement, such unity never existed except in moments of political conflict, such as the distribution of food and supplies during the strike launched by the business sector against the government in October 1972.

This key feature of the movement explains the findings of the field-work study we conducted on 25 *campamentos* in 1971.[28] The social world we discovered did not present any major social or cultural innovation. The only exception was the organization of police and judicial functions,[29] which due to the absence of state legal institutions within the squatter settlements allowed (and forced) the *pobladores* to take a series of measures representing a beginning of popular justice. Yet, concerning the urban issues, the *pobladores'* massive mobilization made it possible for hundreds and thousands to obtain, in a few months, housing and services, against the prevailing logic of capitalist urban development. The urban system was deeply transformed by the *campamentos*. But experiences aimed at generating new social practices were limited by the political institutions where the old order was still the strongest force. A good example of such a situation was the Christian Democrats' congressional veto in 1971, blocking Allende's project to create Neighbourhood Courts (*Tribunales Vecinales*) based on existing experiences of grass-roots justice.

The dependency of the *campamentos* upon the political parties opened the door to their use by each party for its particular interest, lowering the level of grass-roots participation. The most conclusive demonstration on this subject is the careful case-study by Christine Meunier of Nueva la Habana, one of the most mobilized and organized of all *campamentos*, under the leadership of MIR, where she lived and worked between 1971 and 1973, until the military coup.[30] We cross-checked her information with our own observations and interviews in Nueva La Habana in 1971 and 1972, as well as with the demographic and social research conducted on the same *campamento* by Duque and Pastrana in 1970 and 1971.[31] All the findings by the three independent research teams converge towards a similar picture, the significance of which impels us to explore in some detail the social universe of a *campamento* in order to analyse as conclusively as possible the complex relationship between squatters and parties in the midst of a revolutionary process.

Nueva La Habana was one of the most active, well-organized, and politically mobilized *campamentos*. And it certainly was the most highly publicized, both by the media and by the observers of the Chilean socio-political evolution, the main reason being that it was considered the 'model' *campamento* under the leadership of MIR. The Ministry of Housing expedited its settlement in November 1970 by relocating 1,600 families (10,000 people), with their consent, from the previous MIR-led land invasions (*campamentos* Ranquil, Elmo Catalán, and Magaly Monserato). MIR accepted the relocation of the three *campamentos* in a new 86-hectare urban unit as a challenge that would demonstrate its capacity to organize, ability to obtain housing and services, and effectiveness in transforming the squatters into a revolutionary force. If there was potential for an urban social movement in the *campamentos* of Santiago, we could expect to see it emerge from the mud and shacks of Nueva La Habana.

The strength of Nueva La Habana came from its tight grass-roots organization and militant leadership. All *pobladores* were supposed to participate in the collective tasks of the *campamento*, as well as in the decisions about its management. All residents were included in a territorial organization on the basis of *manzanas* (blocks) that delegated one of their members to a board that elected an executive committee (*jefatura*) of five members. At the same time, the most active *pobladores* were invited to form a functional structure, the 'fronts of work', both at the level of each block and in the *campamento* as a whole, to take care of the different services that had to be provided for the residents on the basis of resources made available by the government: health, education, culture, police and self-defence, justice, sports, and so on. As a matter of fact, the capacity of the MIR to agitate and the deliberate commitment of Allende's Government to limit confrontations with the revolutionary left, led to the paradox that Nueva La Habana received preferential treatment for housing and social services compared to the average squatter settlement.[32] On the basis of the legitimacy acquired by its very effective delivery of services, particularly in the field of health care, MIR frequently asked the *pobladores* to show support for its policies outside the *campamentos*, and it was usual, in all major political demonstrations, to see buses and trucks from Nueva La Habana loaded with *pobladores* waving the red-and-black flag of the Movimiento de Pobladores Revolucionarios. A few dozens of Nueva La Habana's residents were dedicated MIR militants under the leadership of a charismatic and thoroughly honest *poblador*, Alejandro Villalobos, nick-named El Mike.[33] For the majority of residents, though they were sincere supporters of left-wing politics, involvement in the political struggle depended upon issues like the access to land, housing, and services.

The ideological gap between the political vanguard and the squatters was the cause of continuous tension inside the *campamento* during the three years of its life.[34] This tension was expressed, for instance, in the resistance of the residents to the efforts for a cultural revolution in the children's schools, set up by MIR using old buses as classrooms. When the young teachers tried to change the traditional version of Chilean history or to recast the teaching to follow Marxist themes, many parents threatened to boycott the school, forcing the staff to preserve the 'official' teaching programme. The reason was not that they were necessarily anti-Marxist, but rather that they did not want their children to become exceptional by virtue of receiving a different education from the rest of the city. The *campamento*, with its revolutionary folklore, its popular theatre group and its 12-metres-high Che Guevara portrait, was clearly seen by most residents as a transitory step towards a more 'normal' neighbourhood, a neighbourhood where one could receive visits from friends and relatives from the outside world, who for a long time had been scared to come and visit the squatters living in areas reported as 'dangerous' by the press and proclaimed as 'revolutionary' by their leadership.

The careful observations by Meunier about the social use of space and housing by the squatters provides a striking illustration of the individualism of the majority of squatters. Most houses, though tiny (with ground measurements of six by five metres), tried to enclose a piece of land, to mark a front yard as a semi-public space, while refusing space for common yards. The shack itself was divided between the main room, where the man could receive visits, and the kitchen-toilet, the private domain of the woman. Only the more enlightened leadership tried to make some space available for public use, but this practice led to spatial segregation: the shacks of the leaders tended to be concentrated towards the centre of the *campamento*, close to the shacks used for public purposes. The discrepancy between the level of involvement and consciousness thus became expressed in the spatial organization of the settlement. Individualism was even more pronounced when the residents were called to decide upon the design of their own houses. While asking for architectural diversity (three types of houses were built to fit the different sizes of families), they emphasized the desire for a standard design, utterly rejecting high-rise buildings. They also asked for the individual connection of each house to the water and electricity supply, restated the convenience of individual yards, and specified that the conventional domestic equipment, including television sets and individual electric appliances, would have to have enough room in the new houses. The real dream of most *pobladores* was that Nueva La Habana would one day cease to be a *campamento* and become an average working-class *población*.

Yet it should not be deduced that cultural conservatism and political opportunism were the reasons for this attitude. In fact the residents of Nueva La Habana were ready to mobilize in defence of their houses and in defence of their political beliefs each time it was required. They invaded land against police repression during the difficult months of 1970. They worked hard to dig sewerage-trenches, connect electricity, provide water, build shacks, set up public services, administer their 'city', and help one another when required. When, in October 1972, the economic boycott from external and internal capitalist forces halted the distribution of basic foods, the entire *campamento* mobilized to obtain supplies from the factories and the fields, and to distribute a basket to each family for weeks, without requiring any payment from those who could not afford it. They also established a new popular morality, banning prostitution and alcohol in the *campamento*, protecting battered women, and taking care of one another's children when the parents were working or involved in political activity. In sum, Nueva La Habana did not refuse its share of mobilization or cultivate a hypocritical attitude towards socialist ideals in exchange for urban patronage. But what was clear to every observer was that such a struggle was a means, and not a goal, for the great majority of the *pobladores*, that Nueva La Habana was an introspective community, dreaming of a peaceful, quiet, well-equipped neighbourhood, while MIR's leadership, conscious of the sharpening of the political conflict, desperately wanted to raise the level of militancy so that the entire *campamento* would become a revolutionary force. Their efforts in this direction proved unsuccessful.

On the basis of 20 focused interviews with residents, Meunier hypothesized the existence of three types of consciousness in the *campamento*:

1. The *individual*, focused upon the satisfaction of urban demands through the participation in the illegal occupation of land.

2. The *collective*, whose goals were limited to the success of the *campamento* as a community through the collective effort of all residents, closely allied to the government's initiatives.

3. The *political*, emphasizing the use of the *campamento* as a launching platform for the revolutionary struggle.

Although their sample is too limited to be conclusive, similar observations can be drawn from the survey by Duque and Pastrana, as well as from our own study. It would seem that the political level was only reached by MIR's cadres, that the mainstream of residents had some kind of collective consciousness, while a strong-minded minority maintained an individualistic attitude, though sympathetic to left-wing politics. It is crucial for our analysis to try to understand some of the reasons behind each level of consciousness, since Meunier's study concludes with the connection between level of consciousness and social mobilization. While the 'collective consciousness' appears to have been randomly distributed among a variety of social characteristics, the two other types tended to be connected with a few significant variables: the 'political consciousness' seemed more common among men than among women, among individuals with lower income, and among unemployed workers (although the fact of being unemployed in 1971's Chile might be a *consequence* of being a revolutionary worker). The 'individualist consciousness' seems to have been associated with higher income, better than average housing conditions before coming to the *campamento*, and women.

On the basis of these observations, two complementary themes can be noted:

1. The mainstream of the working class in Nueva La Habana probably followed the same pattern found elsewhere, collectively defending their living conditions but leaving the task of general political leadership to the government. A minority group of higher-income families joined the invasion to solve their housing problem without further commitment. The radical vanguard of MIR was composed of unskilled workers whose political leanings could surface more easily through the *pobladores'* movement, given the tight political control by communists and socialists in the labour movement (the CUT). This argument, specific to Nueva La Habana, confirms one of our basic general theses on the *pobladores* movement in Chile. Support for this interpretation can be found in the comparison of the occupational structure between Nueva La Habana and the *campamento* Bernardo O'Higgins, the model squatter settlement organized by the Communist Party. Table 19.3, constructed by us on the basis of the census that Duque and Pastrana took in four *compamentos*, shows that the Nueva La Habana's residents had a lower proportion of well-educated people, a lower proportion of workers in the 'dynamic sector' (modern industry), a much higher proportion of unemployed (33.2 per cent against 19.5 per cent in Bernardo O'Higgins), and a much lower proportion of unionized workers.

TABLE 19.3   Social Composition of Four *Campamentos*, Santiago de Chile, 1971.
(Percentage of Residents over the Total of each *Campamento* who have the Listed
Characteristics.)

	Campamentos			
	Fidel Castro	26 de Julio	Nueva la Habana	Bernardo O'Higgins
Low income	39	29	18	n/a
High level of education (primary completed)	15	20	18	24
Self-employed workers	11	13	15	17
Manufacturing workers	39	35	48	53
Service workers	21	27	24	36
Workers in modern industrial companies	16	25	25	28
Workers in large companies (over 50 employees)	36.6	46.8	42.9	37.4
Unemployed workers	37.6	22.3	33.2	19.5
High level of urban experience	56.7	65.9	64.6	66.5
Urbanized workers	30	43	35	44

n/a not available, but other sources indicate that the income levels in Bernardo O'Higgins
were noticeably higher than in the other three *campamentos*.

Source: Joaquin Duque and Ernesto Pastrana, *Survey of Four Campamentos* (Santiago de
Chile: Facultad Latinoamericana de Ciencias Sociales, 1971).

The apparent contradiction that Nueva La Habana came higher in the
proportion of workers it had from large factories is a simple statistical
artefact: a substantial number of those from Bernardo O'Higgins were
skilled and well-paid bus-drivers who could not be counted as working in
factories. In sum, Bernardo O'Higgins and the Communist Party seem to
have relied on the support of the organized working class while MIR and
Nueva La Habana seem to have been more successful among the workers
of the informal urban economy. We will develop this argument at a more
general level, once the profile of Nueva La Habana is complete.

2. Another major factor in Nueva La Habana was that women appear to
have been the group most reluctant to follow MIR's revolutionary ideology
and the ones who emphasized the satisfaction of basic needs before general
political commitment. In fact, this is the main reason advanced by Meunier
for explaining the gap in consciousness and mobilization between the
vanguard and the majority of the squatters. Meunier lists many examples
about the absence of any real transformation in women's roles and lives.
She describes how women cooked and ate in the kitchen while serving their
husbands and friends in the main room. She describes the sexual
domination by some cadres and the difficulty that women faced because
their participation in the running of the *campamento* gave rise to suspicions

of infidelity. She goes on to describe the difficulties women had in taking advantage of contraception services provided by MIR because the men saw them as a threat to their virility. As a result, in 1972, a majority of the women in Nueva La Habana were pregnant. Furthermore, unable fully to participate in the political mobilization, women saw the absence of the men from the house, their unemployment, and their political commitment as threats to family life. Separations were common among political leaders and their women as a result of these tensions. This, in turn, widened the gap between single men, who became full-time political activists, and the majority of families, dominated by women's fears and pragmatic feelings. The situation was paradoxical if we consider that MIR, given its strong student basis, was perhaps the one Chilean party that tried the hardest to liberate women and to integrate them fully into politics. But in Nueva La Habana the form that this liberation took deepened the divide between MIR's militant women and the majority of residents. MIR organized a women's militia that took care of a variety of tasks, particularly those to do with health but also in matters of self-defence. But this initiative was not supported by a change of attitude of men towards 'their women', who were still unable to participate in the collective activities. So it further isolated the few political women and exposed them to the criticism and distrust of the housewives. Such a dramatic contrast can be illustrated by two events:

(a) The general blame put by almost the entire *campamento* on a woman whose unguarded child drowned while she was working at the health centre.

(b) The rejection by women of MIR's proposal to close down the mothers' centres (an inspiration of the Christian Democrats) where women met to learn domestic skills, and to replace them by women's centres which would emphasize women's militant role. Most women felt that such a change would politicize their free space, depriving them of their capacity to autonomously decide how to use these centres. Thus, the mothers' centres continued to function in the heart of revolutionary squatters' settlements.

Although they were unable to challenge the machismo prevalent in the settlements, most resident women rejected MIR's heavy-handed politicization of women's issues and became the prime movers for the use of urban mobilization strictly for the improvement of their conditions.

So Nueva La Habana lived entirely under the shadow of MIR's initiatives. New housing was built, urban infrastructures were provided, health and education services were delivered, cultural activities were organized, goods were supplied, prices were controlled, moral reform was attempted, and some form of democratic self-management was implemented, although under the unchallenged leadership of the *miristas* (MIR militants). Yet the social role of the *campamento* shifted according to the political tasks and priorities established by MIR at the national level. In the first year, MIR supported urban demands as a means of consolidating its position in the squatter movement so as to reinforce its militant power. In May 1971, at the congress of Nueva La Habana residents, the leadership announced that top priority should be given to the penetration of the organized working class by MIR, to counteract reformist forces in the labour movement. So most of the cadres of the *campamento* were sent to other political duties, leaving the *pobladores* in a support role for the main struggle being fought in the work places. The working-class ideology of MIR came into open contradiction with its militant presence among the squatters, and there followed in 1971–2 a period of disorientation in the *campamento*, leading to demobilization and in-fighting.

The general mobilization in October 1972 against the conservative offensive in Chile led to a new role for the *campamento*, first, in the battle over distribution and, later on, in the support of the construction of *cordones industriales* (industrial committees) and *comandos comunales* (urban unions) as centres of revolutionary, popular power. Militants from Nueva La Habana tried on 3 April 1973 to occupy the National Agency of Commercial Distribution (CENADI) as a gesture in favour of this strategy: they built barricades in Santiago's main avenue, Vicuña Mackena, and clashed fiercely with the police for a whole day. Nueva La Habana subsequently kept its subordinate role as a branch of a political party, adopting a variety of tactics, corresponding to the different directions taken by MIR's political activity. So if the 'model *campamento*' was an expression of the militant squatters' capacity to build their city and to try new communal ways of life, it was also, above anything else, an organizational weapon of a revolutionary party.

In our own research with the CIDU team we came to a similar conclusion: for all *campamentos*, whatever their political orientation, the practice of the squatters was entirely determined by the politics of the settlement, and the political direction of the settlement was, in turn, the work of the dominant party in each *campamento*.

The same finding was obtained by studies on mobilization in the *conventillos* (central-city slums)[35] and in the neighbourhood-based demand organizations.[36] Perhaps the only exception consisted of the committees organized to control prices and delivery of food, the *Juntas de Abastecimientos y Precios*, where a high level of popular autonomy was observed, cutting across the political membership. Even so, the practice of these committees was very different: they either aimed to reinforce the socialist government or to structure a dual power, depending upon the political tendency of their leadership.[37]

The political stand of the *pobladores* was a decisive element in enabling the formerly passive popular urban sectors to join the crucial political battle, initiated by the working-class movement, for the construction of a new society. Thus we could speak, in this case, of an urban social movement because the popular masses were mobilized around urban issues and made a considerable political contribution to the impetus for social change.

Nevertheless, the partisan and segmented politicization of the *pobladores* made it impossible for the left to expand its influence beyond the borders of the groups it could directly control. The different sympathies held in each *campamento* hardened into political opposition among different groups, a situation quite unlike that of the trade unions, where Christian Democratic workers often backed initiatives from the left such as the protest against the boycott of Chile by the international financial institutions. Furthermore, the separating of political forces in the squatter settlements led each group to seek support in the administration, splitting the whole system into different constellations of state officials, party cadres, and *pobladores*, each with its own political flag.

The social effects of this development became evident in the changing relationships between the *pobladores* and Allende's Government. During the first year (November 1970 to October 1971) the difficulties of putting the new construction industry to work made it impossible for the government to satisfy the *pobladores'* demands, and the only thing it could do was to accept the land invasion and provide some elementary services by putting the squatters in touch with public agencies. In spite of the government's inability, the *pobladores*, including Christian Democrats, collaborated actively with the administration. In contrast, at the end of the second year, when 70,000 housing units were under way and when health, education, and other services started to be delivered, some serious signs of unrest appeared among the *pobladores*. Finally, after October 1972, when the political battle became inevitable, each sector of the *pobladores* aligned with its corresponding political faction, and the squatters' movement disappeared as an identifiable entity.

To understand the evolution of this attitude towards the government, we must consider the social class content of the political affiliations in Unidad Popular as well as the social interests represented by the squatter settlements. In the first year the government, taking advantage of the political confusion in business circles, successfully implemented a series of economic and social reforms which substantially improved the level of production and standard of living. This policy obtained important popular support, as did the preservation of political freedom and social peace. The political debate was kept within institutional boundaries, and the opposition of the Christian Democrats was moderate.

During the second year, however, the economy deteriorated rapidly, due to the sabotage of the economy by the Chilean business sector and landlords, the international boycott, and the end of the benefits of using formerly idle industrial capacity. The political alliance between the centre and the right isolated Unidad Popular. The radicalization of some popular sectors exacerbated the situation and was used as a pretext for political provocation. International pressure against Allende was reinforced. There was one popular working sector that was particularly sensitive to political alienation, especially in these conditions: workers in small companies. Although not included in the nationalized sector, these workers were asked to restrain their own demands to preserve the alliance with small business. This sector of workers was actually non-unionized, since Chilean laws before 1970 prevented union membership in companies with less than 25 employees, with the result that they tended to be politically less conscious than the unionized working class. So the Unidad Popular government asked for a more responsible effort by this less organized and less conscious segment of the working class. The reaction was a series of errant initiatives, sometimes very radical, sometimes conservative, and generally out of tune with mainstream policy of the popular left. Now all surveys show that this sector of small companies' working class—the workers of the 'traditional sector' or 'informal economy'—represented in Chile the most important share of the squatter settlements (see Table 19.3), and therefore the *pobladores'* movement organized the main expression of this social group. But as we have just noted, the squatters' movement was split into political factions and so reflected the disorientation of this group. Instead of becoming a movement and so a focus for organization and mobilization, the factions only enabled them to manœuvre occasionally against the Unidad Popular government.

During the third year, the left of the Unidad Popular, as well as MIR, tried to build up a people's power base strong enough to oppose the ruling classes by developing territorially based organizations which combined both the *comandos comunales* and *cordones industriales*.[38] But because many people actually leant towards the centre-right, and most workers were following the government, this grass-roots movement gathered only a vanguard of industrial workers in some sectors of the big cities, particularly in the area of Cerrillos-Maipu in Santiago. In fact, the squatters' movement disappeared as an autonomous entity in the decisive moments of 1973, and was less than ever the unified movement around which the left might have organized some popular sectors; people in the end supported the centre-right as much as the left. Not only did the political influence of the *pobladores* movement wane but, as part of the right-wing offensive, the far right organized some middle-class neighbourhoods along socio-political lines (the Proteco organizations), linking the provision of local services to preparations for the military coup. The disappearance of the *pobladores'* movement, then, in 1972–3 was the consequence of the logic of party discipline replacing the search by the left to establish political hegemony.

The only moments of mass participation in the squatters' movement were those where political parties of the left gathered around a clear-cut common cause: the first year of Allende's Government, and the mass response to the business strike of October 1972. In both situations, the squatters' movement started to provide a new urban system corresponding to the political transformation of the state by Allende's Government. On both occasions, we could observe its potential as a mass movement and as a social movement. As a mass movement, it gathered and organized a larger proportion of people than the left-wing parties could. As a social movement it started to produce substantial changes to urban services and the local state institutions because of its capacity to mobilize people. Both moments were exceptional but too short-lived: the unity and the cultural influence of the squatters' movement did not survive the polarization of the political opposition existing inside it.

The squatter movement in Chile was potentially a decisive element in the revolutionary transformation of society, because it could have achieved an alliance of the organized working class with the unorganized and unconscious proletarian sectors, as well as with the petty bourgeoisie in crisis. For the first time in Latin America, the left understood the potential of urban movements, and battled with populist ideology on its own ground, planting the possibilities of political hegemony among urban popular sectors. But the form taken by this political initiative, the over-politicization from the beginning, and the organizational profile of each political party within the movement, undermined its unity and made the autonomous definition of its goals impossible. Instead of being an instrument for reconstructing people's unity, the *pobladores'* movement became an amplifier of ideological divisions. Yet its memory will last as the most promising attempt by Latin American urban masses to improve their social condition and achieve political liberation.

Notes

1. See the collection of data and analyses presented in Janet Abu-Lughod and Richard Hay (eds), *Third World Urbanization* (Chicago: Maaroufa Press and London: Methuen, 1977).
2. See Anthony and Elizabeth Leeds, 'Accounting for Behavioural Differences in Squatter Settlements in Three Political Systems: Brazil, Peru and Chile', in Louis Massotti and John Walton (eds), *The City in a Comparative Perspective* (Beverly Hills, Calif.: Sage Publications, 1976).
3. Etienne Henry, *Urbanisation dépendante et mouvements sociaux urbains* (Paris: Université de Paris, Ph.D. thesis, 1974); and E. Henry, *La escena*

urbana (Lima: Pontífica Universidad Católica del Perú, 1978); David Collier, *Squatters and Oligarchs* (Baltimore: Johns Hopkins Press, 1976).

4. They always start in the periphery of the city, but with the expansion of urban space, some of the early *barriadas* are now located in the core of the metropolitan area.

5. Jacqueline Weisslitz, *Développement, dépendance, et structure urbaine: analyse comparative de villes péruviennes* (Paris: Université de Paris, Ph.D. thesis, 1978).

6. Manuel Castells, 'L'urbanisation dépendante en Amérique Latine', *Espaces et Sociétés* 3 (1971); also Helen Safa (ed.), *The Political Economy of Urbanization in Third World Countries* (New Delhi: Oxford University Press, 1982).

7. Etienne Henry, 'Los asentamientos urbanos populares: un esquema interpretativo', *Debates* (Lima) 1, 1(1977), 109–38.

8. David Collier, *Squatters and Oligarchs* (1976).

9. Alfredo Rodriguez and Gustavo Riofrio, *Segregación social y movilización residencial: el caso de Lima* (Buenos Aires: SIAP, 1974).

10. Leeds and Leeds, 'Accounting for Behavioural Differences', 1976.

11. Luis Unikel, *El desarrollo urbano de Mexico* (Mexico, DF: El Colegio de Mexico, 1976).

12. Jorge Montaño, *Los pobres de la ciudad en los asentamientos espontáneos* (Mexico, DF: Siglo XXI, 1976). And Equipo Pueblo, *Surgimiento de la coordinadora nacional del movimiento urbano popular: las luchas urbano-populares en el momento actual* (Mexico DF: Conamup, 1982); B. Navarro and P. Moctezuma, 'Clase obrera, ejército industrial de reserva y movimientos sociales urbanos de las clases dominadas en Mexico', *Teoría y Política*, 2, (1981).

13. See Susan Eckstein, *The Poverty of Revolution: The State and the Urban Poor in Mexico* (Princeton: Princeton University Press, 1977).

14. See Julio Labastida, 'Los grupos dominantes frente a las alternativas de cambio', *El Perfil de México en 1980* (Mexico, DF: Siglo, XXI, 1972), vol. 3.

15. Diana Villarreal, *Marginalité urbaine et politique de l'état au Mexique: enquête sur les zones résidentielles illégales de la ville de Monterrey* (Paris: Université de Paris, Ph.D. thesis, 1979).

16. For sources on Chile, see Manuel Castells, *The City and the Grassroots: A Cross-Cultural Theory of Urban Social Movements* (London: Edward Arnold; Berkeley and Los Angeles: University of California Press, 1983), 362–5. For an analysis of the overall process, see Manuel Castells, *La lucha de clases en Chile* (Buenos Aires: Siglo XXI, 1975); and Barbara Stallings, *Class Conflict and Economic Development in Chile, 1958–1973* (Stanford: Stanford University Press, 1978).

17. See Manuel Castells, 'Movimiento de pobladores y lucha de clases en Chile', *Revista Latinoamericana de Estudios Urbanos y Regionales*, 7 (1973), 9–36; also, Ernesto Pastrana and Monica Threlfall, *Pan, techo, y poder: el movimiento de pobladores en Chile, 1970–3* (Buenos Aires: Ediciones SIAP, 1974).

18. Cecilia Urrutia, *Historia de las poblaciones callampas* (Santiago: Quimantu, 1972).

19. Jaime Rojas, *La Participation urbaine dans les sociétés dépendantes: l'expérience du mouvement des pobladores au Chili* (Paris: Université de Paris, Ph.D. thesis, 1978).

20. Franz Vanderschueren, 'Significado político de las juntas de vecinos', in *Revista Latinoamericana de Estudios Urbanos y Regionales*, 2 (1971).

21. According to the careful research monograph by E. Santos and S. Seelenberger, *Aspectos de un diagnóstico de la problemática estructural del sector vivienda* (Santiago: Escuela de Arquitectura, Universidad Católica de Chile, 1968).

22. Rosemond Cheetham, 'El sector privado de la construcción: patrón de dominación', *Revista Latinoamericana de Estudios Urbanos y Regionales* 3 (1971).

23. Jaime Rojas, *La Participation urbaine.*

24. Luis Alvarado, Rosemond Cheetham, and Gaston Rojas, 'Movilización social en torno al problema de la vivienda', *Revista Latinoamericana de Estudios Urbanos y Regionales*, 7 (1973).

25. Jose Bengoa, *Pampa irigoin: lucha de clases y conciencia de clases* (Santiago: CESO, Universidad de Chile, 1972).

26. For quantitative data on the evolution of the squatter movement, see Pastrana and Threlfall, *Pan, techo, y poder*, 60–6.

27. Joaquin Duque and Ernesto Pastrana, 'La movilización reivindicativa urbana de los sectores populares en Chile, 1964–72', in *Revista Latinoamericana de Ciencias Sociales*, 4 (1973).

28. See Equipo de estudios poblacionales del CIDU, 'Revindicación urbana y lucha política: los campamentos de pobladores en Santiago de Chile', *Revista Latinoamericana de Estudios Urbanos y Regionales*, 7 (1973).

29. See Equipo de estudios poblacionales del CIDU, 'Experiencia de justicia popular en poblaciones', *Cuadernos CEREN* (Santiago), 8 (1971). The most complete analysis on the subject is provided by Franz Vanderschueren, *Mouvements sociaux et changements institutionnels: l'expérience des tribunaux populaires de quartier au Chili* (Paris: Université de Paris, Ph.D. thesis, 1979).

30. See the extraordinary study by Christine Meunier, *Revendications urbaines, stratégie politique et transformation idéologique: le campamento Nueva La Habana, Santiago, 1970–73* (Paris: Université de Paris, 1976, Ph.D. thesis).

31. Duque and Pastrana, 'La movilización reivindicativa urbana', 1973.

32. According to one of the officials in charge of housing policy under Allende, Miguel Lawner. See Miguel Lawner and Ana-María Barrenechea, 'Los mil días de Allende: la política de vivienda del gobierno popular en Chile' (unpublished Research Report, 1978).

33. Alejandro Villalobos actively resisted the military dictatorship in Chile; he was murdered in the street by the junta's political police in 1975, but his death was officially announced as 'a clash with guerrillas'.

34. After the military coup Nueva La Habana suffered fierce repression and became an impoverished shanty town renamed Amanecer (Dawn) by the junta.

35. François Pingeot, *Populisme urbain et crise du centre-ville dans les sociétés dépendantes: Santiago-du-Chili 1969–73* (Paris: Université de Paris, 1976, Ph.D thesis).

36. Luis Alvarado, Rosemond Cheetham, and Gastón Rojas, 'Movilización social'.
37. See Cristina Cordero, Eder Sader, and Monica Threlfall, *Consejo comunal de trabajadores y cordón cerillos-maipu 1972: Balance y perspectivas de un embrión de poder popular* (Santiago: CIDU, 1973, Working paper 67).
38. See Rosemond Cheetham, Alfredo Rodríguez, Jaime Rojas, and Gastón Rojas: 'Comandos urbanos: alternativa de poder socialista', *Revista Interamericana de Planificación* (July 1974).

20

Urban Labour Movements under Authoritarian Capitalism in the Southern Cone and Brazil, 1964–1983*

Paul W. Drake

FOUR of the most developed, most industrialized, and most unionized countries in Latin America have produced four of the most significant labour movements. From the early 1960s onward, scholars flocked to analyse the rise of those working-class organizations. After the breakdown of populist compromises between capital and labour based on import-substituting industrialization, class conflict was escalating. Then the high hopes for proletarian advances aroused by the presidency of João Goulart in Brazil, the election of Salvador Allende in Chile, the emergence of the Frente Amplio in Uruguay, and the return of Juan Perón in Argentina were all dashed. The authoritarian governments of Brazil (beginning in 1964), Chile (1973), Uruguay (1973), and Argentina (1976) targeted urban workers, their unions, and parties as the principal victims of political repression, social demobilization, and capital reaccumulation. The inability of those potent labour movements to consummate their reform projects prior to the *coups d'état* or to resist reactionary assaults thereafter prompted Latin Americanists to reevaluate the political roles and prospects of urban workers. Now that authoritarian capitalism is receding in the hemisphere, it seems time to synthesize and assess that literature on the situation of labour from the commencement of those military dictatorships in 1964 to their decay in 1983.

Introduction

Analyses of labour movements in South America have passed through three broad stages. Prior to the radicalization of workers in the 1960s and 1970s, most observers stressed that structural limitations and institutional weaknesses kept Latin American trade unionists reliant on moderate

* This essay was written with the support of the Social Science Research Council. I am also grateful for the help of my research assistants, Baldomero Estrada and Marcus de Carvalho.

political mechanisms. Some scholars even suggested that cultural propensities —inherited from the rural past—oriented labourers toward paternalistic caudillos. Despite nationalistic and socialistic rhetoric, most working-class movements, so it seemed, dedicated themselves to practical and prudent bargaining and benefits. The workers' eagerness to participate in populist coalitions rather than revolutionary adventures provided ample evidence for such interpretations. Although objectively and absolutely disadvantaged, the minority of labourers enjoying unionized employment in the industrial sector at least profited from privileges denied the masses in the urban shanty towns and rural shacks. Therefore, it was argued, most organized workers were content to confine their activities within the boundaries of the existing economic and political system.

That school of thought was called into question, however, by spiralling militance and even attacks on private property just prior to the *coups d'état*, especially in Chile and Argentina. Observers began emphasizing the taking of consciousness and radicalization of the working class. The unraveling of populist alliances and programmes, the outbreak of autonomous proletarian activism, and the upsurge of ultra-left alternatives supplied ammunition for these new interpretations. In response, authors wrote about an impending revolution in Brazil, the irreversible gains of the Marxists in Chile, the corrosion of the élitist two-party monopoly in Uruguay, and the socialization of Peronism in Argentina. Then the stunning military take-overs plunged sympathizers of those labour movements from high optimism to deep pessimism.

The subsequent enforced silence of the workers motivated analysts to try to reconcile the above two contrasting images. By the end of the 1970s, gloom descended over portraits of the labour movement and its potential, particularly in the Southern Cone. The inability of unionists to counter-attack the Chilean, Uruguayan, and Argentine dictatorships resurrected older visions of the underlying debilities and caution of the working class. Just as the pre-coup advances inspired some exaggerated estimates of labour's strengths, so the post-coup regression sometimes evoked overblown assessments of labour's weaknesses. Two lines of thinking came to prevail. On the one hand, a few scholars suggested that almost everything had changed, that the draconian transformations wrought by authoritarian capitalism had completely demolished and demoralized the labourite groups. On the other hand, some writers argued that almost nothing had changed, that the dictatorships were merely holding the lid on unrepentant 'popular' forces. Especially in Chile, one of those two explanations seemed necessary to account for the glaring difference between an aggressive, articulate, irresistible labour movement in 1973 and a cowed, controlled, subdued one in 1980.

As those movements re-emerged in the early 1980s, it became evident that they had preserved many continuities with their past, despite maximum efforts by their adversaries to erase those traditions. At the same time, the authoritarian experiences had reshaped those labour organizations, which became more independent from their increasingly moderate party allies. To understand those continuities and changes, this essay surveys the character of working-class politics prior to the military take-overs, and then asks how those labourite movements were contained by the armed forces, how they survived that onslaught, and how they were transformed as a result.

Looking back at the 1960s and 1970s, scholars asked, why did all these labour movements fail to achieve their objectives and instead fall prey to conservative dictatorships? Obviously, the primary reason was the awesome power of the domestic and foreign forces arrayed against them. Those élites erected enormous barriers in the path of these movements and then unleashed ferocious repression against them.

To understand fully the defeats weathered by the workers and their party allies, however, also requires an examination of their vulnerability. That focus becomes essential to understand how repression took advantage of the weak spots in these movements, and how it exacerbated problems which existed before the coups. To varying degrees, these movements all suffered from imposed limitations and engrained handicaps which rendered any victories monumental triumphs against staggering odds. Their experiences in these four countries both before and after the *coups d'état* were shaped by (a) the economic and social structures which placed industrial workers at a disadvantage, (b) the institutional restraints on trade-union growth, solidarity, and action, and (c) the political debilities of their party allies. Although these movements also possessed great assets and made mammoth strides forward, this study will highlight the underlying structural, institutional, and political causes of their pre- and post-coup frustrations up to 1983, prior to any full-scale redemocratization.

This article will compare these four experiences by building upon the crucial Chilean episode as a bedrock. The analysis will concentrate on organized industrial labourers, and, to a lesser extent, construction workers and miners. That is because most research has shown that urban manufacturing workers in unions tied to progressive parties constituted the politicized core and vanguard of the labour movement. 'Labour parties' is, itself, a relative term, referring to those major organizations which, compared to other vehicles, have defined themselves as exceptionally devoted to the working class and have been so categorized by other power contenders. To varying degrees, labourite parties featured leaders who cast themselves as spokespersons for the proletariat, programmes and propaganda which emphasized the concerns of that social stratum, and membership and electoral recruitment campaigns heavily targeted at unions and workers. Even that loose conceptualization obviously fits the Marxists in Chile and the Peronists in Argentina far better than the more multi-class entities in Uruguay and Brazil, where groups such as the Broad Front and the Labour Party aspired with less success to represent blue-collar constituents. Indeed, the differing availability, capabilities, organizations, compositions, ideologies, platforms, strategies, styles, and histories of these parties within particular national political systems constituted key variables in the evolution of these four country cases. Nevertheless, this essay will only be able to skim those party characteristics as they interacted with other structural and institutional factors.

Before the Coup

Structural Constraints on Labour

Within Latin America, Chile, Argentina, Uruguay, and Brazil boasted exceptional levels of industrialization and unionization. Nevertheless, long before the coups, economies of scarcity with large underutilized labour surpluses and relatively small manufacturing sectors severely limited the political strength of the proletariat. Moreover, in Chile, Argentina, and Uruguay, the passage of the high tide of import substituting industrialization, the break-up of populist alliances, and the balloooning of the service sector had further reduced the relative weight of industrial workers in political equations, even before the coups accelerated those trends.

In all the countries, the largest single segment of organized labour came from manufacturing and spearheaded the union movement. Although the most assertive workers typically belonged to industrial unions connected to reformist parties, that proletariat only accounted for a minority of trade unionists, since many of them came from artisanal, white-collar, or rural occupations. For example in Chile 41 per cent of all manufacturing workers enrolled in unions, and they accounted for 32 per cent of all trade union members; in Argentina 60 per cent of all industrial labourers joined unions and comprised 34 per cent of all union members. That minority was encircled by vast numbers of non-industrial unionists, unorganized urban workers, unemployed and semi-employed labourers, rural workers, and salaried employees, many of whom were normally much less favourable to proletarian political projects.

Institutional Constraints on Labour Unions

The ways in which industrial workers, their unions, and parties operated in national politics prior to the military interruptions shaped their subsequent situation. To over-simplify, in Chile these workers mainly pressed their grievances through national parties, in Argentina through powerful unions, in Uruguay as a pressure group on the government bureaucracy, and in Brazil as a supplicant within the state corporatist network. In all four cases, with the partial exception of Chile, unions, as in most of the world, pursued essentially incremental objectives far more than grand political projects.

The likelihood of labour unions and parties achieving more ambitious national transformations was reduced by government regulations, by schisms within these heterogeneous labour movements, and by deficiencies in their organizational capacities. Moreover, many of the individual unions remained tiny and frail. It thus proved difficult to consolidate the labour movement internally, to forge unified ties with any single party, or to weave together coalitions between the proletariat and its potential allies in the white-collar sector, the shanty towns, and the countryside.

In Chile, small, localized unions with scant finances and insecure leaders possessing tenuous ties with national federations and parties constituted the rule. For example one survey of Chilean union presidents in the 1960s found only 10 per cent who thought unions by themselves had much impact on national issues, an attitude that might be expected among tamed union leaders in Brazil but not among their independent counterparts in Argentina. Nevertheless, approximately 41 per cent of the economically active population enrolled in unions. Those Chilean unions divided sharply among blue- and white-collar and urban and rural groups, as well as among competing labour parties. Through various occupational and national federations, some two-thirds of the unions tried to promote labour unity by establishing connections with the over-arching Central Workers' Federation (CUT). This loose amalgam of white- and blue-collar organizations had no legal standing prior to Allende and thus depended heavily on party spokesmen to exert national influence.

Chilean unionists relied most on the Marxist parties at the national level of the CUT, but the Socialists (PS) and Communists (PC) also helped at the plant level by providing leadership. Because the labour code denied union heads special security, salaries, and perquisites, only party inspiration usually convinced workers to take those burdensome leadership posts. This resulted in a dualistic system where union chiefs dedicated themselves to the particularistic needs of the rank and file at the local level but also to the political aspirations of their parties at the national level.[1]

These party connections also benefited organized labour because the 1924 national code left unions disadvantaged in bargaining with employers and reliant upon the state to intervene on their behalf. Only Brazil had more restrictive labour laws, and only it and Uruguay created more dependence on the state. The real key to the impact of such legislation on unions was its implementation by the government and its manipulation by the workers and their parties.[2]

After years of interdiction of their Peronist party, Argentine labour unions had developed much more autonomous power and bureaucracy than their Chilean counterparts. Approximately 34 per cent of Argentina's economically active population belonged to trade unions. They boasted unparalleled pre-coup experience at coping with conservative military regimes. Repeated persecution also drove most union leaders and members closer together, though the bosses brooked little dissent from the rank and file. Their situation epitomized 'political unionism', in which the union and party fused. Indeed, the unions took the place of the outlawed party as the mobilizing agent for the working class on concrete as well as 'political' issues.

Argentine labour's political effectiveness derived not from any radical commitment but from unity behind a single party banner—albeit tattered by intramural squabbling. That movement also profited from organizational solidarity within the General Confederation of Labour (CGT). Although workers faced a less constraining labour code than in Chile, existing laws gave Argentine administrations great potential control over elections, finances, and activities.[3]

Uruguay's organizations for the manufacturing proletariat never wielded great power. Industrial labourers were scattered in tiny factories, where over 90 per cent of the firms employed fewer than 20 workers apiece (by contrast, only some 20 per cent of Chile's manufacturing proletariat toiled in establishments with fewer than 25 workers). Only 30 per cent of Uruguay's economically active population belonged to trade unions.

Normally pursuing highly defensive demands, Uruguayan unions promoted explicitly political causes much less often than did their counterparts in Chile and Argentina. Although some 90 per cent of Uruguay's trade unions had affiliated with the National Workers' Federation (CNT) by the time of its consolidation under Communist leadership in 1966, it remained a young and unimposing peak organization. Moreover, labour usually supported Communists in union elections but the populist wing of the Colorado Party in national contests. Those politicians in Congress served as 'ombudsmen' between unionists and government bureaucrats but seldom as national labour spokesmen. Neither a professional union bureaucracy nor political parties provided the primary channel for working-class grievances. Instead, workers, like other pressure groups, mainly negotiated directly with the clientelistic state. Labour laws promised ample access and protection to both white-collar employees and blue-collar workers. Given the efficacy of those institutionalized networks for resolving most conflicts, no intricate, restrictive labour code played a part.[4]

On its own, the Brazilian labour movement suffered from relatively small size, unusually frail independent organizations and leadership, feeble party allies, and corporatist control by the state. Barely 9 per cent of the economically active population enrolled in trade unions. The movement expressed itself through generally anemic municipal-level unions and their federations. They tried to resolve disputes with employers by attracting governmental attention and support. More important to a labour leader's success than any links to political parties was access to and influence with the state.

The labour movement remained very disunited, and only affiliated through five separate national federations. Although the newly formed General Command of Workers (CGT)—under the leadership of the Brazilian Communist Party (PCB) and the Brazilian Labour Party (PTB)—claimed to speak for organized labour as a whole by the early 1960s, it remained an illegal organization with minimal grass-roots strength. Traditionally, the pro-labour parties, including the PCB, reconciled themselves to operating within the paternalistic legal system.[5]

Labour Parties

In each of the four countries, parties supplied workers with union and political leaders, legal aid, legislative victories, influence with government agencies, media outlets, political education, and allies from other social sectors, such as professionals and students. Vast segments of the working class, however, were not necessarily controlled by those parties nor dedicated to their broader programmes. All studies of working-class opinions and voting behaviour—whether scientific samples or smaller sets of interviews—found the largest and most enduring attachment to labour parties in Chile and Argentina rather than Uruguay, or especially Brazil. Of all the parties, the Peronists attracted the most solid backing from the proletariat and the largest electoral percentages. Hardly any correlation between social class and voting patterns showed up in Uruguay and Brazil, although research was not extensive.

No electoral or attitude surveys anywhere turned up a majority of organized workers in favour of socialist or Marxist alternatives. The highest propensities cropped up in Chile and Uruguay prior to the coups, revealing respectively 46 and 23 per cent of urban workers in favour of socialist party coalitions. Some slight leftward movement in working-class electoral preferences took place everywhere in the years immediately preceding the coups. This trend could be seen in the rise of the Peronist left in Argentina, the drainage away from the historical two parties by the Frente Amplio in Uruguay in 1971, and the rising percentage for the Popular Unity (especially the Socialist Party) in Chile in 1973. The greatest radicalism, however, seemed to be associated not with the industrial unionists of long standing but rather with a minority of younger workers. They were newer to politicization and often in smaller industries or even outside the industrial sector; they sometimes pressed the traditional labour parties to be more daring and identified with innovative, more revolutionary political options.

The most extensive studies of working-class political orientations were carried out in Chile. One measure, albeit flawed, of party strength within that union movement came from the elections of directors of the CUT. Significantly, those returns showed persistent massive support for the leftist parties but a noteworthy decline under the Allende Government. That slippage was reflected in a surge for the Christian Democrats (DC). Thus Augusto Pinochet would confront a labour movement with approximately two-thirds of the organizations loyal to the Popular Unity (UP) or kindred forces but with many fractures within those ranks, no evidence of rising union support for the Marxists, and indeed some signs of shrinking strength. Moreover, in the decade prior to the coup, whenever a survey sample was taken of how many people politically supported the Communists, Socialists, and Allende, the result always fell just short of a majority among workers, and of course dwindled off dramatically among higher social sectors.[6]

A few ecological voting studies in Argentina suggested that workers there even more solidly identified with and voted for Peronism than workers in Chile did for the Socialists and Communists. Apparently great continuity prevailed in the electoral base of Peronism from 1946 to 1973, with a clear majority of urban labourers casting ballots for the party in that latter election. But almost half the Peronist votes in 1973 came from non-workers, which gave that movement a much larger following than the UP and much more reliance on other social strata, especially white-collar employees.[7]

In Uruguay, the Frente Amplio grew primarily at the expense of the Colorados. From 1966 to 1971 in Montevideo, the Communists, Socialists, and Christian Democrats rose from 16 to 30 per cent of the electorate, while the Colorados sagged from 51 to 30 per cent. Despite this leftward drift, the Colorados and Blancos (National Party) only declined nationally from 89 per cent of the votes to 81 per cent, while the Frente climbed from 10 to 18 per cent (contrasted with a peak of 44 per cent for the UP in Chile and 63 per cent for the peronists in Argentina). Not only were clear-cut labour parties much weaker in Uruguay than in Chile or Argentina, so too was any linkage between social class and party preference. Polls on the eve of the 1971 election showed the Broad Front attracting a very mixed social following.[8]

In Brazil, no powerful, well-structured, programmatic, mass labour machine appeared. The Communist Party developed strength in the unions but not in the electorate, and endured repeated persecution. The Brazilian Labour Party and the populists offered only patchwork, regionalized, opportunistic vehicles with scant vigour, roots, or class content.

Little pre-coup evidence existed of strong Brazilian working-class attachment to party democracy in general or any party or ideology in particular. The few electoral and attitudinal studies suggested that working-class loyalties were widely spread among political options, though the industrial proletariat exhibited a slight tendency to vote for paternalistic populists. Very few workers openly favoured drastic socio-economic transformations, even among industrial labourers affiliated with the Communists.[9]

Labour Activism prior to the Coup

A picture emerges of handicapped working-class movements which, to varying degrees, were not fervently committed to or capable of fundamental socio-economic changes advocated by powerful labour parties. If accurate, how can that portrait of relative moderation and weakness be reconciled with scores of accounts of pre-coup mounting worker radicalism which elicited savage anti-labour reactions from the frightened upper and middle classes through the armed forces? If, with the partial exception of Chile, labour unions and parties presented no uncontainable threat to the established order, why were they viewed as such dangerous enemies to be curtailed?

One reason may have been that, in conservative societies with scarce resources, galloping inflation, industrial stagnation, and precarious manu-facturing élites heavily dependent on the state, foreign capital, and low labour costs, even relatively mild challenges from the organized working class and its party partners appeared inordinately threatening. A narrow margin for capital accumulation rendered non-revolutionary pressures quite menacing. A second reason could be that the mobilization, politicization, and radicalization of the worker movements were exaggerated not only by panicky opponents but also by idealistic proponents, who were eager to embellish the strength of their parties and followers. One example was the so-called 'Festive Left' in Brazil. A third explanation would note that, except in Uruguay, pro-labour governments took power before the coups. Those administrations permitted and encouraged a very unusual level of labour activism, which sometimes went farther than the governments intended. The fourth and final answer might be, with the partial exception of Brazil, that the breakdown of populist alliances, the deteriorating position of the traditional proletariat, and the rise of newly activated worker groups did engender some unprecedented working-class radicalism prior to the coups. That is, the relatively pragmatic standard portrait of Latin American labour probably still held true for a majority of unionists, but a militant minority of worker activists had emerged whom the ruling élites reacted to by punishing all labour unions and parties.

In Brazil, Chile, and Argentina, the presence of a labourite government—promising advances while calling for sacrifices—led to accelerating divisions and politicization among the workers. Those cleavages carried on past the coups. Although many labourers loyally heeded directives from their parties in power, both so-called 'economistic' and 'radical' groups caused their governments severe anxieties. First, the labour parties in office mainly encountered disobedience and resistance from worker groups demanding 'bread-and-butter' benefits for themselves. After years of supporting politicians because of their promises of instrumental gains, these workers wanted immediate material rewards rather than exhortations for continued dedication to long-range ideological goals. The striking copper-miners in Chile, backed by the Christian Democrats, offered a case in point. Second, these labour governments found it equally hard to discipline smaller worker groups who supported promises of ultimate revolutionary change for the proletariat and who believed their parties were not pursuing those ideological objectives avidly and rapidly enough. The workers forming Chile's 'industrial cordons' provided an example. The dual role of the labour parties of providing their followers with both clientelistic services and ideological missions proved more manageable out of office than in power.

All four coups took place against labour union and party movements which, to varying degrees, had already lost a great deal of thrust, cohesion, purpose, and optimism prior to the military intervention, especially in Argentina and Uruguay. Only in Chile, where labour and its party allies had been the most integrated and radical and where some had experienced and still had high hopes for a leap towards socialism, did any significant working-class resistance to the coup arise. But that was quickly, brutally snuffed out.

After the Coup

After seizing power, all the military regimes deployed enough repressive measures—indeed, more than enough—to quell the labour unions and parties. The voluminous records of murders, arrests, kidnappings, tortures, beatings, dismissals, exiles, and other abuses need not be repeated here. Nor need it be reiterated that these coercive methods provide the fundamental explanation for the inability of the proletariat and its political allies—already handicapped in many ways—to lash back. Within the confines of those police states, however, it is worth noting certain variations.

In all four countries, the anti-communist fervour of the armed forces directed the harshest tactics at the Marxist parties and unions. This selectivity within the general onslaught against working-class organizations proved most damaging in Chile, where the Communists and Socialists played a predominant role. The Chilean movement also suffered somewhat more than its counterparts because the military regime there delivered more severe political and economic shocks to the proletariat. Anti-labour measures intensified gradually in the other three cases.[10]

All four countries adopted broadly similar policies against all labour unions and parties, but different emphases emerged. To over-simplify, the Chilean military directed most of its fire-power against the leftist parties, the Argentines primarily focused on the labour bureaucracy, the Uruguayans simply closed down the entire representative system while maintaining some direct Ministry of Labour dealings with workers, and the Brazilians mainly strictly enforced long-standing corporatist controls. These strategies all reflected the varying strengths and weaknesses of those labour movements prior to the coups. The dictatorships intensified the structural, institutional, and political handicaps which had held back those worker organizations for many years.

Structural Constraints on Labour

After the political shock treatment of the coups and their aftermath, the military regimes subjected the workers to the economic shock treatment of neo-capitalist, free-market policies. Although their economic programmes varied, all contained some ingredients which would have undercut the proletariat even under a civilian government. The main exception was Brazil, which launched its restructuring project earlier.

All four regimes slashed the leverage of workers producing for domestic consumption and of workers in any one export industry by promoting, liberalizing, and diversifying foreign trade. This policy had the most impact in an enclave export economy like Chile. There, the reduction of copper from 77 to 50 per cent of the value of all exports shrank the bargaining power of the miners.

Campaigns by all four regimes against triple-digit inflation also dampened worker activism by defusing one of the long-standing rallying points for protests. The military's anti-inflationary measures also accelerated redistribution from workers to capitalists, concentrating income in the hands of the wealthiest 20 per cent of the population. That deterioration in labourers' standard of living left them scratching for subsistence rather than crusading for new gains. Along with declines in real wages, workers were afflicted by government austerity programmes which gutted social welfare provisions, leaving labour far more dependent on employers and secure employment. The workers' vulnerability and isolation also mounted in proportion to the growth of the service sector and underemployment.

In the three Southern Cone cases—especially Chile—deindustrialization and record-breaking unemployment further eroded the position of the proletariat. To weaken industrial workers, it proved necessary to weaken industry by lowering its protected position and concomitantly its labour costs. Although reliable figures remain unavailable, it has been estimated that by the early 1980s unemployment had reached at least 25 per cent in Chile, 10 per cent in Argentina, and 13 per cent in Uruguay. Thus unions had to emphasize job security rather than enhancement.[11]

After the political shock treatment of 1973–4, Pinochet instituted his economic shock treatment in 1975. The most extreme free-market approach of all the cases, that programme removed most tariffs and subsidies for industry, price controls for consumers, and welfare safety nets for the poor. Total imports and non-traditional exports approximately tripled in value from 1970 to 1978.

Through deflationary austerity and wages policies, Pinochet by 1978 had reduced the rate of inflation from 500 to 39 per cent and workers' real wages by approximately 30 per cent. That widening income gap between labourers and the middle social strata diminished the likelihood of political alliances across class lines. The elimination or privatization of most social services also punished workers and left many of them too concerned with individual survival in the market-place to engage in spirited collective action.

The sectoral profile of the Chilean economy showed by 1978 that the percentage of the economically active population employed in industry had fallen from 19 per cent in 1970 to 13 per cent and in construction from 6 to 4 per cent, while the service sector bulged from 27 to 35 per cent. In absolute terms, the national labour force in industry plummeted from 776,000 in 1972 to 585,000 in 1981. The agricultural work-force also shrank from 25 per cent of national employment in 1970 to 20 per cent by 1978, following Chile's transformation into a service economy.

As welfare cushions were removed, unemployment and underemployment soared, especially among younger urban wage-workers in industry and construction. Total national unemployment exploded from 98,000 in 1972 to 417,000 in 1981. When that open, vulnerable economy reeled under the force of the international recession by 1982, the downward spiral became worse than during the Great Depression of 1929–32.[12] Reacting to those structural changes, some leaders of the leftist parties realized by the early 1980s that their resurrection would have to rely less on the classic proletariat and more on 'popular' elements among the service-sector workers, the self-employed, the unemployed, the slum-dwellers, and the white-collar employees.[13]

Progressive movements in Argentina were also tempted to replay populist programmes and coalitions—long the leitmotif of Peronism—while broadening their appeal to underemployed workers outside the manufacturing sector. By paring down protective barriers against externally manufactured goods, the Jorge Videla Government forced domestic industrialists to meet foreign competition by lowering their prices and real wages. Trade promotion and currency devaluation transferred income from urban workers to agro-exporters. Those measures formed part and parcel of the anti-inflation campaign, which became neither so extreme nor so successful as in Chile. The annual growth in consumer prices dropped from 335 per cent in 1975 to 140 per cent in 1979. After rising from a base index of 100 in 1970 to 105 in 1975, real wages plunged to 63 by 1978. During 1975–80, industrial output fell by 3 per cent and industrial employment by 26 per cent.

From 1978 on, however, that military regime backed away from highly recessionary policies discouraging industry and encouraging unemployment. Much more than in Chile, industrialists in Argentina prodded the government to reverse its free-market policies. Those manufacturers were larger and more powerful than their Chilean counterparts, who had been undercut by expropriations under Allende. Although discontented with post-coup policies, Chilean industrialists remained far more fearful of the workers and their Marxist parties and therefore less willing and able to restrain anti-labour programmes under the dictatorship.

Moreover, the government in Argentina only wanted to discipline, not destroy, industry and its workers. Many in that regime still hoped to lure labour away from Peronism and the left through state paternalism, and so Videla did not want to create massive unemployment which might radicalize the workers. Regardless of the government's desires, the global recession in 1981–2 spurred falling productivity and employment, which soon resembled the more intentional Chilean situation.[14]

Uruguay proved even less avid than Argentina in its adherence to free-market economics. Metamorphosis away from an emphasis on industry towards a predominantly service-sector economy had been under way there for two decades. From the eve of the military take-over through the late 1970s, the percentage of the economically active population employed in manufacturing went from 21 to 19 per cent and in services from 27 to 29 per cent. Favouring only piecemeal liberalization, Uruguay's dictatorship proved much more reluctant to erase protectionism, to trim down the central government, or to foment unemployment.

That regime did expand foreign trade, which rose from 31 per cent (1970–2) to 41 per cent (1976–7) of GDP. At the same time, austerity measures and wage controls lowered the rate of inflation from 104 per cent in 1973 to 45 per cent in 1977, private consumption from 77 to 63 per cent GDP, and the real income of workers by 29 per cent. The share of national income accruing to employers increased by 27 per cent and that going to wage-earners plummeted by 34 per cent. Following in the footsteps of Pinochet, the government compounded this deprivation of workers by cutting back many social services and terminating employer contributions to social security and fringe benefits. Unemployment rose from 8 to 13 per cent, cresting during the 1981–3 international recession.

Since manufacturing for export continued to prosper under the Uruguayan military, industrialists there remained more satisfied with the regime's policies than did their counterparts in Chile and Argentina. Except for some dismayed producers for domestic consumption, most Uruguayan manufacturers seemed pleased with anti-union policies which held down their labour costs. Any effective response by labour unions and parties would have to rely upon workers in the manufacturing export sector and in white-collar occupations.[15]

By contrast, Brazil's military rulers—starting with a more agrarian, less industrialized economy—stimulated the growth of the modern urban sector. Especially during the 1967–73 'miracle', the export–import sector blossomed, as did manufacturing sales abroad. While GDP grew an average of 10 per cent per year, inflation fell from 27 to 17 per cent. Income distribution became increasingly regressive.

The spectacular growth of Brazilian industry—especially its more sophisticated sectors linked to foreign corporations and capital—generated no constituency for Chilean-style free-market policies. Instead, it produced a larger, more skilled, and more literate proletariat. The absolute number of industrial labourers more than doubled under the military government, rising from 9 to 11 per cent of the economically active population, while the service sector climbed from 22 to 24 per cent. Working-class leverage increased particularly in the most modern manufacturing branches, such as metal-working, where higher skill levels tightened the labour market somewhat.

Although the regime did not promote unemployment, the labour movement still suffered from vast underemployment and surplus manpower, especially in the mushrooming tertiary sector and squatter communities. The proletariat remained small, young, and compacted regionally (nearly half of all industrial workers lived in the state of São Paulo and had yet to reach the age of 30). A very high proportion of Brazilian workers in the cities still lacked steady jobs, marketable skills, literacy, organization, or political mobilization. Far more than in the Southern Cone, working-class parties would have to reach out from the narrow industrial base to recruit followers in the *favelas* and the countryside.[16]

Institutional Constraints on Labour

All these regimes initially and massively injured the labour movement through (a) widespread denial of independent union rights to organize, meet, bargain, or strike; (b) surgical removal of the most militant leaders and organizations, especially at the level of federations and confederations; and (c) sweeping elimination of party allies. Although uniformly harsh, these blows descended most heavily on industrial, blue-collar, larger, and Marxist unions. By nullifying national leaders and organizations, the military tried to molecularize the working class, so that the labour movement would at most survive at the plant or local level.

These regimes could get rid of politicians but not the working class. Therefore they soon sought new institutional arrangements for the permanent containment of the proletariat without the need for constant surveillance. These schemes to restructure labour policies, organizations, and systems, however, proved much less successful than crude repression. For example attempts to create surrogate unions by fiat or inducements foundered. Although the dictatorships also imposed or encouraged new union leaders, many older representatives survived and many newer ones proved equally unbending in their defence of worker interests. Even union spokesmen opposed to socialism or populism had to advocate the practical needs of their membership. The ejection of many leftist unionists from their posts and jobs did increase the strength of more conservative labour leaders, but that transformation sometimes appeared superficial and temporary. Because of their fear and enmity towards workers as well as the workers' own recalcitrance, the regimes made only minimal efforts to tame labour through co-optation, least of all in Chile. When governments extended small incentives for co-operation, they usually went to white-collar, non-Marxist, and smaller unions, some of which reciprocated with support.

Although all of the dictatorships considered creating full-fledged corporatist systems for encapsulating labour, only the Brazilians adopted this model and allowed unions to grow under state auspices. Realizing the barriers to really recasting a deep-seated labour movement, none of the regimes except the Chilean actually tried to devise an entirely fresh legal system of labour–industrial relations. Even there that full-blown new code mainly just rationalized the anti-labour policies of the junta. Whereas the dictatorships proved skilled at destruction, the workers themselves carried out the reconstruction of the labour movement.

In Chile, the military regime's fundamental labour policy aimed to deprive all unions of any effective role in society so that workers would subsist at the mercy of the market-place. The junta's initial decrees outlawed any national federation and 75 per cent of the organizations affiliated with the CUT, denied all unions and their leaders any associational or bargaining rights, and forbade any collective or political activities. Simultaneously, the military executed, exiled, detained, and otherwise removed or terrorized those unionists sympathetic to the UP.

Pinochet tried to spawn a new, non-Marxian generation of labour leaders by banning most UP sympathizers from office. Those loyalties proved so deeply engrained, however, that, beneath the surface, labour's political orientations apparently changed very little. Shortly after the coup, the junta, having abolished union elections, simply ordered former Popular Unity leaders replaced by the oldest union members. But those elder workers usually turned out to have equally strong Marxist leanings, particularly in Communist unions. In other cases, Marxists, especially Socialists, willingly ceded their union posts to Christian Democrats so as to alleviate government harassment of the union. Consequently the DC and even more conservative unionists made substantial gains as official and public leaders of the labour movement.

To what extent that shift really represented a massive political conversion by the workers, however, remained open to doubt. As the new labour code emerged in 1978–9, the government confidently allowed some plant-level elections with only 48 hours' notice. Neither formal partisan candidates nor workers with former political or union leadership experience were permitted to run. Nevertheless, the best evidence suggests that the balloting almost replicated the pre-coup party alignments of unionists. The Communists, Socialists, and Christian Democrats each accounted for about one-third of the newly chosen leaders.

Especially in its first three years, the junta also tried to coax some non-Marxian unions over to its side. The government won support from a few anti-Allende, Christian Democrat, and white-collar groups, for example among truck-owners and copper-miners. Those efforts to encourage pro-government unions soon encountered frustration, however, because the regime refused to give them significant rights or benefits and maintained economic policies inimical to workers. Therefore it increasingly came into conflict with these more co-operative unions and drove most of them into the opposition.

Instead of trying to envelop labour in a new corporatist structure. Chile's dictatorship sought to exacerbate its existing weaknesses by further encouraging the atomization and localization of unions. By 1978–9, however, the regime felt it necessary to codify this new *laissez-faire* system. Because the revamped labour code hamstrung union organizations and activities while removing the central government from labour–management disputes, it abandoned most workers to the caprice of owners. Although Chilean unions uniformly condemned the legislative package, they also tried to take advantage of the small space opened up to at least express opinions about the lot of workers.[17]

Like Pinochet, Videla in Argentina first unleashed naked coercion, including similar physical savagery and erasure of all significant trade-union legal rights. In response to ferment by radicalized factory workers, numerous interventions, arrests, and disappearances took place at the plant level as well as among the militant Peronists in the CGT. Unions suffered devastating losses, but they survived because of their powerful bureaucratic structures, their resilience from previous experiences with military regimes, and their immunity against anti-communism as a rationalization for wide-scale atrocities against workers.

After eliminating many intractable labour leaders, especially at the middle and lower levels, the military regime sought a *modus vivendi* with more pliable unionists. Since Peronism permeated virtually the entire movement, however, it proved harder in Argentina than in Chile to divide labour politically. Even moderate or conservative Peronists normally resisted abandoning that political umbrella to endorse the regime's anti-worker economic and social policies.

The new 1979 labour code tried to dilute the hegemony of monolithic Peronist unions, curb their bargaining power, and roll back their benefits. For example it prohibited the closed shop, long terms for union officials, political affiliations, and numerous state and employer welfare provisions. This legislation did not go as far towards individualization and emasculation as the Chilean plan. Nor did it endure as a blueprint for labour–industrial relations once external economic and geo-political shocks in 1981–2 upset the military's project.[18]

The Uruguayan generals wreaked havoc on labour by decapitating its top leadership, shearing off its more radical sectors, and abolishing all its legal rights to exist or bargain. Lacking strong party or bureaucratic mechanisms, the union movement caved in within a month after the 1973 coup. The repression weighed heaviest on the Tupamaros and the CNT at first and then bore down on the communists from 1976 on. Soon Uruguay hosted the largest number of political prisoners per capita in the world, most of them unionists.

As in Chile and Argentina, the military found it easier to hush Uruguayan labour than to reinvent it. Immediately after the coup, the Ministry of Labour called for all workers to renew their union affiliations, expecting fresh organizations and leaders to arise in place of the previously predominantly Communist apparatus; when 90 per cent of the workers reaffiliated with their old organizations, the regime cancelled all such democratic union processes. Another effort by the government to replace the CNT with an official General Confederation of Uruguayan Workers also encountered rejection by labourers and by the International Labour Organization.

In contrast to Chile and Argentina, the Uruguayan rulers extended labour a few more inducements along with constraints. They halted the CNT general strike against the coup not only with mass arrests but also with small salary increases. Even more than before the military take-over, the Ministry of Labour became active in addressing the needs and complaints of individual workers about wages and working conditions; the number of cases handled by that office quintupled from 1973 to 1980. Despite the near total illegality of labour organizations, parties, and activities, the regime soon quietly allowed a few non-communist unions to reorganize and bargain locally. Nevertheless, a state of siege rather than any new institutional rules and rights continued to govern labour up to 1983.[19]

After exorcizing the populists and Communists following the 1964 coup, the Brazilian authorities principally suppressed large federations and big unions rather than small, local, less political organizations. Instead of having to muzzle most workers as in Chile, the Brazilians only found it necessary to pulverize the small, autonomous, radical segments. A second wave of repression in 1968, responding to student and labour unrest, consolidated the continuing system of corporatist controls.

More than in the Southern Cone, the new Brazilian rulers succeeded in wholesale replacement of previous working-class leaders with compliant servants of the regime. As in past raids on labour, the government mainly aimed to drive out the communists, whose party and union leaders had to go into hiding or jail. The government took control of all the major unions and placed them under the command of state representatives, the so-called '*pelegos*', while barring former PCB or PTB union leaders from holding office.[20]

Responses by Labour Unions and Parties

Denied most democratic opportunities for expression, resistance, and advance, labour unions and parties did not succeed in undermining or ambushing these regimes up to 1983. They did, however, devise creative ways to survive, to parry some of the worst abuses of these governments, and to prepare for future liberalization or democratization. Enforced demobilization seemed to lead to the 'Argentinization' of the labour movements in Chile and, to a lesser degree, Uruguay and Brazil. Whereas unions in Chile—and somewhat in the other countries—worked through political parties before the coups, thereafter the parties started working through the unions. Like other civic organizations, the trade unions came to play roles previously monopolized by the parties as representative mechanisms, as opposition spokespersons, as providers of policy alternatives. In many cases, persecution apparently increased worker reliance on unions, solidarity between union leaders and followers, unity among unions, and identification of unionists with their martyred party colleagues. The unions also became more independent from party control and took the initiative in pressing the dictatorship for changes.[21]

Organized labourers in all the countries primarily stressed particular material objectives rather than grand social or ideological conquests, mainly because they had little choice. Economic and institutional constraints under the dictatorships accentuated the defensive, incremental orientation of the labour movement. Rather than seeking improvements in wages and working conditions, unions had to concentrate on job security in the face of declining incomes, rising unemployment, and government animosity. But both they and their parties realized that pursuit of intrinsically unionist demands—basic organizational rights, contracts, and benefits—inevitably also furthered the long-range broader goals of democratization and working-class advances.

Although forced to lower expectations and exerting less leverage than before the coups, labourers could still endanger capital accumulation in economies on the periphery of the world system. As a consequence, working-class activism, however restrained, helped produce not only the previous implantation but also the subsequent disintegration of the dictatorships in countries with narrow margins for economic manoeuvring. More than any other factor, the deepening global recession in the early 1980s reignited trade union activities in protest against government policies which jeopardized the very survival of their rank and file.

In the absence of open proletarian party activities, workers came to rely more not only on their own organizations but also on surrogate allies, notably international labour entities and the Roman Catholic Church. In Chile, where labour parties were most circumscribed but had the highest international visibility, unions most benefited from these foreign connections. By contrast, the Peronist unions had virtually no international ties, the Uruguayan CNT only wove thin strands overseas, and Brazilian federations were not allowed to join internationals. And only in Chile and Brazil did the Church supply significant backing to unionists.

Labour unions became more effective than labour parties at resisting these regimes and their assaults. The unions proved most resilient and most crucial to survival of their parties in Argentina and Chile, where those bonds had been strongest before the coup. In Brazil and Uruguay, parties became more effective at acquiring room for manoeuvre because they were less tied to labour and programmes for structural change, and were therefore more acceptable to the military. Labour's party allies usually managed to operate only in the spaces granted them by the ruling élites;

they rarely displayed any ability to create or force those openings themselves. Those least capable of counter-attacking these regimes were the parties most persecuted—the Marxists—whereas the Peronists, the Colorados, and even the PTB were allowed more leeway to engage in guarded political activity.[22]

The dictatorships unintentionally promoted greater unity of the parties as well as the unions. Up to 1983, this co-operation was most impressive in Argentina, where the *multipartidaria* coalesced all major parties behind a push for democratization, and in Uruguay, where the two traditional parties joined forces in promoting a return to electoral democracy and also called for the readmittance of the Frente members. In contrast with the Argentine Radicals, the Uruguayan Blancos, and the Brazilian Democratic Movement (MDB), however, the Chilean centrist and rightist parties proved very reluctant to align with the labour parties behind projects for redemocratization. Although the Christian Democrats and even a few leaders of the National Party increasingly moved into opposition to Pinochet, they were slow to bridge the bitter divisions which preceded the coup. Even the forging of the Democratic Alliance in 1983 failed to encompass the communists or to insure unity among social democratic sympathizers. Multi-party friction retarded the resistance movement in Chile, but the presence of non-Marxist parties with labour components— mainly the DC—also offered workers a less ostracized shelter until the return of normality.[23]

Even under severe restrictions, labour parties retained significance for the workers, particularly in Chile and Argentina, where they had historically exerted more sway over unions. At the same time, party ties which had proved beneficial under democratic systems now became liability for unions linked to the political enemies of the dictatorship. Therefore party influence withered somewhat and, in some cases, shrivelled down to the hard-core loyalists. Nevertheless, most unions held fast to their party affiliations. The new regimes offered them few attractive alternatives, their historical memories, class-consciousness, socialization, and ideological proclivities still drew labourers to the parties, and those vehicles still supplied them support services.[24]

Among previous militants, the parties continued to provide group identity and networks, keeping alive historic referents and beliefs as well as hopes for the future. Martyrdom, sacrifices, and struggles by party leaders also reinforced bonds between them and the oppressed workers. In closed societies, those politicians' clandestine criticisms of the ruling group suggested alternative programmes. They provided the outlines of shadow governments. Haunting the dictatorships, they continued to imply the illegitimacy and transience of those regimes. Their implicit presence and explicit protests or subversive activities sometimes limited more extreme government initiatives and carved out small niches for pro-labour manifestations. They continued to furnish allies—such as students and intellectuals—and assistance—such as legal aid. Through exile politics, they mobilized international pressure on the dictatorships, channelled support to unions, and demonstrated their survival and potential. Moreover, failure and exclusion prompted these parties to undergo profound soul-searching and self-reappraisal which proved salutary for future reactivation.

Up to 1983, the Chilean case showed the least progress towards any democratic openings for the labour movement. After initial suicidal resistance by a minority of workers, the unions reacted to the junta's bombardment from 1973 to 1975 by lying low in an effort to at least survive. The regime's atomization of labour organizations promoted decentralization and disunity but also allowed anti-government unions to carry on at the plant level. As the pogrom against Marxist unionists became more systematic and permanent, however, more and more unions converted to DC leadership. By the time the labour movement resurfaced in 1976, perhaps 80 per cent of the public leadership positions had passed into the hands of Christian Democrats.

During the second half of the 1970s, Chilean labour unions reasserted themselves from their local bases and reorganized into broad federations to speak out on government policies and worker needs at the national level. Although not allowing such associations, neither did the regime prevent their emergence. The two most important alliances became the Group of Ten and the National Union Co-ordinator (CNS). Partly replicating pre-coup divisions within the labour movement, the Ten primarily identified with the Christian Democrats, while the larger CNS, although also led by a Christian Democrat leader expelled from the country in 1982, tended to include most of the older, stronger, blue-collar unions historically linked with the leftist parties. After initially supporting the junta, the Ten joined the Christian Democrat Party in aligning with the opposition. As the junta's anti-labour economic, social and political policies rippled out, former DC and UP unions—although still quarrelling—managed to develop more co-ordinated positions and activities than did the parties themselves, eventually forming the National Labour Command.

Those labour organizations recaptured a voice in national affairs mainly through their own efforts, but also with significant assistance from international organizations, the Church, and the forbidden parties. The international arms of US, European, and other Latin American unions applied pressure on the junta to respect union rights, delivered financial and advisory assistance to Chilean labour, and squeezed concessions from the government by on-site visits. The exiled CUT, the UP parties, and the Christian Democrats helped forge these connections between external and internal unionists.

The Roman Catholic Church, often working in conjunction with the Christian Democrats, also helped rally international backing for organized labour under Pinochet. It shouldered some of the functions of the proscribed parties in speaking for and protecting unions. The Church provided workers with legal defence and social services, and protested on their behalf.

Organized workers had to develop more autonomous skills because the military so thoroughly dismembered their political allies. While the previous awesome electoral and parliamentary strengths of the UP parties proved useless against a dictator, their prior weaknesses—especially disunity—remained. By closing down their electoral and congressional arenas, the government did not eliminate the opposition parties. It did, however, reinforce their factionalism and polarization, freeze their leadership and ideas, and keep them on the defensive for a decade.

Despite towering obstacles and their consequent inability to challenge the regime's existence, the labour parties did register some accomplishments important to the working class. During 1974–82, the members of the UP managed to (a) preserve their identity and key leaders, usually in exile; (b) mount an effective international campaign to condemn and isolate Pinochet; (c) maintain some communication, contact, and impact with their social bases; (d) exert influence through non-party channels, such as unions, universities, the Church, and clandestine media; and (e) reassess their own ideological, strategic, and tactical positions for the past, present, and future. Because the regime did not adopt corporatist, populist, or state party instruments to take over the groups and functions previously monopolized by the political parties, it left those areas open to them—albeit subterraneously for the time being.

That the parties lost some influence but by no means all was revealed through union activities and the 1980 plebiscite. Although unions operated much more autonomously after the coup, they still heeded party cues. For example copper unions linked to the Christian Democrats did not mobilize against Pinochet's policies until the DC itself moved solidly into opposition by 1976.

Although the government rigged the 1980 plebiscite against any valid showing by adversaries, it conceded that at least 30 per cent of the voters responded negatively to the constitution that was to allow Pinochet to stay in office for 17 more years. Those 'no' votes mainly came from the time-honoured strongholds of labour and the left, such as the far north and south, the mining zones, and the working-class districts of Santiago. This suggested that the Marxist parties, with their long history of a social base largely defined by class and ideology, might be preserving their hard core. Even in a fair election, Pinochet might have won at least a plurality at the height of the short-lived economic boom, but the fragmented centrist and leftist opposition failed to mount a concerted test of his strength.

The plight of Chilean labour parties after the coup reflected not only the government's repression but also their own long-standing characteristics. Both before and after the end of democracy, the Socialists failed to discipline their ideologically and socially diverse ranks. As many of its upper and middle-level leaders fled or died, the party lost contact with its nuclei. The PS splintered repeatedly until it and the UP essentially dissolved in 1979.

Gradually, however, this fragmentation began yielding some positive results. The virtual dissolution of the old PS and UP opened the way to a renovation of the party and its coalitions. By the early 1980s, some leaders began trying to construct new unity within the PS, new alliances between it and other progressive movements, and new understandings with non-leftist parties, all in the name of 'socialist convergence'. Despite the Socialists' devastation, it remained possible that the enduring popularity of many of their past programmes, symbols, and leaders, their martyrdom since 1973, and their very factionalism, diversity, and elasticity could give them remarkable ability to rebound during any future democratic openings.

Predictably, the Communists survived more intact because of their disciplined organization, deep roots in the labour movement, prior underground experience (1948–58), and foreign backing. From 1973 to 1983, the party was reportedly retaining most of its union militants, especially in mining, textiles, and construction, while adding numerous adherents among shanty-town dwellers. Beneath the surface, the PC may have influenced more blue-collar workers than any other single party during the dictatorship.

After the 1980 plebiscite undergirded Pinochet's continuation, the historically gradualist PC discussed violent insurrection as one approach to toppling the tyrant. Although the party did not instigate armed uprisings by the working class, it did signal the left's willingness to entertain a multiplicity of tactics against the regime, as the Argentine opposition did during 1969–72. Like the Socialists, however, most Communists mainly counted upon international pressures, labour demands, mass demonstrations, multi-party coalitions, and the regime's own contradictions and errors to bring down Pinochet.

Meanwhile, the Christian Democrats, with less intimidation from the government and more protection from the Church, became much more of a labour party, although to what extent that was a marriage of convenience remains to be seen. Like the former UP parties, the DC underwent agonizing self-reappraisal about its social base, current immobility, and future agenda after the 1980 plebiscite. The intrinsic debilities and dilemmas of a centrist party in a polarized, authoritarian situation further raised questions about its continuing strength and scope. Nevertheless, the DC remained the major hope for providing a transition to a more democratic tomorrow.[25]

The Argentine labour movement mounted no resistance to the 1976 coup. Compared to previous confrontations with antipathetic governments, it was effectively demobilized from the take-over until the recession and Malvinas War of the early 1980s. During 1976–80, union activism shrank to its lowest point since 1943. Labour remained divided among the factions of Peronism and over tactical issues, but it also clung loyally to the legacy of Perón and to past union leaders, even though many of them had to operate extra-legally and clandestinely. Only a few unions reached an accommodation with the military government. Others, while unable to muster the kind of mass counter-attack which exploded in 1969–71, still responded with work slow-downs and halts, plant occupations, sabotage, and public demonstrations. When the regime and its economic programme lost momentum, general strikes erupted in 1979, 1981, and, most effectively, 1982.

Now that Perón had passed from the scene, the myth of his return could no longer galvanize the labour movement against the military or hold aloft the dream of a successful alternative. Although largely ineffectual, the Peronists continued to dominate the labour movement and to speak for it to the government and to other opposition parties with much greater freedom than did the PS and PC in Chile. The Argentine military regime hunted down Marxist groups and the Peronist left but allowed the mainline Peronist, Radical, Christian Democrat, Socialist, and even Communist parties to maintain most of their leaders and structures so long as they suspended activities. Once the defeat in the Malvinas discredited the armed forces, most observers assumed that the other major power in Argentine politics—the labour-based Peronists—would recapture the presidency.[26]

In Uruguay, the military swiftly flattened the labour unions and their parties, who played no significant role after the coup. The CNT sustained an impressive 15-day general strike against the army take-over in 1973, but then succumbed to banishment and lethal repression. With significant support from neither the Church nor international organizations, unions thereafter functioned at most illegally and inaudibly at the factory level, even though many began to recompose themselves by the end of the 1970s.

Working-class parties fared little better. The Uruguayan military outlawed all Marxist organizations and Frente members, except the Christian Democrats. The most important labour vehicle—the Communist Party (PC)—suffered the worst persecution, enduring banishment for the first time since its birth in 1921. Although it had never fetched more than 6 per cent on its own in national elections (1966), the PC had controlled approximately 80 per cent of the leadership positions in the CNT. Not surprisingly, in light of their great strength in past Uruguayan politics and their much lesser oppression under the dictatorship, it was the traditional multi-class parties—rather than labour or the left—which proved most capable of reinvigorating some democratic activities.

Without declaring them illegal, the dictatorship ordered the Colorados, Blancos, and Christian Democrats to suspend all activities. Although numerous Colorados and a few Blancos collaborated with the authoritarian regime, many of those traditional politicians joined the Christian Democrats and other centre-left and leftist parties in the opposition. By the end of the 1970s, most of the Blancos, the Christian Democrats, the Socialists, the Communists, and even some Colorados came together in the Grupo de Convergencia Democrática en Uruguay (CDU). Similar to the *multi-partidaria* in Argentina, the CDU represented a far broader opposition coalition than anything stitched together up to that time in Chile. Since Uruguayan parties had been less polarized and the left less menacing, the traditional machines reacted to the dictatorship by becoming more co-operative with one another and their competitors.

Uruguayan opposition forces—unlike their Chilean counterparts—defeated the government in the 1980 plebiscite, 57 to 43 per cent. The parties succeeded in that effort because a broad anti-dictatorial front transcended socio-ideological cleavages and included a democratic right-wing. Gallup polls indicated that a majority of members of all parties, especially those affiliated with the Frente, voted against the military's constitutional proposal. Just as before the coup, however, those opinion surveys did not turn up any sharp social correlations with voting patterns, indicating that roughly two-thirds of both the upper and lower classes in Montevideo cast negative ballots. By 1983, the Uruguayan military was withdrawing, while permitting and policing two moderate parties, behind which the labour movement would ostensibly have to channel its desires.[27]

In Brazil, any widespread working-class resistance to the military's 1964 take-over and subsequent reign was slow to take shape. Because unions became even more subservient social welfare agencies for the government, it usually required the development of new labour organizations and leaders—often outside the official structures—to revive any worker activism. By shackling legal unions, the military unintentionally encouraged the growth of more genuine extra-legal movements. Whereas survival of the labour movement was the crucial question in the other three countries, the recreation of such a movement was what was needed in Brazil.

After economic growth tapered off, recession set in, inflation revived, and the government relaxed controls on opposition political and social expressions, labour burst forth at the end of the 1970s. Especially at the plant level in the new modern industries of São Paulo, a grass-roots movement had grown up outside the official unions and parties. Spearheaded by metalworkers in the auto factories closely tied to foreign capital and by their charismatic leader Luis Inácio da Silva ('Lula'), massive strikes did not achieve many of their concrete objectives. But they did resurrect labour politics. Like their counterparts in Argentina, these Brazilian metalworkers constituted a novel challenge because of their strategic employment and because of their lack of loyalty to old parties and corporatist agencies. The Church, which had been vociferously defending labour and human rights, backed this movement. Because the government remained intractable, the worker campaign linked up with the broader national drive for democratization, also supported by some domestic industrialists. From the late 1970s onward, unions and parties gained ground simultaneously.

Although the indirect influence of greater party activity helped the Brazilian labour movement reassert itself, direct ties between parties and unions remained lean and problematic. Worker connections with progressive parties had been frail prior to the coup, after which those populist vehicles disappeared. By the late 1970s, Lula and other new labour leaders remained suspicious of the old opposition parties as well as the government and official unions. They promoted greater autonomy of the workers' movement to avoid populist or corporatist traps. Therefore they launched, along with some intellectuals, their own small (3 per cent of the votes in the 1982 elections) but innovative Party of Workers (PT).

Many other workers, especially those still beholden to the Communist Party, continued to vote for the larger official opposition vehicle, the Brazilian Democratic Movement, which had absorbed many elements of the defunct PTB. The PCB itself remained outlawed and reportedly lost some strength in the labour movement to newcomers like Lula and the PT. The Communists normally argued for working-class moderation so as not to derail democratization. Other worker votes and sympathies scattered among a variety of lesser parties, such as Leonel Brizola's rejuvenated Brazilian Labour Party. Although politicization of more skilled, organized workers seemed to be rising, their national alternatives remained mainly fragile patron–client parties heavily dependent upon non-ideological, multi-class regional machines and personalities.

By 1983, Brazil stood out among the four cases for the relatively wide latitude permitted labour union and party activity under military rule. That limited political opening occurred not because the labour movement there was more powerful than in the Southern Cone but rather because it had been historically weaker. The Brazilian Government apparently tolerated liberalization because labour unions and parties never constituted a fundamental threat. None the less, working-class vehicles now held out more promise than previously in Brazil and operated much more effectively than their enchained counterparts in Chile. That did not mean, however, that those Brazilian organizations yet exhibited any commensurate potential to extract significant concessions for the workers from the established order.[28]

Redemocratization

Up to 1983, labour unions and parties had displayed little capacity to influence or dethrone the authoritarian regimes. How these proletarian movements fared before and after the *coups d'état*, however, is no sure guide to their future role in rebuilding a durable democratic system. Under the adverse conditions of authoritarian capitalism, the survival and revival of labourite groups seems far more impressive than their deterioration.

Given the constraints and weaknesses besetting working-class parties and unions, they could oust the dictatorships only in concert with more numerous 'popular' sectors and with more powerful groups from the middle and upper classes, and indeed the international arena. After all, these regimes were faltering by the early 1980s mainly because of relatively exogenous factors, especially the global recession and the Malvinas war. The crisis in the world economy had helped foster and now undermined those dictatorships. Equally important, that crisis was compounded in South America by the workers' resistance to further immiseration in economies with narrow latitude for compromises.

Parties seldom create political space; they fill it. Subordinated groups rarely establish the rules of the game; they take advantage of them. Under these repressive governments, the capacity for action by workers and their organizations had been largely determined by the ruling élites, and labour unions and parties had seized every opportunity. The dictatorships had hammered many labour leaders and politicians into moderate positions, but once these regimes began to crack and fall, there was no doubt that parties speaking for the workers would again rush into the breach.

In the long run, then, perhaps authoritarian capitalism had failed. During the 1960s and 1970s, it was the labour parties and unions who fell short of their more ambitious dreams of power and redistribution. Rather than dislodging the foreign and domestic capitalists, they had enraged those dangerous foes. By the early 1980s, however, it was the élites and armed forces who had stumbled in their crusade to excise working-class populism and socialism once and for all. Both workers and their antagonists had encountered limits on their political projects.

As their authoritarian regimes and their economies teetered on the brink of bankruptcy, the guardians of the status quo saw labour unions and parties in Brazil reassert themselves even more energetically than in 1964. While the Argentine dictatorship fell victim to its own internal and external misadventures, the Peronists strode forth not unscathed but undaunted. After being repudiated in their own plebiscite, the rulers of Uruguay discovered that they had not succeeded in suffocating the past political habits of their citizenry. Even in Chile, where Pinochet remained, ensconced and steadfast, the sturdy labour movement had not been stamped out.

While demonstrating their survival and potential, these working-class movements also reflect the debilitating changes of the past two decades. Although scaling new heights, the Brazilian proletariat remains a minor force in national politics, as suggested by the PT's poor showing in the 1983 elections. In that same year, the weakened Peronists surprised most experts by losing the presidency—and many working-class votes—to the resurgent Radical Party, following a broader Latin American trend toward leadership by social democrats. Uruguayan leftists and unionists gained inclusion in the process of limited liberalization, but they had to play a muted role, secondary to the two traditional parties. And in Chile, the snowballing protests against the dictatorship began, predictably, with the copper-miners, but they wielded less leverage and encountered less labour solidarity than in the pre-1970 years. Although other unions hailed the daring initiatives of the copper-miners, working-class vulnerability discouraged general strikes. Therefore that movement had to gather momentum in 1983 by spreading out to other groups from the lower class and even from above worker ranks. Only gradually did the mobilization come under the partial co-ordination of the outlawed parties. That remarkable reawakening of the Chilean opposition, however, still has to contend with the social and political fragmentation aggravated by years of tyranny.

By 1983, the workers and their parties still needed to overcome those regimes and their legacies. These tenacious labour movements also must submit to or transcend the intensified structural, institutional, and political constraints which had hampered previous reform efforts. Understanding those continuing and, in many ways, deepening constraints may influence expectations for the near future. In the absence of social revolution and the presence of economic austerity, South American workers will confront very difficult political calculations. Now that their nemeses have failed in maximum attempts to cripple working-class organizations and progress, it remains for labour and its allies to craft inventive coalitions, programmes, and strategies which can not only initiate but also secure economic, social, and political democratization.

Notes

1. Alan Angell, *Politics and the Labour Movement in Chile* (London, 1972);
   J. Samuel Valenzuela, 'The Chilean Labour Movement: The Institutionalization
   of Conflict', in Arturo Valenzuela and J. Samuel Valenzuela, *Chile: Politics
   and Society* (New Brunswick, 1976), 135–71.
2. Ruth Berins Collier and David Collier, 'Inducements versus Constraints:
   Disaggregating "Corporatism",' *American Political Science Review*, 73 (4)
   (Dec. 1979), 967–86.
3. Juan Carlos Torre, 'The Meaning of Current Workers' Struggles', *Latin
   American Perspectives*, 1 (3) (Fall 1974), 73–81; Marcelo Cavarozzi, *Autoritarismo y democracia (1955–1983)* (Buenos Aires, 1983); Daniel James,
   'Power and Politics in Peronist Trade Unions', *Journal of Interamerican Studies
   and World Affairs*, 20 (1) (Feb. 1978), 3–36.
4. Alfredo Errandonea and Daniel Costabile, *Sindicato y sociedad en el Uruguay*
   (Montevideo, 1969); Howard Handelman, 'Labor-Industrial Conflict and the
   Collapse of Uruguayan Democracy', *Journal of Interamerican Studies and
   World Affairs*, 23 (4) (Nov. 1981), 371–94; Gustavo Cosse, 'Notas acerca de la
   clase obrera, la democracia y el autoritarismo en el caso uruguayo', mimeo
   (1983).
5. Kenneth Erickson, *The Brazilian Corporative State and Working-Class Politics*
   (Berkeley, 1977); Thomas Skidmore, 'Politics and Economic Policy Making
   in Authoritarian Brazil, 1937–71', in Alfred Stepan, *Authoritarian Brazil*
   (New Haven, 1973), 3–46; Leoncio Martins Rodrigues, *Trabalhadores,
   sindicatos e industrialização* (São Paulo, 1974).
6. Manuel Castells, *La lucha de clases en Chile* (Buenos Aires, 1974), Francisco
   Zapata S., 'The Chilean Labor Movement under Salvador Allende, 1970–1973',
   Latin American Perspectives, 3 (1) (Winter 1976), 85–97; Peter Winn, 'Loosing
   the Chains: Labor and the Chilean Revolutionary Process, 1970–1973', *Latin
   American Perspectives*, 3 (1) (Winter 1976), 70–84; James Petras, 'Nationaliza-
   tion, Socioeconomic Change, and Popular Participation', in Valenzuela and
   Valenzuela, *Chile: Politics and Society*, 172–200; J. Zylberberg and C. Lalive,
   'Corporatism–Populism–Socialism: The Political Culture of the Chilean
   Workers', *Canadian Journal of Latin American Studies*, 4 (9) (1979), 172–90;
   Henry A. Landsberger, 'The Labor Elite: Is It Revolutionary?', in Seymour M.
   Lipset and Aldo Solari, *Elites in Latin America* (New York, 1967), 256–300;
   Brian H. Smith and José Luis Rodriguez, 'Comparative Working-Class
   Behavior: Chile, France, and Italy', *American Behavioral Scientist*, 18 (1)
   (Sept. 1974), 59–96; Kenneth P. Langton and Ronald Rapoport, 'Social
   Structure, Social Context, and Partisan Mobilization: Urban Workers in
   Chile', *Comparative Political Studies*, 8 (3) (Oct. 1975), 318–44; James W.
   Prothro and Patricio E. Chaparro, 'Public Opinion and the Movement of
   Chilean Government to the Left, 1952–72', in *Chile: Politics and Society*,
   67–114; Robert Ayres, 'Unidad Popular and the Chilean Electoral Process',
   ibid. 30–66.
7. Dario Canton and Jorge R. Jorrat, 'Occupation and Vote in Urban Argentina:

The March, 1973 Presidential Election', *Latin American Research Review*, 13 (1) (1978), 146–57; Manuel Mora y Araujo and Ignacio Llorente, *El voto peronista* (Buenos Aires, 1980).

8. Ronald H. McDonald, 'Electoral Politics and Uruguayan Political Decay', *Inter-American Economic Affairs*, 26 (1) (Summer 1972), 25–45; Juan Rial, 'Notas sobre el sistema de partidos políticos en el Uruguay, 1904–1971', *CIESU/DT* 42 (1982).

9. Martins, *Trabalhadores, sindicatos e industrialização*, 88–124; José Alvaro Moises, 'Current Issues in the Labor Movement in Brazil', *Latin American Perspectives*, 6 (4) (Fall 1979), 51–70.

10. Karen L. Remmer, 'Political Demobilization in Chile, 1973–1978', *Comparative Politics*, 12 (3) (Apr. 1980), 277–82.

11. Rosemary Thorp and Laurence Whitehead, *Inflation and Stabilization in Latin America* (New York, 1979); Alejandro Foxley, *Latin American Experiments in Neoconservative Economics* (Berkeley, 1983).

12. Ricardo Ffrench-Davis, 'Políticas de comercio exterior en Chile: 1973–1978', *Working Papers* (Latin American Program: The Wilson Center), 67 (1980); Guillermo Campero and José A. Valenzuela, *El movimiento sindical chileno en el capitalismo autoritario (1973–1981)* (Santiago, 1981); Roberto Zahler, 'Recent Southern Cone Liberalization Reforms and Stabilization Policies: The Chilean Case, 1974–1982', *Journal of Interamerican Studies and World Affairs*, 25 (4) (Nov. 1983), 509–62; Philip J. O'Brien, 'The New Leviathan: The Chicago School and the Chilean Regime, 1973–1980', *Occasional Papers* (Institute of Latin American Studies, University of Glasgow), 38 (1982); Cristina Hurtado-Beca, 'Chile, 1973–1981: Desarticulación y reestructuración autoritaria del movimiento sindical', *Boletín de Estudios Latinoamericanos y del Caribe*, 31 (Dec. 1981), 91–117; Foxley, *Latin American Experiments in Neoconservative Economics*.

13. Enzo Faletto, 'Algunas características de la base social del Partido Socialista y del Partido Comunista, 1958–1973', *FLACSO* (Santiago), 97 (Sept. 1980); Guillermo Campero, 'Las nuevas condiciones en las relaciones del trabajo y la acción política en Chile', *Revista Mexicana de Sociología*, 41 (2) (Apr.–June 1979), 481–93; Francisco Rojas Aravena, *Autoritarismo y alternativas populares en América latina* (San José, 1982).

14. Adolfo Canitrot, 'Teoría y práctica: Liberalismo, política anti-inflacionaria y apertura económica en la Argentina, 1976–1981', *Estudios Cedes*, 3 (10) (1980); Walter Little, 'Argentine Labour Crisis since 1976', mimeo (1982).

15. Juan Rial, 'Los partidos tradicionales y la búsqueda de una nueva alternativa para la democracia', mimeo (1982); Howard Handelman and Thomas G. Sanders, *Military Government and the Movement toward Democracy in South America* (Bloomington, 1981); M. H. J. Finch, *A Political Economy of Uruguay since 1870* (New York, 1981); James Hanson and Jaime de Melo, 'The Uruguayan Experience with Liberalization and Stabilization, 1974–1981', *Journal of Interamerican Studies and World Affairs*, 25 (4) (Nov. 1983).

16. Kenneth S. Mericle, 'Corporatist Control of the Working Class: Authoritarian Brazil since 1964', in James M. Malloy, *Authoritarianism and Corporatism in Latin America* (Pittsburgh, 1977), 303–38; John Humphrey, *Capitalist Control*

and Workers' Struggle in the Brazilian Auto Industry (Princeton, 1982); Moises, 'Current Issues in the Labor Movement in Brazil'.

17. Campero and Valenzuela, *El movimiento sindical chileno en el capitalismo autoritario*; Hurtado-Beca, Chile, '1973–1981'; Manuel Barrera, 'Political laboral y movimiento sindical chileno durante el regimen militar', *Working Papers* (Latin American Program: The Wilson Center), 66 (1980); Gonzalo Falabella, 'Labour in Chile under the Junta', *Working Papers* (Institute of Latin American Studies, University of London), 4 (July 1981); Nigel Haworth and Jackie Roddick, 'Labour and Monetarism in Chile, 1975–1980', *Bulletin of Latin American Research*, 1 (1) (Oct. 1981), 49–62; Janine Miguel, 'La nueva institucionalidad laboral de los regimenes de seguridad nacional: La experiencia chilena', *Research Paper Series* (Institute of Latin American Studies, Stockholm), 33 (Feb. 1982); Arturo Valenzuela and J. Samuel Valenzuela, 'Party Oppositions under the Chilean Authoritarian Regime', *Working Papers* (Latin American Program: The Wilson Center) 125 (1983).

18. Little, 'Argentine Labour Crisis', Foxley, *Latin American Experiments in Neoconservative Economics*, 104–23.

19. Rial 'Los partidos tradicionales'; Martin Weinstein, *Uruguay: The Politics of Failure* (Westport, 1975); Edy Kaufman, *Uruguay in Transition: From Civilian to Military Rule* (New Brunswick, 1979).

20. Erickson, *The Brazilian Corporative State*; Mericle, 'Corporatist Control of the Working Class'; Youssef Cohen, 'The Benevolent Leviathan: Political Consciousness among Urban Workers under State Corporatism', *American Political Science Review*, 79 (1982), 46–59; Philippe Schmitter, 'The "Portugalization" of Brazil?', in Stepan, *Authoritarian Brazil*, 197–232; Angela Mendes de Almeida and Michael Lowy, 'Union Structure and Labor Organization in the Recent History of Brazil', *Latin American Perspectives*, 3 (1) (Winter 1976), 98–119.

21. Donald C. Hodges, *Argentina, 1943–1976: The National Revolution and Resistance* (Albuquerque, 1976).

22. Douglas Chalmers and Craig Robinson, 'Why Power Contenders Choose Liberalization: Perspectives from South America', *International Studies Quarterly*, 26 (1) (Mar. 1982), 3–36.

23. Remmer, 'Political Demobilization Chile'.

24. Valenzuela and Valenzuela, 'Party Oppositions under the Chilean Authoritarian Regime'.

25. Campero and Valenzuela, *El movimiento sindical chileno*; Valenzuela and Valenzuela, 'Party Oppositions under the Chilean Authoritarian Regime'; Falabella, 'Labour in Chile under the Junta'; Arturo Valenzuela, 'Six Years of Military Rule in Chile', *Working Papers* (Latin American Program: The Wilson Center), 109 (1982); Manuel Antonio Garretón, 'Institucionalización y oposición en el régimen autoritario chileno', *Working Papers* (Latin American Program: The Wilson Center), 59 (1980); 'Evolución política y problemas de la transición a la democracia en el régimen militar chileno', *FLACSO* (Santiago), 148 (June 1982); Francisco Zapata S., 'Los mineros del cobre y el gobierno militar en Chile entre 1973 y 1981', *Boletín de Estudios Latinoamericanos y del Caribe*, 32 (June 1982), 39–47; Paul E. Sigmund, 'The 1980 Constitution and

Political Institutionalization in Chile', (mimeo, 1982); Carmelo Furci, 'The Chilean Communist Party (PCCH) and its Third Underground Period, 1973–1980', *Bulletin of Latin American Research*, 2 (1) (Oct. 1982).

26. Cavarozzi, *Autoritarismo y democracia*; Little, 'Argentine Labour Crisis'.
27. Handelman, 'Labor Politics and Plebiscites: The Case of Uruguay', *Working Papers* (Latin American Program: The Wilson Center) 89 (1981); Rial, 'Los partidos tradicionales'; Kaufman, *Uruguay in Transition*; Cosse, 'Notas acerca de la clase obrera'; Luis E. Gonzalez, 'Uruguay, 1980–1981: An Unexpected Opening', *Latin American Research Review*, 18 (3) (1983), 237–72; Ronald H. McDonald, 'The Rise of Military Politics in Uruguay', *Inter-American Economic Affairs*, 28 (4) (Spring 1975), 25–43.
28. Erickson, *The Brazilian Corporative State*; Chalmers and Robinson, 'Why Power Contenders Choose Liberalization'; Moises, 'Current Issues in the Labor Movement'; Humphrey, *Capitalist Control and Workers' Struggle*; Mericle, 'Corporatist-Control of the Working Class'; Martins, *Trabalhadores, sindicatos e industrializacao*; Mendes and Lowy, *Union Structure and Labor Organization*; Fernando Henrique Cardoso and Bolivar Lamounier, *Os partidos e as eleicoes no Brasil* (Rio de Janeiro, 1975).

21

The Urban Character of Contemporary Revolutions*

Josef Gugler

THEDA SKOCPOL's *State and Social Revolutions*, while indebted to Barrington Moore's (1966) innovative *œuvre*, constitutes a landmark in the study of the origins and outcomes of revolutions. Two emphases in particular distinguish Skocpol's approach: international structures and world-historical developments take a central place in her analysis, and the state is seen as potentially autonomous from—though conditioned by—socio-economic interests and structures. Skocpol's analysis focuses on the French, Russian, and Chinese revolutions, but she suggests that these share certain broad resemblances with all the social revolutions that have taken place since World War II: they occurred in predominantly agrarian countries where peasants were the major producing class; peasant revolts or mobilization for guerrilla warfare played a pivotal role in each revolutionary process; and, without such revolts, urban radicalism in the end was not able to accomplish social-revolutionary transformations (Skocpol, 1979, 112–13, 287).[1] The very year these statements were published revolutionary movements triumphed in Iran and Nicaragua. If they give urgency to a re-evaluation of Skocpol's analysis,[2] there is reason to question her interpretation of contemporary revolutions in general.

The basic thrust of Skocpol's argument fits national wars of liberation well enough if allowance is made for the fact that such movements do not threaten the very existence of the metropolitan power, but rather its more or less important, but in every case circumscribed, overseas interests. National wars of liberation have punctuated like thunderstorms the decolonization process after World War II. That war loosened the grip of the colonial powers on their territories. The process was abrupt and

* This is a revised version of an article published in *Studies in Comparative International Development* 17 (2), 1982. It is reprinted here by permission of the editor. I wish to thank, without implicating, Susan Eckstein, Howard Handelman, Dietrich Rueschemeyer, Marilyn Rueschemeyer, and two anonymous reviewers of SCID for helpful comments. An earlier version of that article appeared under the title 'Der städtische Charakter moderner Revolutionen' in Heine von Alemann and Hans Peter Thurn (eds), *Soziologie in weltbürgerlicher Absicht: Festschrift für René König zu seinem 75. Geburtstag* (Westdeutscher Verlag, 1981).

dramatic where they were displaced by the Japanese. Once Britain had given up its Indian Empire, a new period in world history was marked, and all attempts to set back the clock failed. The Dutch in Indonesia and the French in Indo-China reestablished themselves after the Japanese occupation, but, faced with unending wars, eventually withdrew. By the 1970s, Portugal's attempt to hold on to its African possessions had become an anachronism and undermined a government unprepared to acknowledge the new realities. The settler community in Algeria prolonged the agony, but to no avail.

National wars of liberation have invariably been fought in predominantly rural countries. More than 80 per cent of the country's population lived in rural areas when the Dutch abandoned Indonesia in 1949, as the French left Indo-China in 1954, and when the Portuguese granted independence to Guinea-Bissau, Mozambique, and Angola in 1974–5. Algeria was a more urbanized country but still about two-thirds rural when the French finally made peace in 1962. In every case, the colonial power was challenged by a rural-based guerrilla movement that curtailed the production, processing, and transport of crops for export, thus striking at the very backbone of the colonial economy. If the colonial regime's capability to extract agricultural surplus was thus impaired, it invariably continued to control the cities effectively; this was the case even in Algeria, where the French effectively destroyed the large terrorist organization in Algiers. Colonial regimes withdrew of their own accord, and at a time of their choosing, even if precipitately in the case of the French after their defeat at Dien Bien Phu. The benefits to be derived from maintaining the colonial presence had been reduced, and the expeditionary force, the metropolitan government, and/or powerful sectors of public opinion were led to weigh the costs of protracted warfare. The colonial power eventually accepted that its interests would be better served by a harmonious relationship with a formally independent country. Indeed, most colonies were granted independence before any armed resistance emerged.

Four Revolutions

We propose to distinguish wars of national liberation directed against a colonial regime from revolutionary movements challenging a national government for control. Essential characteristics of the revolutionary movement are that it employs extra-legal means in challenging a national government and that elements outside government and the security forces play the principal role; it is 'popular' in the sense that it bases itself outside the ruling élite. We will argue that contemporary revolutions are largely urban in character.[3]

Revolutionary movements have seized power in four countries since the establishment of the People's Republic of China on 1 October 1949. A brief look at the cases of Bolivia, Iran, and Nicaragua will show that these were essentially urban struggles. The case of Cuba is more ambiguous.[4]

Cuba

If Ernesto 'Che' Guevara inspired revolutionaries throughout the Third World, he exalted rural guerrillas in his *Guerrilla Warfare*. The argument of the primacy of the rural guerrilla *foco* was further developed by Regis Debray (1967). The austere, difficult, and dangerous life experienced by the Cuban *guerrilleros*, and their encounter with rural poverty, profoundly marked Fidel Castro, Che Guevara, and their fellow fighters, many of whom came similarly from the urban middle class. The fact that it was Castro and a nucleus of his comrades from the Sierra Maestra who seized control of the destiny of Cuba added weight to their experience. And, indeed, the Cuban revolution is unique in the last four decades in that attacks by rural-based guerrillas led to the disintegration of the Cuban Army.

Still, a closer look at the Cuban revolution reveals a more complex picture. First, a large proportion of the *guerrilleros*, 60–80 per cent according to one source, were drawn from the urban population. And the urban underground was instrumental in recruiting them and in enabling them to join the fighters in the mountains. Second, the urban underground provided the lifeline for the *guerrilleros*. It supplied them with arms, information, money, and even food, and established the contacts through which they gained national and international recognition.[5] Third, the urban underground carried out a wide range of violent actions and sustained most of the casualties. More than five thousand bombings were reported in 1957 and 1958. Havana's international airport was burned. The Argentine racing driver Juan Fangio was kidnapped and kept prisoner for two days, attracting world-wide attention. The underground in Cienfuegos collaborated with the conspirators of the naval uprising there in 1957. The most spectacular action was carried out by the Directorio Revolucionario, the student organization, when it stormed the presidential palace on 13 March 1957, but failed in its attempt to kill the dictator.[6] In comparison, there were probably never more than three hundred *guerrilleros* in the Sierra Maestra at one time. This was the case even during the largest army offensive, in the summer of 1958, when 40 of their number fell. Finally, when the regime collapsed, rebel troops were too few to seize power. It was the urban underground that policed the streets and took over the administrative machinery. And it was a general strike that signified mass support for the rebels and discouraged attempts to establish a conservative successor regime (Karol 1970, 164–80; Thomas [1971] 1977, 256–63 and *passim*).

Bolivia

The Bolivian revolution may be seen as compressed into three days of intensive fighting when the government was overthrown in 1952. Or it may be traced back to its roots in organized labour and disaffected elements of the middle class. The artisan-labour movement had reached a degree of national coherence at the beginning of the Great Depression. During World War II, organized labour had become politicized and radicalized as its leadership shifted to the tin-miners. In an uneasy alliance with the labour movement, from 1946 on the Movimiento Nacionalista Revolucionario (MNR) increasingly became the rallying point of middle-class opposition committed to revolution. The first major joint effort of the MNR and the labour movement occurred in 1949 when a nation-wide revolt erupted. Rebel forces seized control of every provincial capital and mining-camp but failed in La Paz. Loyal government troops used the capital as a base and succeeded in suppressing the rebels, province by province.

The 1952 insurrection was launched in La Paz by Los Grupos de Honor and the national police, whose commanding officer had agreed to join the MNR in a coup. The Groups of Honour were paramilitary cells, their composition was mainly lower-middle class: artisans, less well-organized workers of small factories, and elements of the *clase popular*. When the insurrection appeared to be doomed, the police general sought asylum in a foreign embassy. However, armed workers from the mines and factories joined the fight in several provincial centers, turned the tide in La Paz, and cut off possible reinforcements for the capital. What had started out as a coup with limited civil participation, ended with Victor Paz Estenssoro, the leader of the MNR, presiding over a government that included three labour ministers, i.e. official representatives of the labour left. Whichever way the Bolivian revolution is delineated, there was never any involvement of rural elements in seizing power (Malloy 1970, 103–6, 127–50, 157, 167–8).

Iran

The success of the revolution against the Shah is the more remarkable for the odds it faced. Iran had a well-established government and enjoyed an economic boom. The Shah boasted a large modern army that was extremely well equipped. SAVAK, the secret police, had established a reign of terror: tens of thousands had been arrested and savagely tortured, hundreds had been executed. Many dissidents had gone into exile, and great numbers of young people had refused to return after completing their studies abroad.

Some elements in the anti-Shah movement stand out, though it is too early to assess their relative contribution to its success. Most conspicuous were the street demonstrations. On 8 January 1978, theological students in the holy city of Qom staged a sit-in. It was broken up by security forces, an action that quickly provoked retaliation, and the security forces started shooting. In two days of disturbances dozens were killed, according to one estimate; at least 70 persons, by another. In February, the Tabriz demonstration of sympathy and solidarity to commemorate those killed in Qom rapidly turned into a vehement protest against the Shah. The local Azerbaijani police refused to intervene; troops were called in and responded violently. The demonstration was transformed into a riot, spearheaded, it is said, by poor recent immigrants and radical students. An estimated one hundred persons were killed. A pattern evolved as demonstrators took to the streets again and again to commemorate the victims of earlier confrontations and to face troops who were shooting to kill. Over 3,000 are believed to have died in the first 11 months of 1978.

A second key element in the overthrow of the Shah was worker protest. Strikes did not gather momentum until mid-September. By the end of October, oil production had dropped by nearly three-quarters; many factories were forced to go on short time or to close for lack of energy supplies. The drying-up of oil exports threatened to create foreign exchange problems. By early November all public services—transport, telecommunications, ports, and fuel supplies—were paralysed or nearly so. Strikes at major banks affected import credits; strikes at customs halted industrial production by shutting off raw materials and spare parts.

A third factor was that large sectors of the middle class, and especially university students, had become increasingly disaffected over the years. Several guerrilla groups sprang from the student milieu abroad and at home, but their impact was circumscribed effectively by SAVAK, and their significance was primarily symbolic until very late in the struggle (Graham 1979, 214–15, 220, 224, 233, 237; Bill 1978). Central to our argument is that neither a rural guerrilla movement nor the rural masses played any role in the Iranian revolution.

Nicaragua

On the Nicaraguan revolution we have even less information. The junction of an organized guerrilla force with an urban insurrection appears to have been crucial to the overthrow of the Somoza regime. The Cuban example inspired the establishment of El Frente Sandinista de Liberación Nacional (FSLN). However, as elsewhere in Latin America, rural campaigns ended in defeat throughout the 1960s and into the 1970s. In 1975, splits developed within the FSLN over strategy: whether to concentrate on rural guerrilla warfare or on organizing among urban workers, and whether to give priority to military action or political activity. But then a number of spontaneous urban uprisings forced the FSLN's hand. They were spontaneous in two senses: triggered by specific events, and erupting independent of the FSLN. In February 1978, an insurrection in Monimbó held out for five days. In August youngsters armed only with pistols, rifles, and home-made contact bombs, their faces covered with black-and-red bandanas, forced the National Guard back to their barracks in Matagalpa. The FSLN, faced with the alternative to stop such insurrectionary tactics or to lead them, launched uprisings in five other cities on 9 and 10 September but eventually had to retreat to the hills. In April 1979, Estelí rose against Somoza once more—accounts differ whether this uprising was initiated by the FSLN or a local initiative the FSLN decided to support (Black 1981, 149; Booth [1982] 1985, 173). In any case, a comprehensive plan of action was already taking shape. A general strike called for 5 June paralysed the country; guerrilla units spearheaded insurrections in the major towns; and within six weeks the dictator was put to flight (Chavarría 1982; see also Booth ([1982] 1985) and Black 1981).

Humberto Ortego [1979] 1980, 4), leader of the Tercerista tendency within the FSLN, summarized the events in an interview:

We always took the masses into account, but more in terms of their supporting the guerrillas, so that the guerrillas as such could defeat the National Guard. This isn't what actually happened. What happened was that it was the guerrillas who provided support for the masses so that they could defeat the enemy by means of insurrection. We all held that view, and it was practice that showed that in order to win we had to mobilize the masses and get them to actively participate in the armed struggle. The guerrillas alone weren't enough, because the armed movement of the vanguard would never have had the weapons needed to defeat the enemy. Only in theory could we obtain the weapons and resources needed to defeat the National Guard. We realized that our chief source of strength lay in maintaining a state of total mobilization: social, economic and political mobilization that would disperse the technical and military resources of the enemy.

Since production, the highways and the social order in general were affected, the enemy was unable to move his forces and other means about at will because he had to cope with mass mobilizations, neighbourhood demonstrations, barricades, acts of sabotage, etc. This enabled the vanguard, which was reorganizing its army, to confront the more numerous enemy forces on a better footing.

An Interpretation

Each of these revolutionary movements was largely urban in character. Even where rural-based guerrillas played a prominent role, in Cuba and

Nicaragua, their leadership, a large proportion of their fellow combatants, and much of their material support were drawn from the urban milieu. It was in the cities that all the confrontations took place in Bolivia and Iran, that the decisive battles were fought in Nicaragua, and that the Cuban *guerrilleros* found crucial support to establish their regime. Urban workers determined the outcome in the battle for the control of La Paz, paralysed the economy in Iran and Nicaragua, and thwarted efforts to snatch the fruits of victory from the hands of the *guerrilleros* who had put the Cuban dictator to flight.

Our definition of revolutions, unlike Skocpol's, focuses on the seizure of power and is not concerned with subsequent transformations. Such a narrow definition would seem advisable in view of the difficulty of securing agreement as to which transformations qualify as revolutionary. Recent analysts such as Skocpol, S. N. Eisenstadt, Jeffrey M. Paige, and Kay Ellen Trimberger are sharply divided on the issue, as Jack A. Goldstone (1980, 450) has pointed out. There is the further difficulty that it is rather early to tell what the outcomes will be in two of the four cases we have discussed. If we were to focus on revolutionary outcomes, our argument would remain unaffected in most, but not all cases. Rural elements played no active role in the transformations wrought after the seizure of power in Cuba, they are unlikely to do so in Iran, and may be expected to remain in a subordinate position in Nicaragua as well. In Bolivia, however, peasants played a critical role in undermining the power of the landed oligarchy, once the MNR had come to power. Additional cases can be argued to constitute revolutions in as much as a military coup was followed by a revolutionary phase. Rural elements did not come to the forefront in that phase in Egypt and Peru, but they were mobilized against the landed aristocracy in Ethiopia. Of course, many wars of national liberation were followed by attempts at revolutionary transformation, but in no case did rural elements play a significant role in that transformation—except in rhetoric.

How to explain the urban character of contemporary revolutions? The four revolutionary struggles successful in the 1950s and 1970s were carried out in countries that varied greatly in their level of urbanization. Little more than a fifth of the population lived in cities in Bolivia at the time of the revolution, slightly less than half in Iran and Nicaragua, somewhat more than half in Cuba. The importance of urban elements in the revolutions of the latter three countries may be seen as a function of their high level of urbanization. In such a perspective, the rapid urban growth experienced by most Third World countries can be taken to presage the age of urban revolutions. As Abraham Guillén (1966, 238), the intellectual mentor of urban guerrillas in Uruguay and beyond, put it:[7]

Strategically, in the case of a popular revolution in a country in which the highest percentage of the population is urban, the center of operations of the revolutionary war should be in the city. Operations should consist of scattered surprise attacks by quick and mobile units superior in arms and numbers at designated points, but avoiding barricades in order not to attract the enemy's attention at one place. The units will then attack with the greatest part of their strength the enemy's least fortified or weakest links in the city. . . The revolution's potential is where the population is.

The demographic observation reflects an economic reality: the surplus is increasingly produced in the urban sector. Even in largely rural Bolivia, it was tin-mining rather than agriculture that constituted the core element in the national economy. The state's ability to extract agricultural surplus is no longer crucial to its very operation. Rather, it is the urban economy that finances the state apparatus. The extreme example is Imperial Iran, which filled its coffers with petro-dollars while importing food.

Because the state is dependent on the urban economy, there are limits to repression in the urban context. Managers, professionals, skilled workers, and even semi-skilled workers cannot be replaced in great numbers at short notice. To imprison them for any length of time, to push them into exile, or to kill them, entails severe economic losses. This means not only a reduction in the resources available to the state, but also a deterioration in living conditions for the population at large that may foster discontent.[8] Rather than using the stick, governments usually deal with the elements of the labour force that control key sectors of the economy such as mining, heavy industry, and transport by offering them privileges of income, fringe benefits, job security, and social security.[9]

Difficulties in extracting surplus provide powerful motivation for colonial powers to reconsider the merits of their direct control over distant lands. For national governments and their supporters the situation is quite different. For one thing, they can appropriate surplus via the money-printing press, though inflation has its political costs. For another, many governments can count on foreign assistance to tide them over a crisis. If patron countries find it in their interest, they may provide substantial economic and/or military support to local governments for many years. Even a bankrupt government is unlikely to abdicate of its own accord. It may fall victim to shifts within its political base, or it may be toppled by sections of the army dissatisfied with the lack of resources. But, confronted with a revolutionary movement, the ruling group will desperately cling to power. If a national war of liberation threatens the tentacles of empire, a revolutionary movement attacks the very existence of political and economic élites. And while members of the élite can make provision to live out their days in comfort in exile, much of the indigenous middle class, unlike colonial civil servants, has nowhere to go. The level of resistance of the élite and the middle class tends to be accordingly high.[10] The MNR, it is true, could appeal effectively to a large part of a middle class that had been deeply divided ever since Bolivia's humiliating defeat in the Chaco War. In Cuba, large sectors of the middle class had been hostile to Batista since 1952, when, after three popularly elected administrations, he seized power in a *coup d'état*; elements of the middle class in Iran and Nicaragua had been similarly alienated from autocratic rulers. But in these three countries it was only when the revolutionary movement made life in the cities insecure and brought the economy to a halt that a full-scale withdrawal of support from government took place. And only when they had thus been quite isolated did Batista, the Shah, and Somoza flee their countries.

Colonial governments withdrew in spite of their success in maintaining control of the cities. But loss of control over rural areas is not a sufficient condition to persuade a national government to relinquish power to a revolutionary movement. For such a movement to succeed it must confront the government in its urban location. Control over the capital city is usually of crucial importance, as the abortive revolution of 1949 demonstrated in Bolivia.

The Chinese communists launched their attack on the cities from the rural areas. The attempt to base a socialist revolution on the urban proletariat had failed in 1927. The Communist Party moved to build up peasant support through land and tax reforms in the areas it controlled, and from guerrilla origins it raised peasant armies that were victorious in a civil war fought on conventional lines. The Chinese experience demonstrated, or so it seemed, that Third World revolutions would have to be based on the peasantry. But the circumstances in which the Chinese communists were able to control entire provinces and to build up their armies were unique. There was no government exercising hegemonic power over China, and the country's vastness inhibited the establishment of such power. The Japanese invasion further undermined any effort at central government, brought the Communist forces recognition as nationalists fighting the invader, and at times forced a truce on the communists' opponents.

Rural areas can shield small mobile guerrilla units even today. But the transition from rural guerrilla activity to peasant army faces impossible odds in contemporary circumstances. Governments everywhere can rapidly dispatch army units across an entire country. Any attempt to move beyond the guerrilla stage in rural areas has to reckon with the immense fire-power, the high mobility, and the efficient communications system that are the hallmark of a modern army.[11] Governments have shown little hesitation in razing rural settlements and relocating peasants to counter rural-based challenges.

Urban guerrillas were revealed to be even more vulnerable than their rural counterparts. The city seemed to hold specific attractions for guerrilla activity. For the students and professionals who invariably predominate among the guerrillas in the early stages, and frequently longer, the city constitutes familiar terrain: they know it and their presence does not attract undue attention. Furthermore, while most guerrillas in rural areas are outsiders, even if they are of rural origin, the urban guerrilla can maintain the cover of a conventional life until he/she arouses the suspicions of authorities. Finally, the urban crowd promises the guerrilla anonymity. But systematic torture revealed the urban guerrillas' vulnerability to security leaks because of their dependency on safe houses that, once uncovered, can be rapidly investigated by vastly superior armed forces. A circle was established as torture—at the beginning often of people arrested quite haphazardly—provided information that led to the capture of participants or sympathizers who under torture provided further information. Torture served also to deter recruits to the ranks of the guerrillas and their sympathizers.[12]

Conclusion

A national government will not surrender to a revolutionary movement unless it is confronted in its urban location by forces it can no longer contain. Such forces are most unlikely to be constituted of a peasant army. They have to be established in an urban insurgency, as shown in Bolivia, in Iran, and most dramatically in Nicaragua. Substantial elements of the urban population must come to reject a regime to the point where they are prepared to confront its armed forces in the streets. The extent and depth of such disaffection will be affected by social and political mobilization. And the organization of the urban insurgency will bear on its success or failure.

Urban insurgency confronts governments with difficult choices. No government has yet responded by relocating urban populations on any scale. And if heavy arms are used in re-establishing control over a city, or major parts of it, the destruction wrought is bound to be huge and support of the government will be alienated. Somoza, though, went so far as to order the destruction of entire small towns and of neighbourhoods in Managua held by the Sandinistas.

Against heavy odds, guerrillas have maintained their presence in a number of countries for many years. They face the even more difficult task of mobilizing an urban mass movement that will confront the fire-power of a modern army. The failure of urban guerrilla movements, while they were operating on a significant scale in Venezuela, Brazil, Uruguay, and Argentina, was precisely their inability to mobilize broad-based support and to gain control of the city streets.

Successful revolutions are exceptional events in world history. Dozens of guerrilla movements began fighting after the victory of the Cuban revolution, but only 20 years later were they successful in one small country, Nicaragua. It is too early to tell whether El Salvador and Guatemala will follow the example of their neighbour. When revolutions have been successful, they have been characterized by unique features. Indeed, revolutionaries have to be innovators, not only to adapt to circumstances that are different in every case and that change even as the struggle proceeds, but also to overcome the initial advantage of their opponents. It is governments that are in the better position to draw lessons from history and to anticipate challenges to their rule.[13]

Such considerations suggest caution in drawing generalizations from the few cases at hand. Still, the largely urban character of contemporary revolutions merits consideration, not only because it contradicts widely held views, but because it suggests a new perspective on revolutionary movements and their adversaries in the contemporary world.

Notes

1. The position that 'only where urban riots have helped trigger rural uprisings have urban tumults ushered in the old regime's demise' is maintained by Goldstone (1982, 200) in his recent review of the literature on revolutions. Dix (1983), on the other hand, recognizes the role of urban elements and urban conflicts in revolutions that have taken place in 'semi-modern societies' with 'semi-modern regimes' such as Cuba and Nicaragua, but characterizes these as following a Latin American pattern distinct from the case of Iran.
2. Skocpol (1982) has begun to address the issues the Iranian revolution raises for her analysis.
3. Welch (1977) has emphasized the importance of the distinction between national wars of liberation and revolutions, and predicted that the African peasantry will not play a revolutionary role in the next several decades.
4. The overthrow of the Lon Nol government in Cambodia in 1975 can be argued to constitute a fifth case. The Khmer Rouge's transition from rural guerrilla to peasant army may then be seen to have been effected in circumstances not unlike those in China a generation earlier—which we will discuss shortly. The success of the guerrilla movement under Yoweri Museveni in Uganda in 1986 may also be taken to contradict the proposition advanced here, but there the government forces were disorganized to an extraordinary degree.
5. The dependence of rural guerrillas on urban support was highlighted when Che Guevara (1968) found himself without such support in Bolivia.
6. For a chronological listing of actions by the urban underground from 1952 to 1959, see Bonachea and San Martín (1974, 338–44). The composition of the first government after the victory of the revolution, in which the urban front

was more strongly represented than the rural, may be taken as one indication of the importance of the urban movement (Thomas [1971] 1977, 283–5).

7. Hodges (1973) provides a biographical account of Guillén, an introduction to his writings, and a discussion of his influence on the armed struggle in Latin America.

8. The crucial role of certain elements of the labour force was demonstrated by the crippling effect that strikes of copper-miners had on the Chilean economy during the Allende regime.

9. For a more detailed discussion of the political role of organized labour in the Third World, see Gilbert and Gugler (1982, 151–5); on the resurgence of organized labour under military rule in Chile, Argentina, Uruguay, and Brazil, see Drake (1988).

10. The position of élites in settler regimes is akin to that of national élites. Their followers tend to be even more committed to resistance than a national middle class: elements of the latter may anticipate accommodation with the new rulers, but for settlers defeat invariably means what they consider exile and destitution.

11. Governments, rather than revolutionary movements, are as a rule the prime beneficiaries of external support. The United States, in particular, quite apart from its open intervention in Vietnam and the Dominican Republic, has assisted many governments against their domestic opponents. Not only were foreign governments provided with equipment on a lavish scale but their troops were given training specifically geared to meet the guerrilla challenge. La Escuela de las Américas, located in the Panama Canal Zone, received its name in 1963 when a new curriculum emphasizing training in counter-insurgency and civic action was introduced. More than 20,000 officers and enlisted men, representing every Latin American country except Cuba, were trained there in the 1960s. In addition, small groups of Green Berets, stationed at Southern Command in the Panama Canal Zone, have worked with the troops of every Latin American nation except Mexico, Cuba, and Haiti. Fifty-two such missions, including parachute drops into guerrilla zones, were reported for 1965 alone. In 1966 and 1967, Green Berets assisted the Guatemalan army and suffered several losses at the hands of the guerrillas. In 1967 they set up a camp in Bolivia where they trained 600 raw recruits of the Bolivian army who later that year tracked down Che Guevara and his comrades (Gott 1971, 450–1, 488–9; Klare 1972, 301, 306, 379–81; Lartéguy [1967] 1970, 195–7). It is also the United States that, with its support for the Contras against the Sandinista government in Nicaragua, provides a significant exception to the rule that governments are the prime beneficiaries of external support.

12. For an account of urban guerrilla warfare in Brazil, Uruguay, and Argentina, and excerpts from a large array of writings by and about the guerrillas, see Kohl and Litt (1974); for a comprehensive annotated bibliography, Russell, Miller, and Hildner (1974); for an outstanding collection of documentary materials on the Tupamaros in Uruguay, drawn from a wide variety of sources, Mayans (1971).

13. Gates (1986) castigates the ahistorical character of most work on revolutions, sketches the history of the changing balance of forces between revolutionaries

and the state since the seventeenth century, and calls attention to the learning process of revolutionaries and governments alike.

References

Bill, James A. (1978) 'Iran and Crisis of '78', *Foreign Affairs*, 57, 323–42.

Black, George (1981) *Triumph of the People: The Sandinista Revolution in Nicaragua* (London: Zed Press).

Bonachea, Ramon L. and Marta San Martín, (1974) *The Cuban Insurrection 1952–1959* (New Brunswick, NJ: Transaction Books).

Booth, John A. (1982) *The End and the Beginning: The Nicaraguan Revolution* (Boulder, Col. and London: Westview Press; 2nd edn. 1985).

Chavarría, Ricardo E. (1982) 'The Nicaraguan insurrection: An Appraisal of its Originality' in Thomas W. Walker (ed.), *Nicaragua in Revolution* (New York: Praeger), 25–40.

Debray, Regis (1967) *Révolution dans la révolution?* Cahiers Libres 98 (Paris: Maspero). Eng. trans. *Revolution in the Revolution? Armed Struggle and Political Struggle in Latin America* (New York and London, Monthly Review Press, 1967).

Dix, Robert H. (1983) 'The Varieties of Revolution', *Comparative Politics*, 15, 281–94.

Drake, Paul W. (1988) 'Urban Labour Movements under Authoritarian Capitalism in the Southern Cone and Brazil, 1964–32', chapter 20 in this volume.

Gates, John M. (1986) 'Toward a History of Revolution', *Comparative Studies in Society and History*, 28, 535–44.

Gilbert, Alan and Gugler, Josef (1982) Cities, Poverty, and Development: Urbanization in the Third World (Oxford: Oxford University Press).

Goldstone, Jack A. (1980) 'Theories of Revolution: The Third Generation', *World Politics*, 32, 425–53.

—— (1982) 'The Comparative and Historical Study of Revolutions', *Annual Review of Sociology*, 8, 187–207.

Gott, Richard (1971) *Guerrilla Movements in Latin America* (Garden City, NY: Doubleday).

Graham, Robert (1979) *Iran: The Illusion of Power* (New York: St Martin's Press).

Guevara, Ernesto Che (1960) *La guerra de guerrillas* (Havana: Departamento del Minfar). English trans. *Guerrilla Warfare* (New York: Monthly Review Press, 1961).

—— (1968) *The Diary of Che Guevara. Bolivia: November 7, 1966–October 7, 1967*. Authorized text in English and Spanish (Toronto and New York: Bantam Books).

Guillén, Abraham (1966) *Estrategia de la guerrilla urbana: Principios básicos de guerra revolucionaria*. Manuales del Pueblo. (Montevideo: Ediciones Liberación, 2nd edn., 1969). English trans. of excerpts from 1st and 2nd edns, in Donald C. Hodges (ed.), *Philosophy of the Urban Guerrilla: The Revolutionary Writings of Abraham Guillén* (New York: William Morrow, 1973), 229–51.

Hodges, Donald C. (1973) 'Introduction: The Social and Political Philosophy of Abraham Guillén', in Donald C. Hodges (ed.), *Philosophy of the Urban Guerrilla: The Revolutionary Writings of Abraham Guillén* (New York: William Morrow), 1–55.

Karol, K. S. (1970) *Les guérilleros au pouvoir: L'itinéraire politique de la révolution cubaine.* L'histoire que nous vivons (Paris: Robert Laffont). English trans. *Guerrillas in power: The Course of the Cuban Revolution* (New York: Hill & Wang, 1970).

Klare, Michael T. (1972) *War without End: American Planning for the Next Vietnams* (New York: Alfred A. Knopf).

Kohl, James; and Litt, John (1974) *Urban Guerrilla Warfare in Latin America* (Cambridge, Mass. and London: MIT Press).

Lartéguy, Jean (1967) *Les guérilleros* (Paris: Raoul Solar) English trans., *The guerrillas* (New American Library, 1970).

Malloy, James M. (1970) *Bolivia: The Uncompleted Revolution* (Pittsburgh: University of Pittsburgh Press).

Mayans, Ernesto (ed.) (1971) *Tupamaros: Antología documental.* CIDOC Cuaderno 60 (Cuernavaca: Centro *Intercultural de Documentación*).

Moore, Barrington, jun. (1966) *Social Origins of Dictatorship and Democracy: Lord and Peasant in the Making of the Modern World* (Boston: Beacon Press).

Ortega, Humberto (1979) 'Nicaragua: La estrategia de la victoria' (Entrevista a Humberto Ortega, commandante en jefe del Ejército Popular Sandinista), *Bohemia*, 71 (52), 4–19. Eng. trans. 'Nicaragua: The Strategy of Victory' (Interview with Humberto Ortega, Commander in Chief of the Sandinista People's Army) in the Eng. language edn. of *Granma*, 27 Jan. 1980, 3–8.

Russell, Charles A., Miller, James A., and Hildner, Robert E. (1974) 'The Urban Guerrilla in Latin America. A Select Bibliography', *Latin American Research Review*, 9 (1), 37–79.

Skocpol, Theda (1979) *States and Social Revolutions: A Comparative Analysis of France, Russia, and China* (Cambridge, London, New York, New Rochelle, Melbourne, and Sydney: Cambridge University Press).

—— (1982) 'Rentier State and Shi'a Islam in the Iranian Revolution', *Theory and Society*, 11, 265–83.

Thomas, Hugh (1971) *Cuba: The Pursuit of Freedom.* Second half reprinted as *The Cuban Revolution* (New York, Hagerstown, San Francisco, and London: Harper & Row, 1977).

Welch, Claude E., jun. (1977) 'Obstacles to "Peasant War" in Africa', *African Studies Review*, 20, 121–30.

Name Index

Subject Index

(Cities are listed by country)